Solutions Manual to Accompany

HEAT TRANSFER

SEVENTH EDITION

J.P. HOLMAN
Professor of Mechanical Engineering
Southern Methodist University

McGraw-Hill Publishing Company
New York St. Louis San Francisco Auckland Bogotá Caracas
Lisbon London Madrid Mexico City Milan Montreal
New Delhi San Juan Singapore Sydney Tokyo Toronto

Solutions Manual to Accompany
HEAT TRANSFER / Seventh Edition

ISBN 0-07-029639-1

3 4 5 6 7 8 9 0 HAM HAM 9 0 9 8 7 6 5 4 3

CONTENTS

FIGURES FROM THE TEXT SUITABLE FOR USE AS

VIEWGRAPHS OR LARGER CALCULATION CHARTS

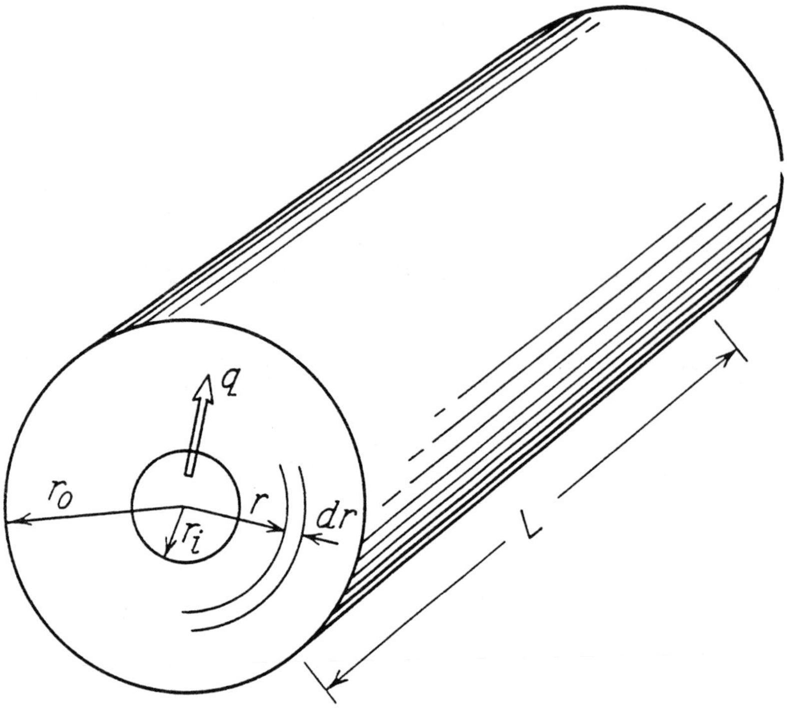

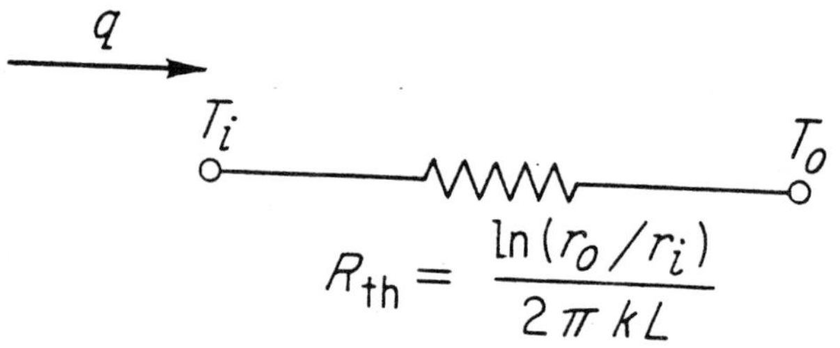

ONE DIMENSIONAL HEAT FLOW THROUGH A HOLLOW CYLINDER AND ELECTRICAL ANALOG. From Holman, J. P.:HEAT TRANSFER, McGraw-Hill Book Co.

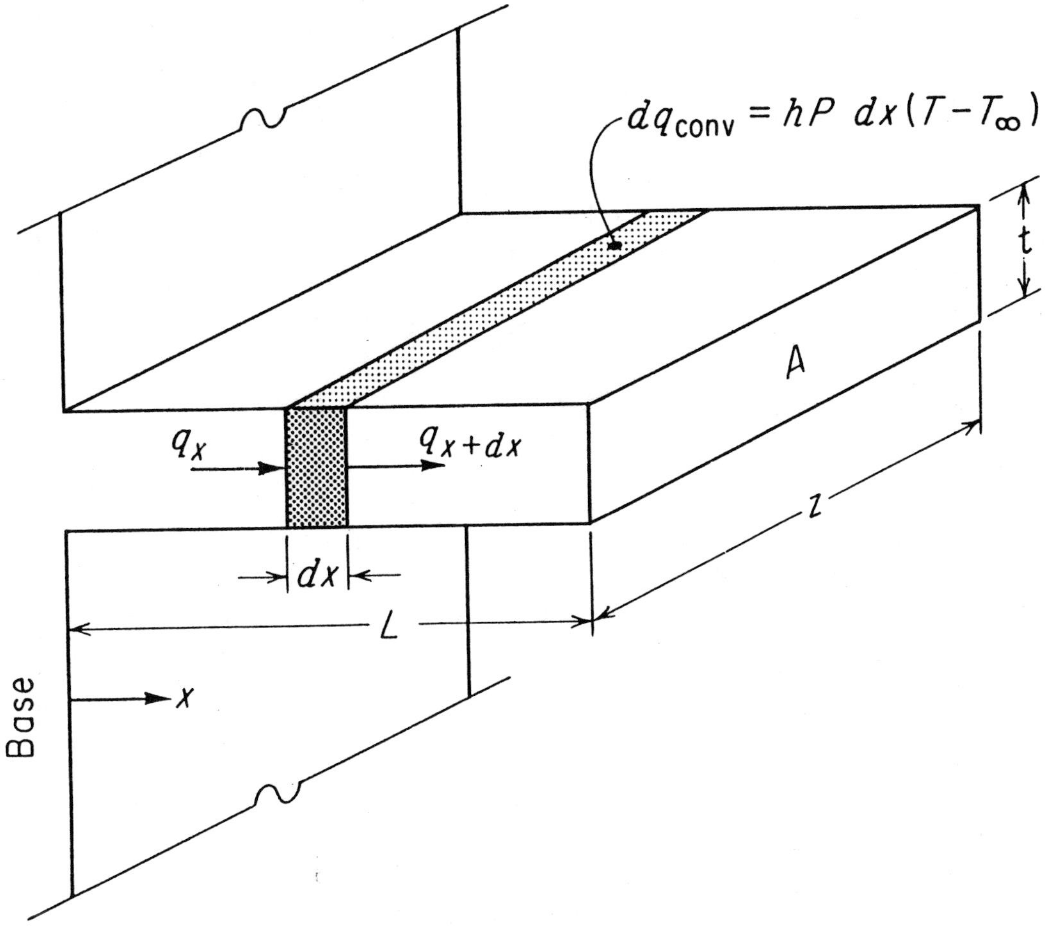

SKETCH ILLUSTRATING ONE DIMENSIONAL CONDUCTION AND CONVECTION THROUGH A RECTANGULAR FIN. From Holman, J. P.: HEAT TRANSFER, McGraw-Hill Book Co.

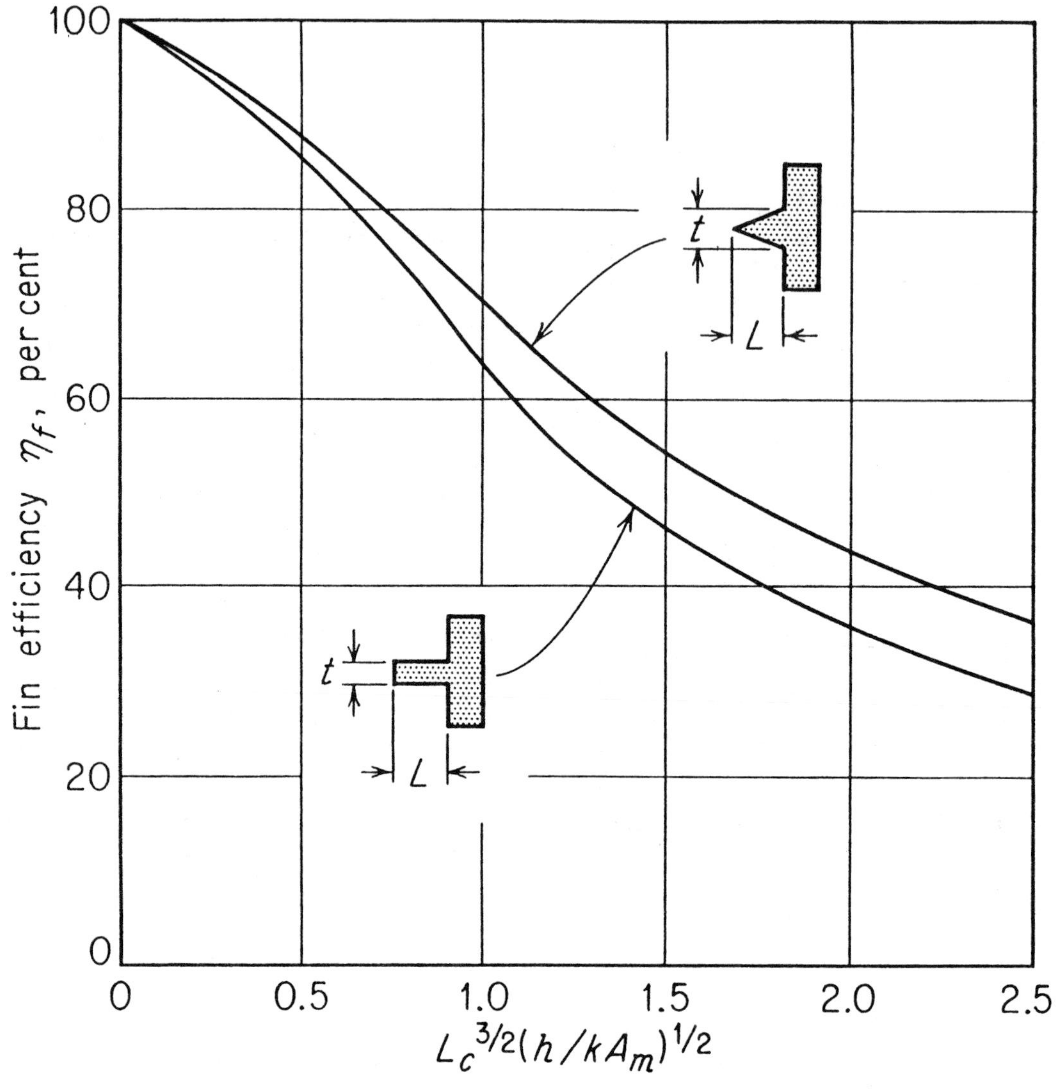

EFFICIENCIES OF RECTANGULAR AND TRIANGULAR FINS. From Holman, J. P.: HEAT TRANSFER, McGraw-Hill Book Co.

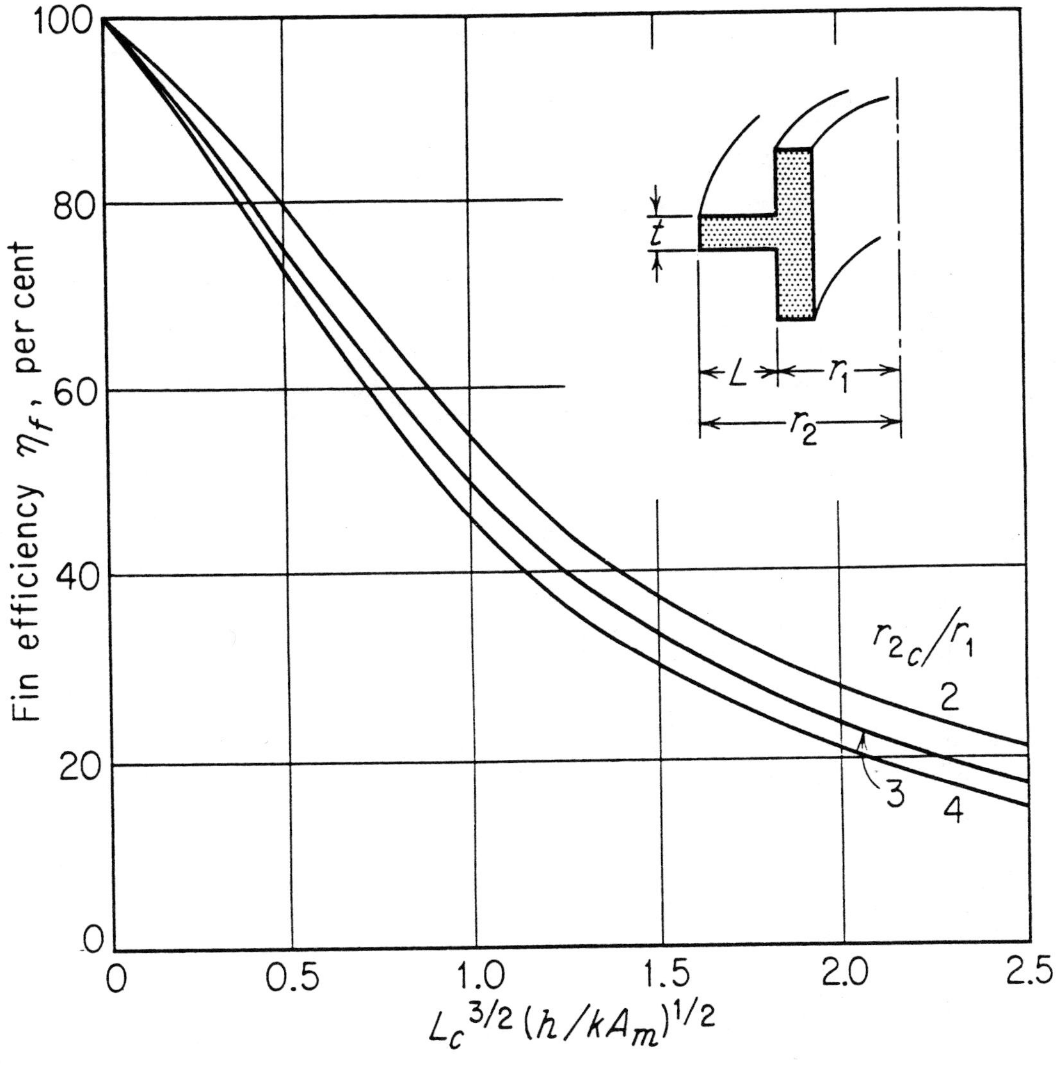

EFFICIENCIES OF CIRCUMFERENTIAL FINS OF RECTANGULAR PROFILE. From Holman, J. P.: HEAT TRANSFER, McGraw-Hill Book Co.

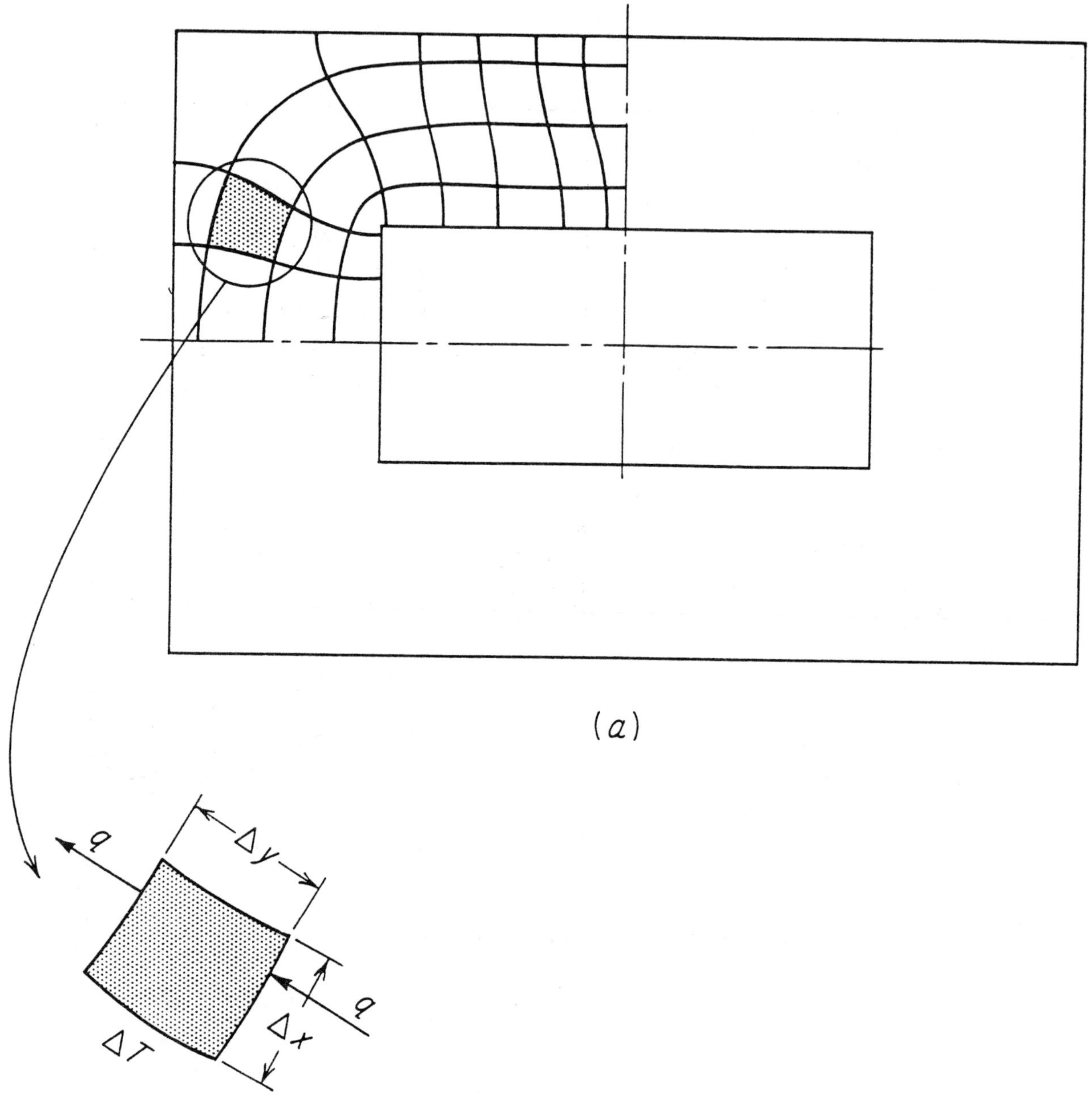

SKETCH SHOWING ELEMENT USED FOR CURVILINEAR-SQUARE ANALYSIS OF TWO-DIMENSIONAL HEAT FLOW. From Holman, J. P.: HEAT TRANSFER, McGraw-Hill Book Co.

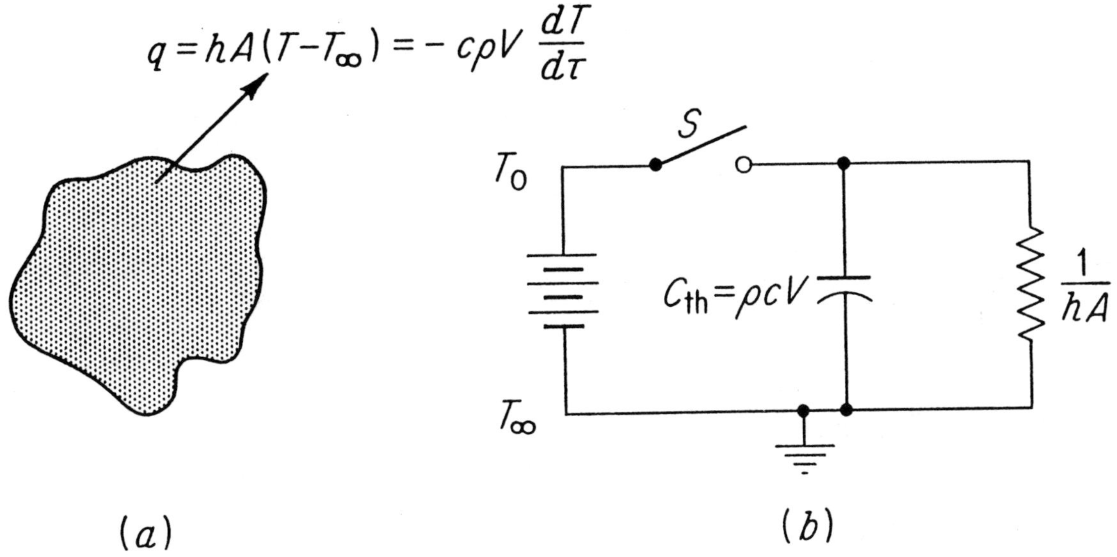

NOMENCLATURE FOR SINGLE-LUMP HEAT-CAPACITY ANALYSIS. From Holman, J. P.: HEAT TRANSFER, McGraw-Hill Book Co.

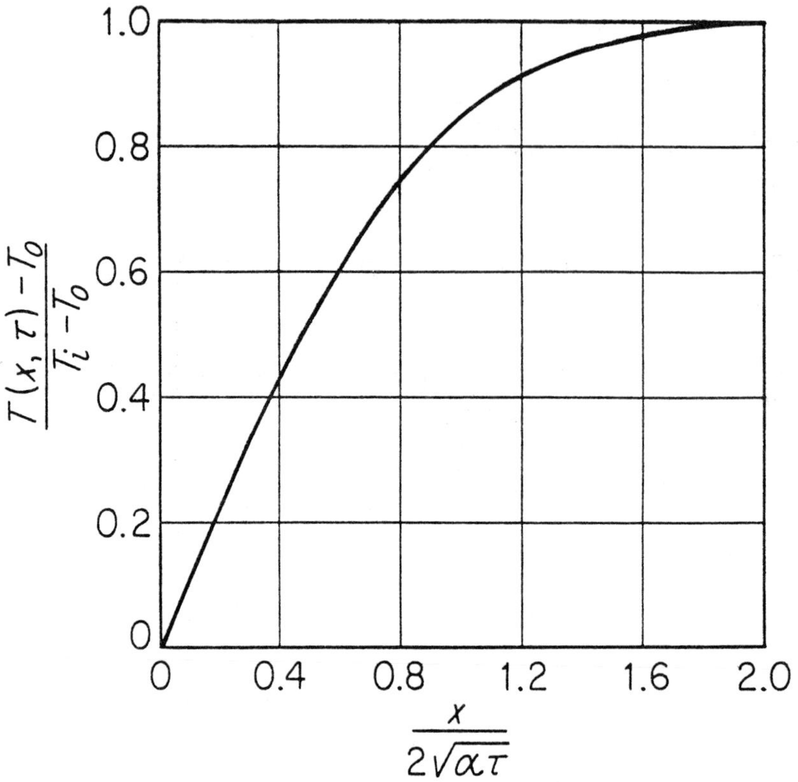

TEMPERATURE DISTRIBUTION IN THE SEMI-INFINITE SOLID. From Holman, J. P.: HEAT TRANSFER, McGraw-Hill Book Co.

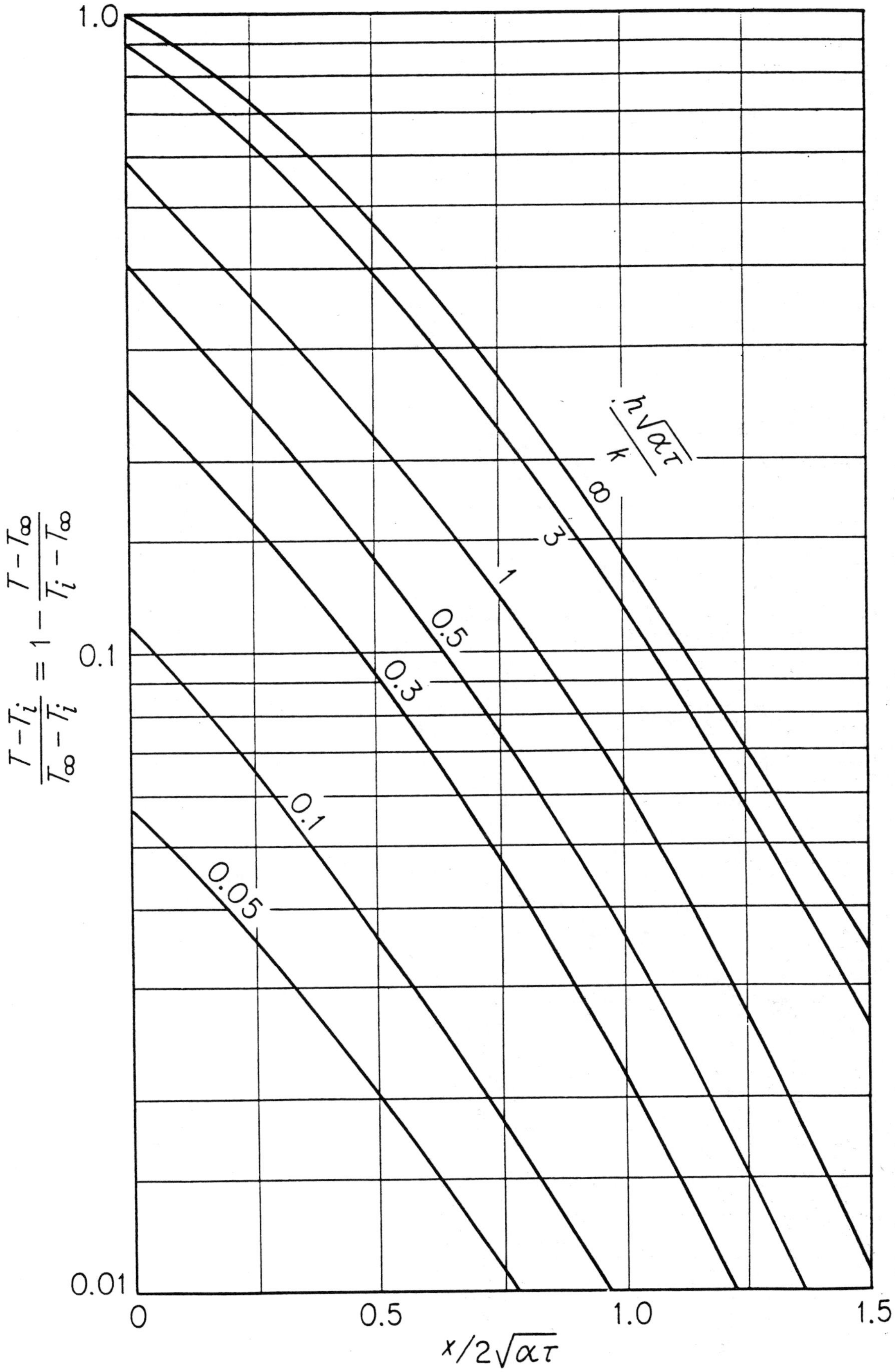

TEMPERATURE DISTRIBUTION IN THE SEMI-INFINITE SOLID WITH CONVECTION BOUNDARY CONDITION. From Holman, J. P.: HEAT TRANSFER, McGraw-Hill Book Co.

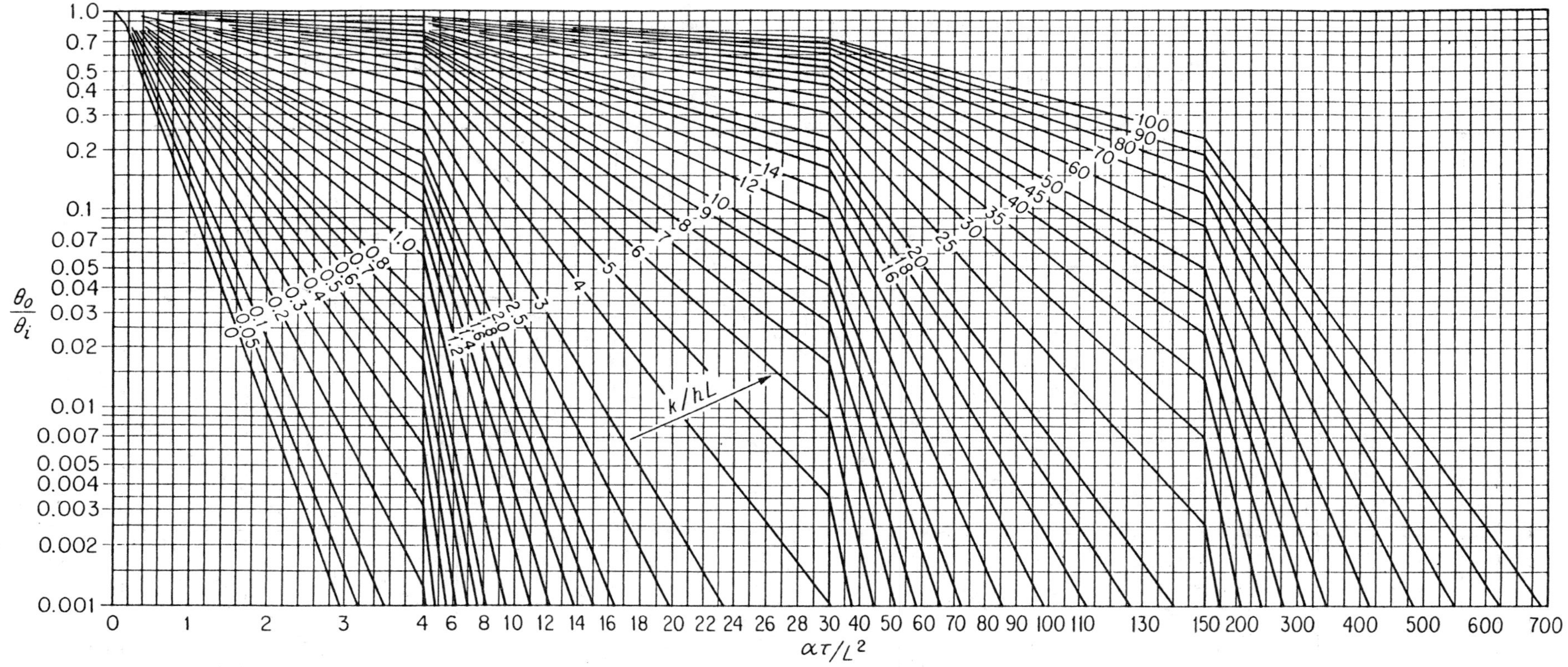

MIDPLANE TEMPERATURE FOR AN INFINITE PLATE OF THICKNESS 2L, FROM HEISLER (2). From Holman, J. P.: HEAT TRANSFER, McGraw-Hill Book Co.

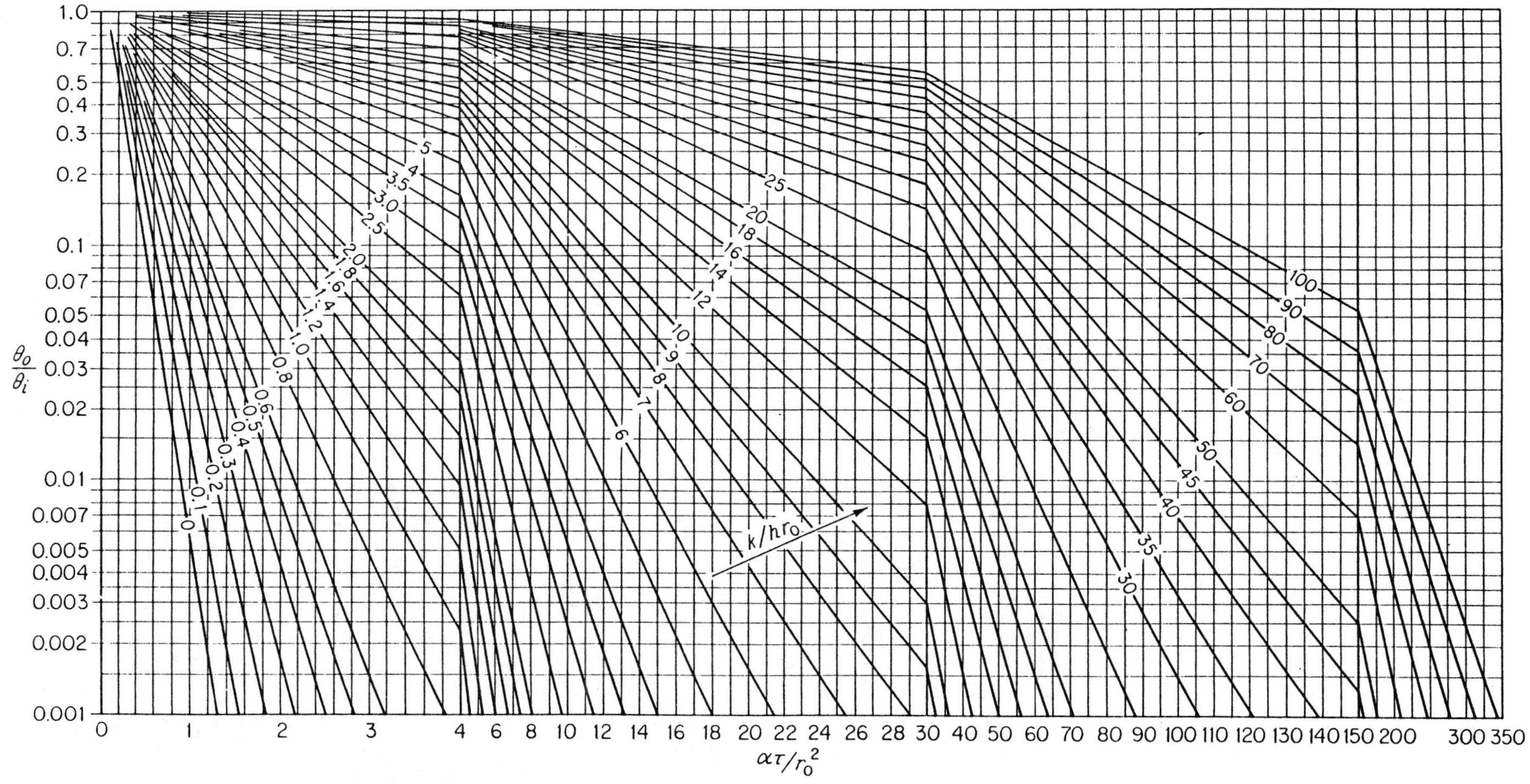

AXIS TEMPERATURE FOR AN INFINITE CYLINDER OF RADIUS r_o , FROM HEISLER (2). From Holman, J. P.: HEAT TRANSFER, McGraw-Hill Book Co.

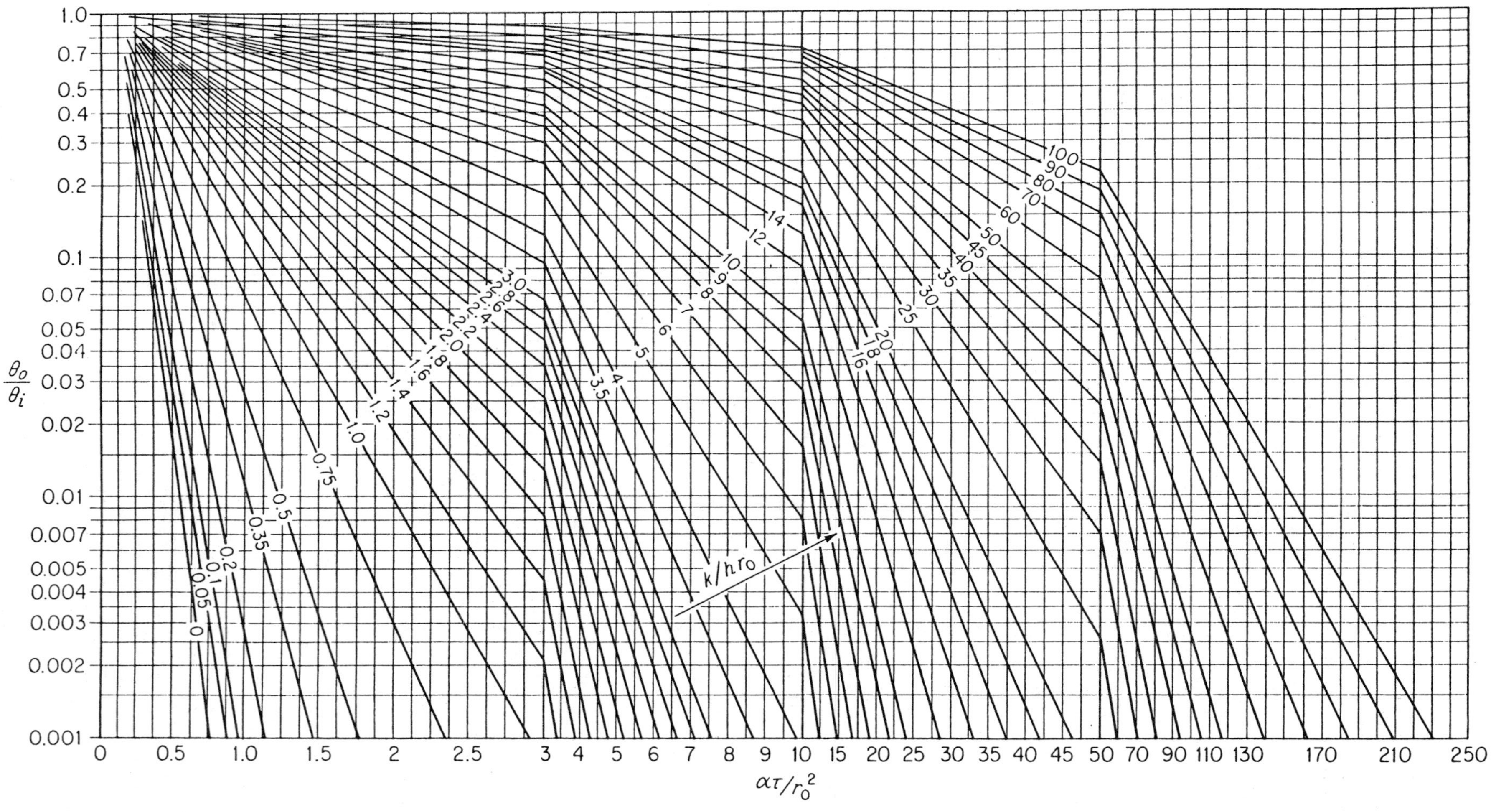

CENTER TEMPERATURE FOR A SPHERE OF RADIUS r_0, FROM HEISLER (2).
From Holman, J. P.: HEAT TRANSFER, McGraw-Hill Book Co.

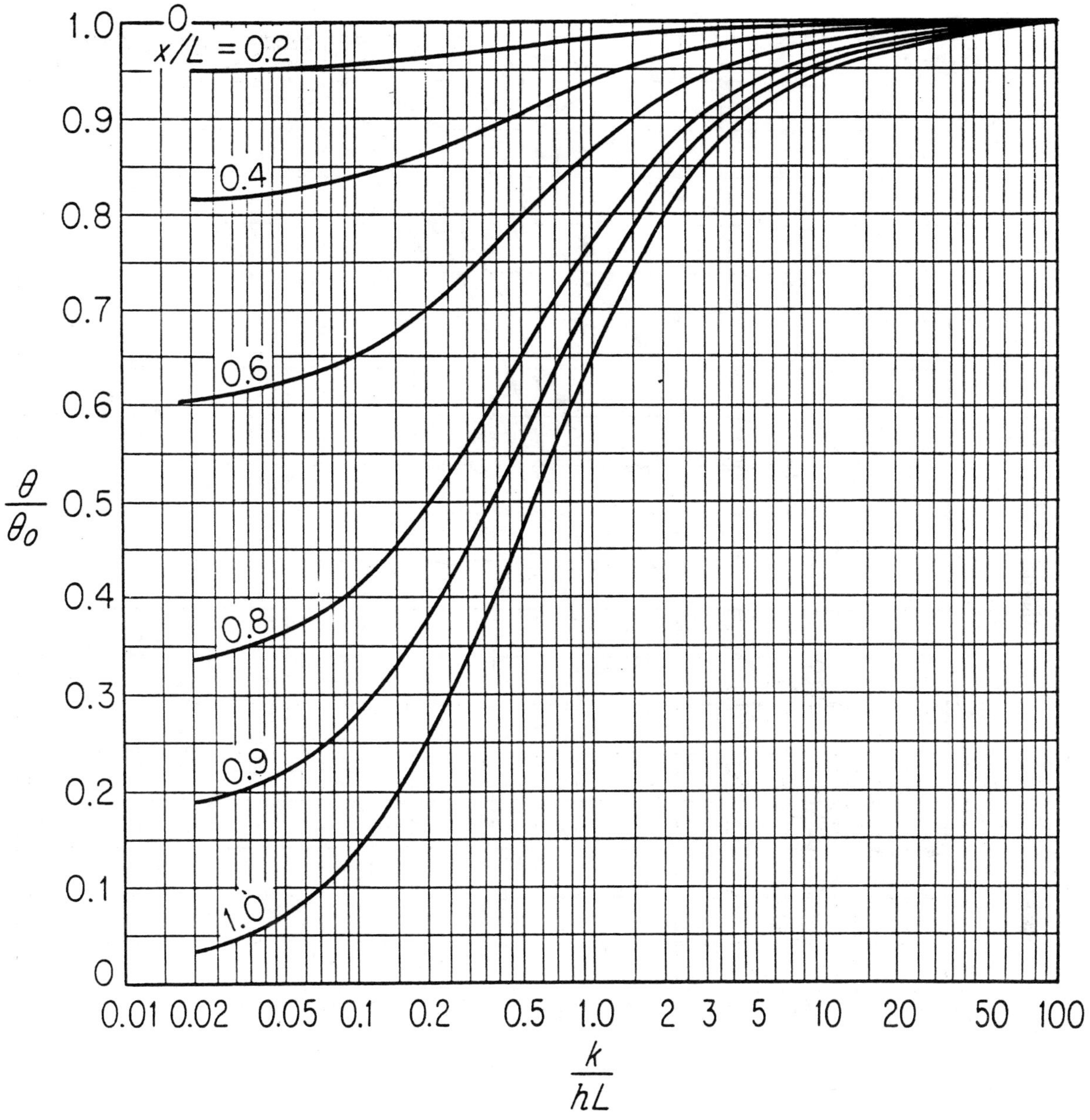

TEMPERATURE AS A FUNCTION OF CENTER TEMPERATURE IN AN INFINITE PLATE OF THICKNESS 2L, FROM HEISLER (2). From Holman, J. P.: HEAT TRANSFER, McGraw-Hill Book Co.

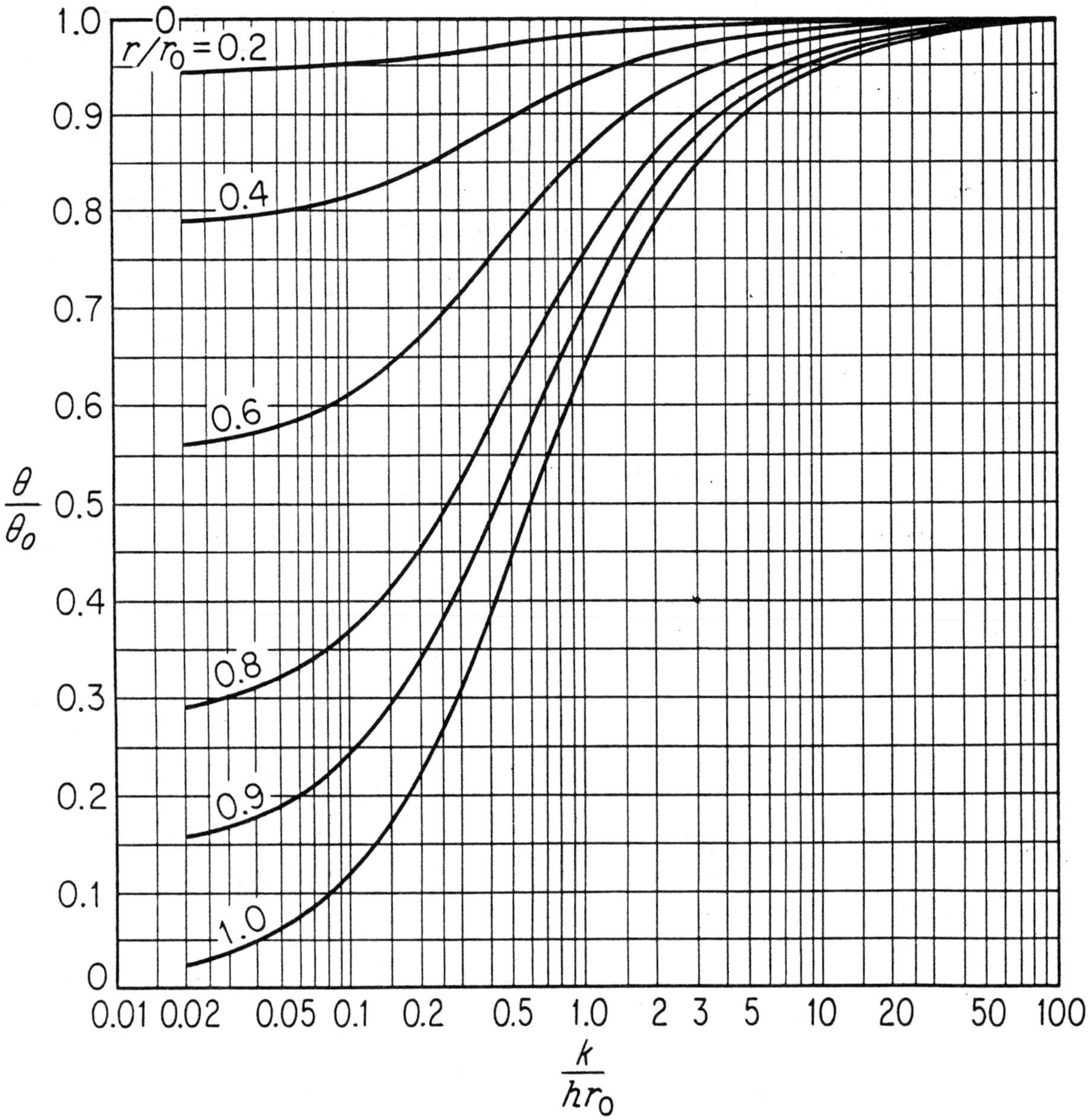

TEMPERATURE AS A FUNCTION OF AXIS TEMPERATURE IN AN INFINITE CYLINDER OF RADIUS r_0 ,FROM HEISLER (2). From Holman, J. P.: HEAT TRANSFER, McGraw-Hill Book Co.

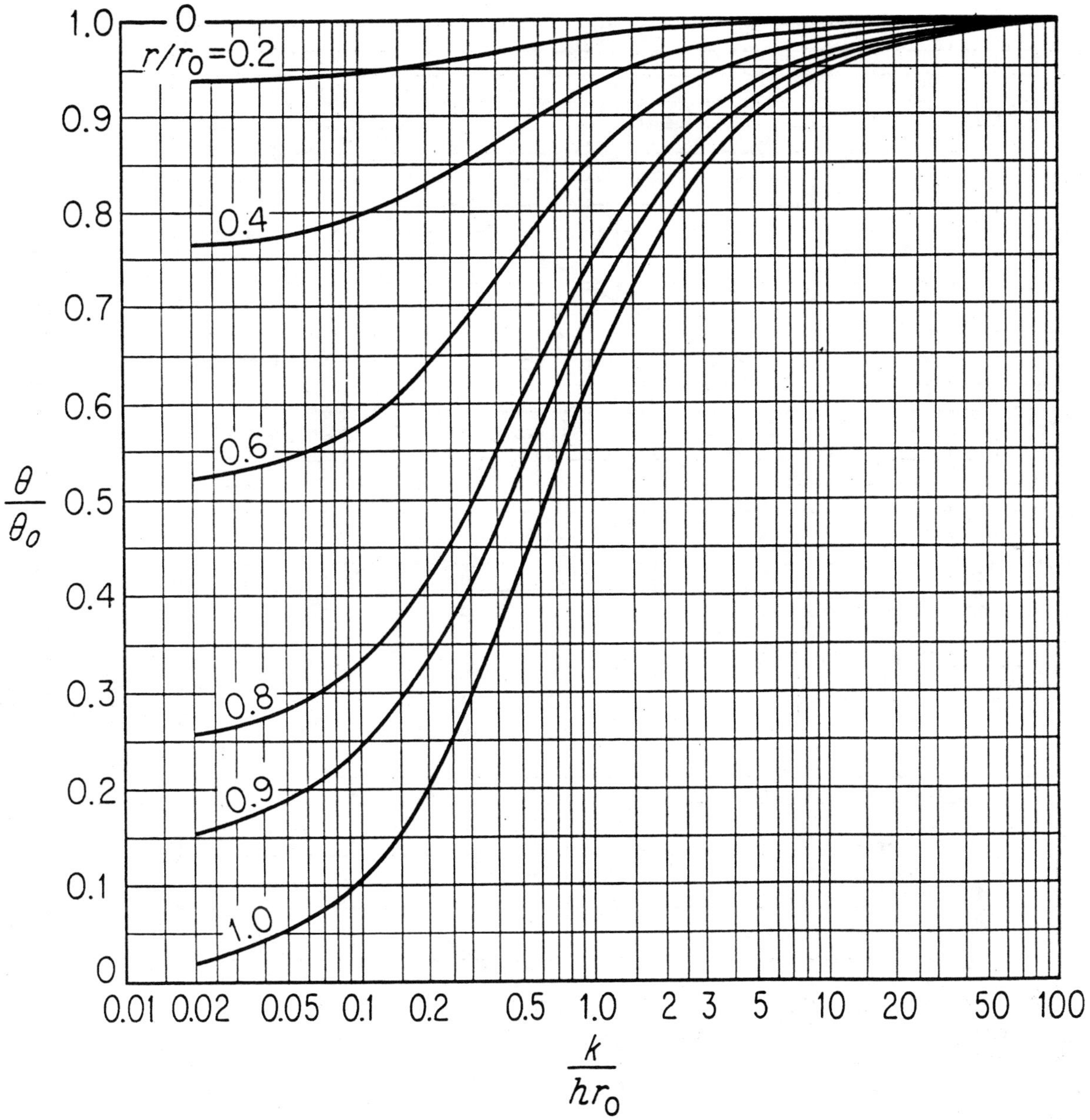

TEMPERATURE AS A FUNCTION OF CENTER TEMPERATURE FOR A SPHERE OF RADIUS r_0 , FROM HEISLER (2). From Holman, J. P.: HEAT TRANSFER, McGraw-Hill Book Co.

Plate: $m = 1$
Cylinder: $m = 2$
Sphere: $m = 3$
Range of these lines:
$\alpha \tau / s^2 > 0.2$
$hs/k < 0.01$

$\dfrac{\theta_i}{\theta_0}$

$\dfrac{m \alpha \tau h}{s k}$

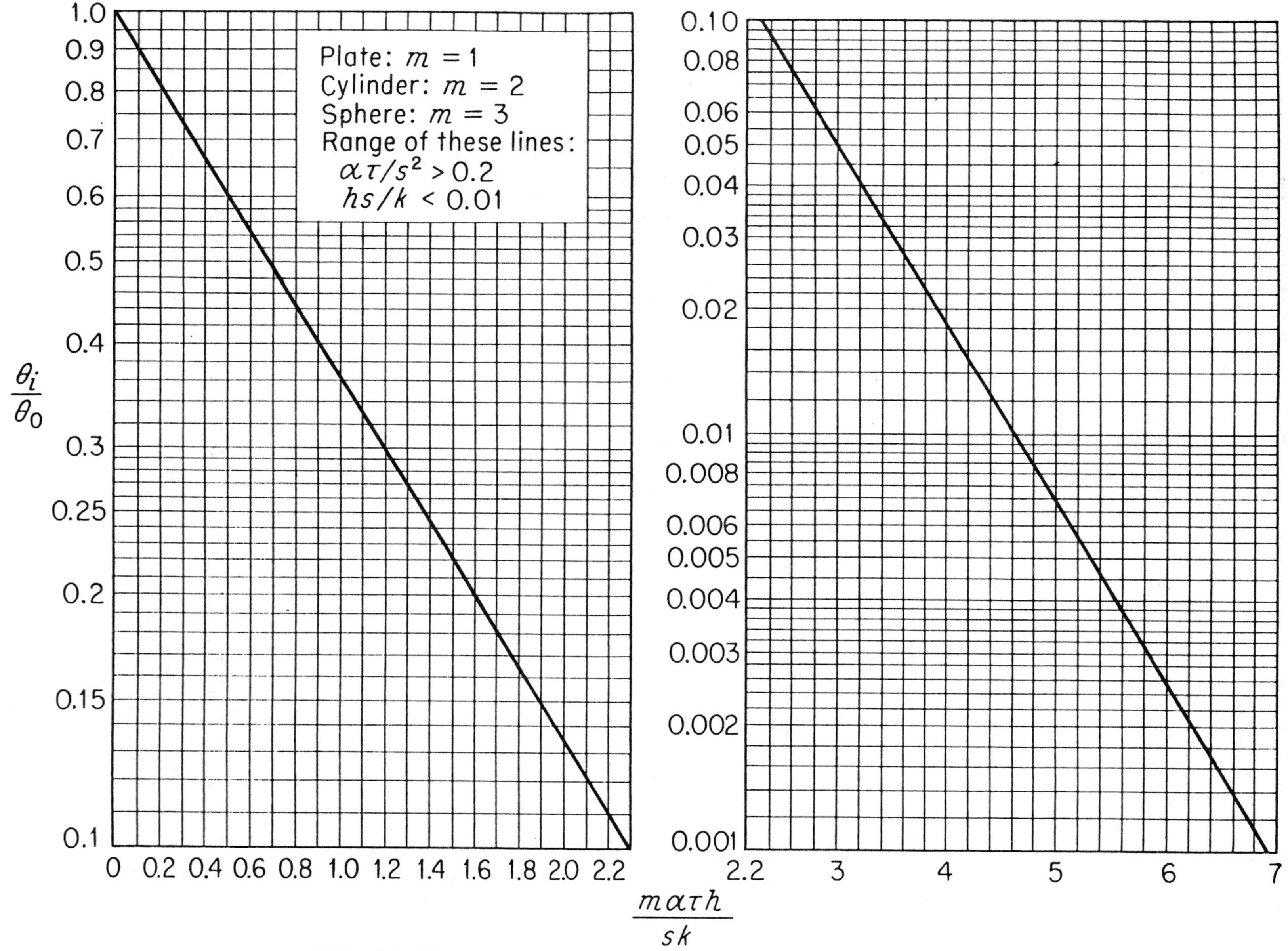

CENTER TEMPERATURES FOR PLATES, CYLINDERS, AND SPHERES FOR SMALL VALUES OF h, ACCORDING TO HEISLER (2). From Holman, J. P.: HEAT TRANSFER, McGraw-Hill Book Co.

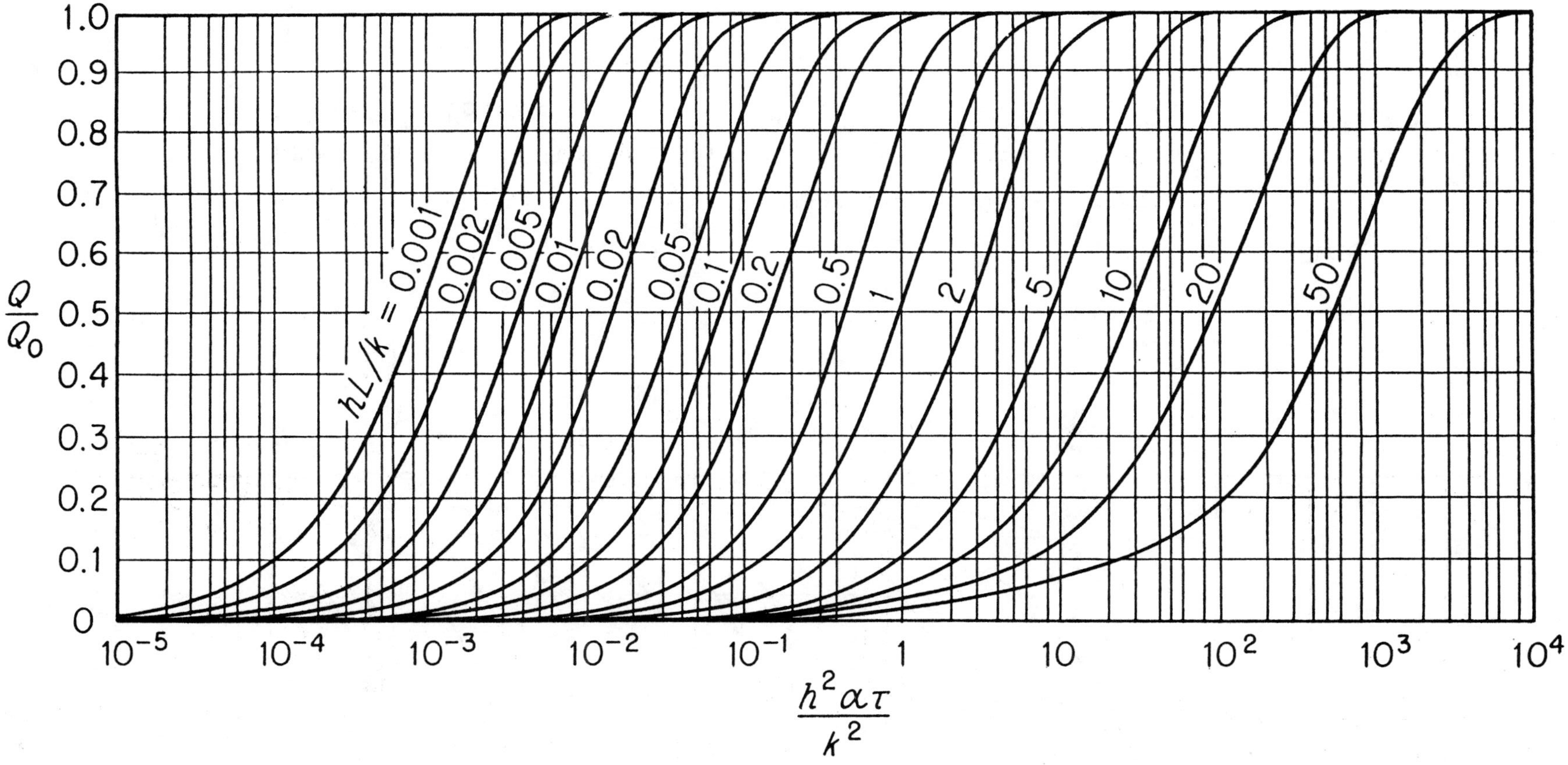

DIMENSIONLESS HEAT LOSS Q/Q_o OF AN INFINITE PLATE OF THICKNESS 2L WITH TIME, FROM GROBER (6). From Holman, J. P.: HEAT TRANSFER, McGraw-Hill Book Co.

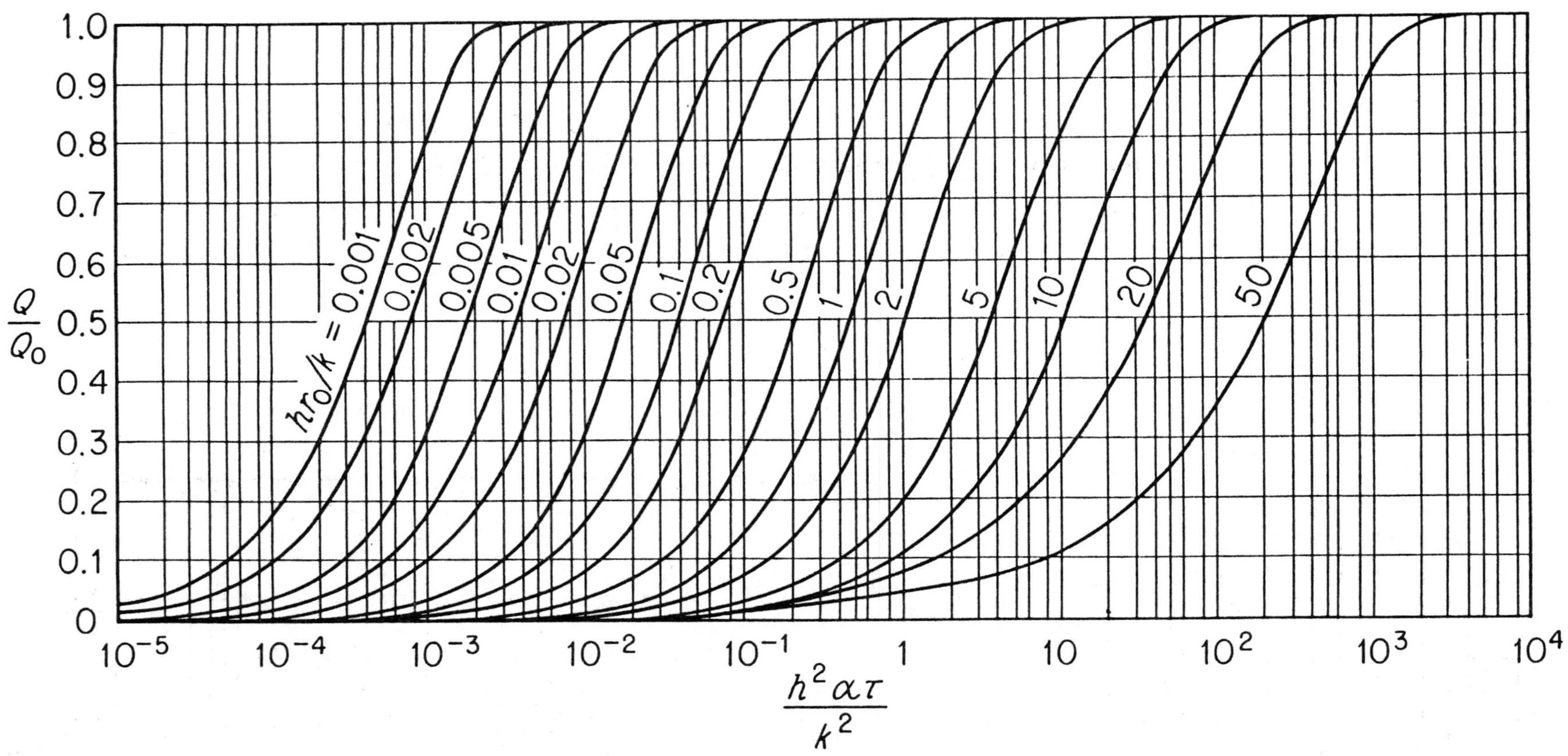

DIMENSIONLESS HEAT LOSS Q/Q_o OF AN INFINITE CYLINDER OF RADIUS r_o WITH TIME, FROM GROBER (6). From Holman, J. P.: HEAT TRANSFER, McGraw-Hill Book Co.

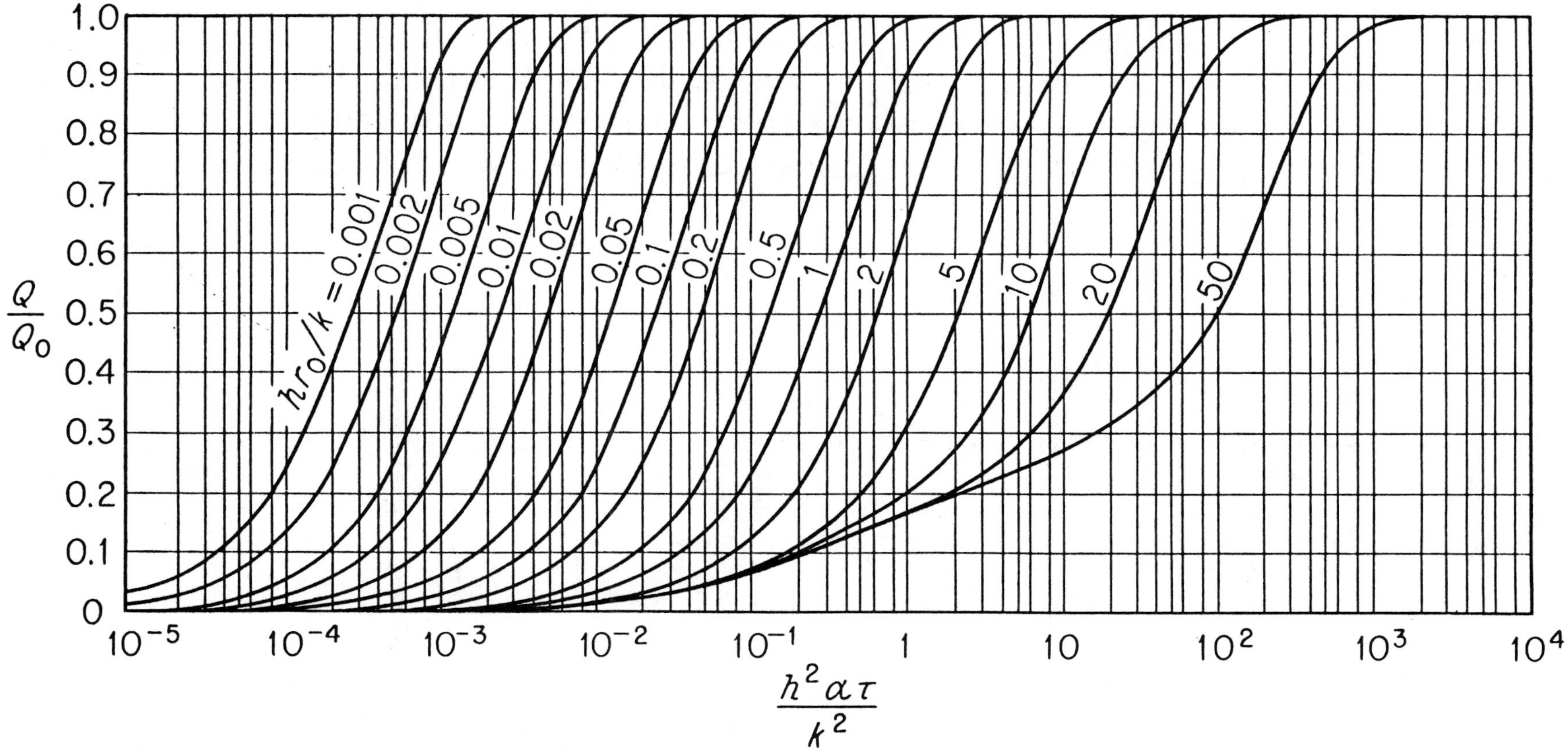

DIMENSIONLESS HEAT LOSS Q/Q_0 OF A SPHERE OF RADIUS r_0 WITH TIME, FROM GROBER (6). From Holman, J. P.: HEAT TRANSFER, McGraw-Hill Book Co.

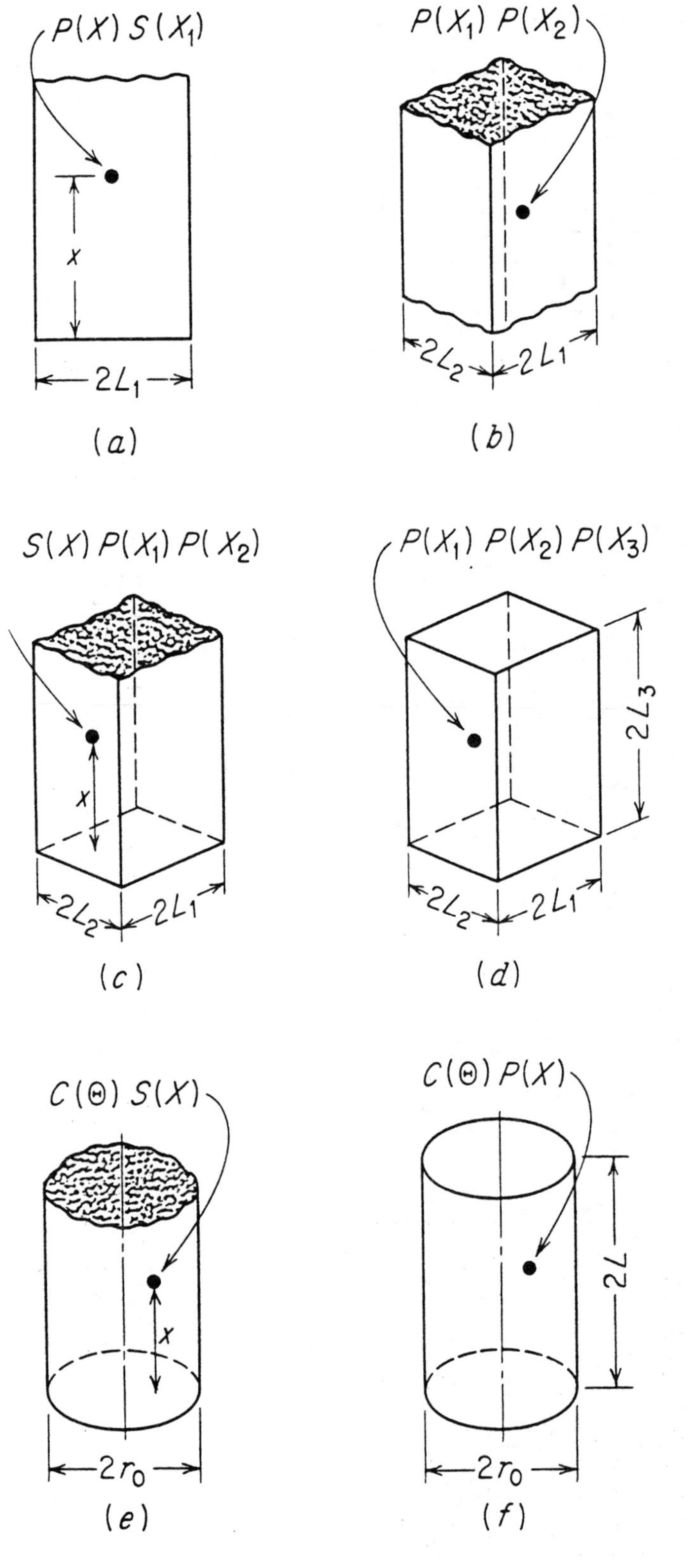

PRODUCT SOLUTIONS FOR TEMPERATURES IN MULTIDIMENSIONAL SYSTEMS. (A) SEMI-INFINITE PLATE; (B) INFINITE RECTANGULAR BAR; (C) SEMI-INFINITE RECTANGULAR BAR; (D) RECTANGULAR PARALLELEPIPED; (E) SEMI-INFINITE CYLINDER; (F) SHORT CYLINDER. From Holman, J. P.: HEAT TRANSFER, McGraw-Hill Book Co.

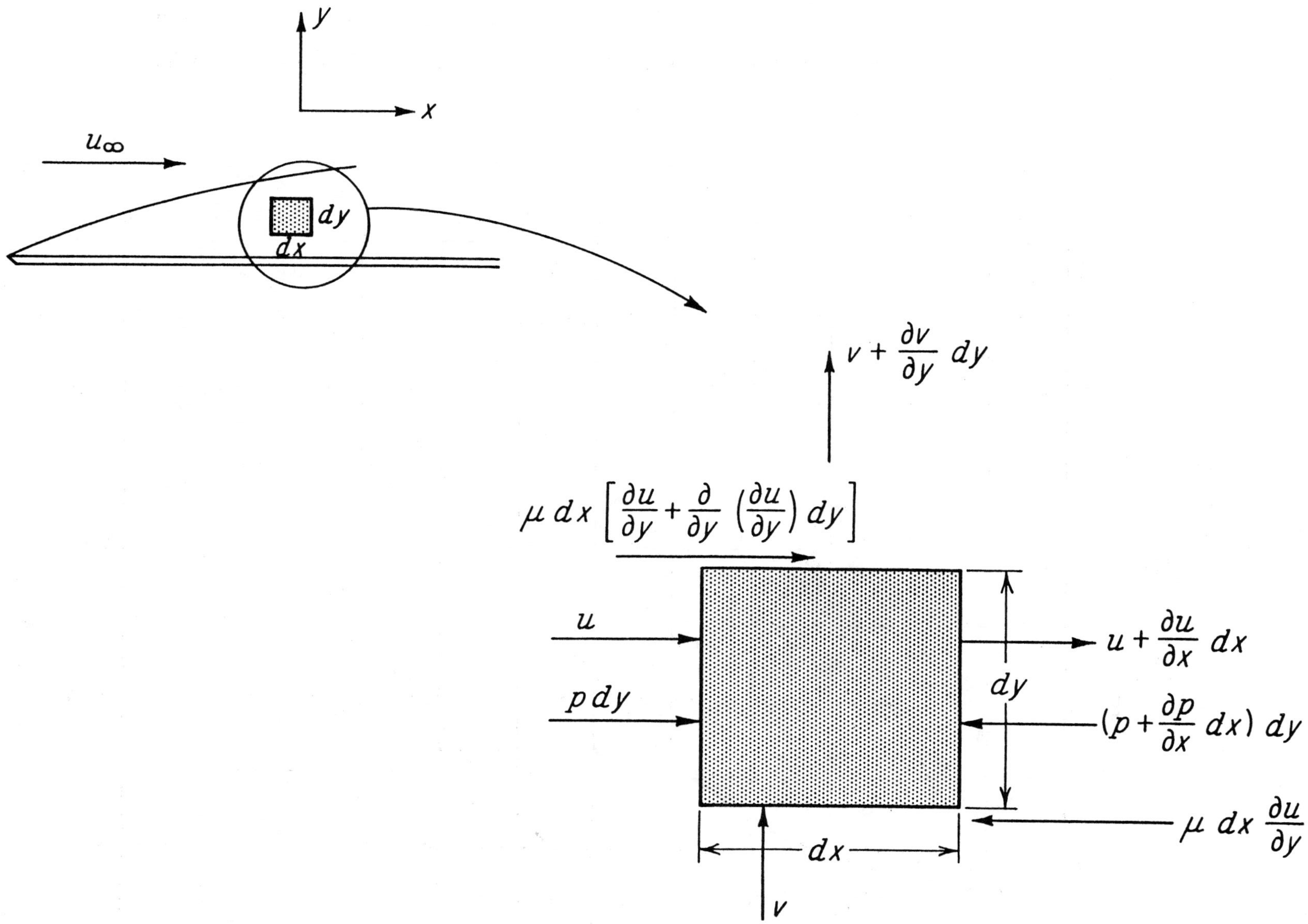

ELEMENTAL CONTROL VOLUME FOR FORCE BALANCE ON LAMINAR BOUNDARY LAYER.
From Holman, J. P.: HEAT TRANSFER, McGraw-Hill Book Co.

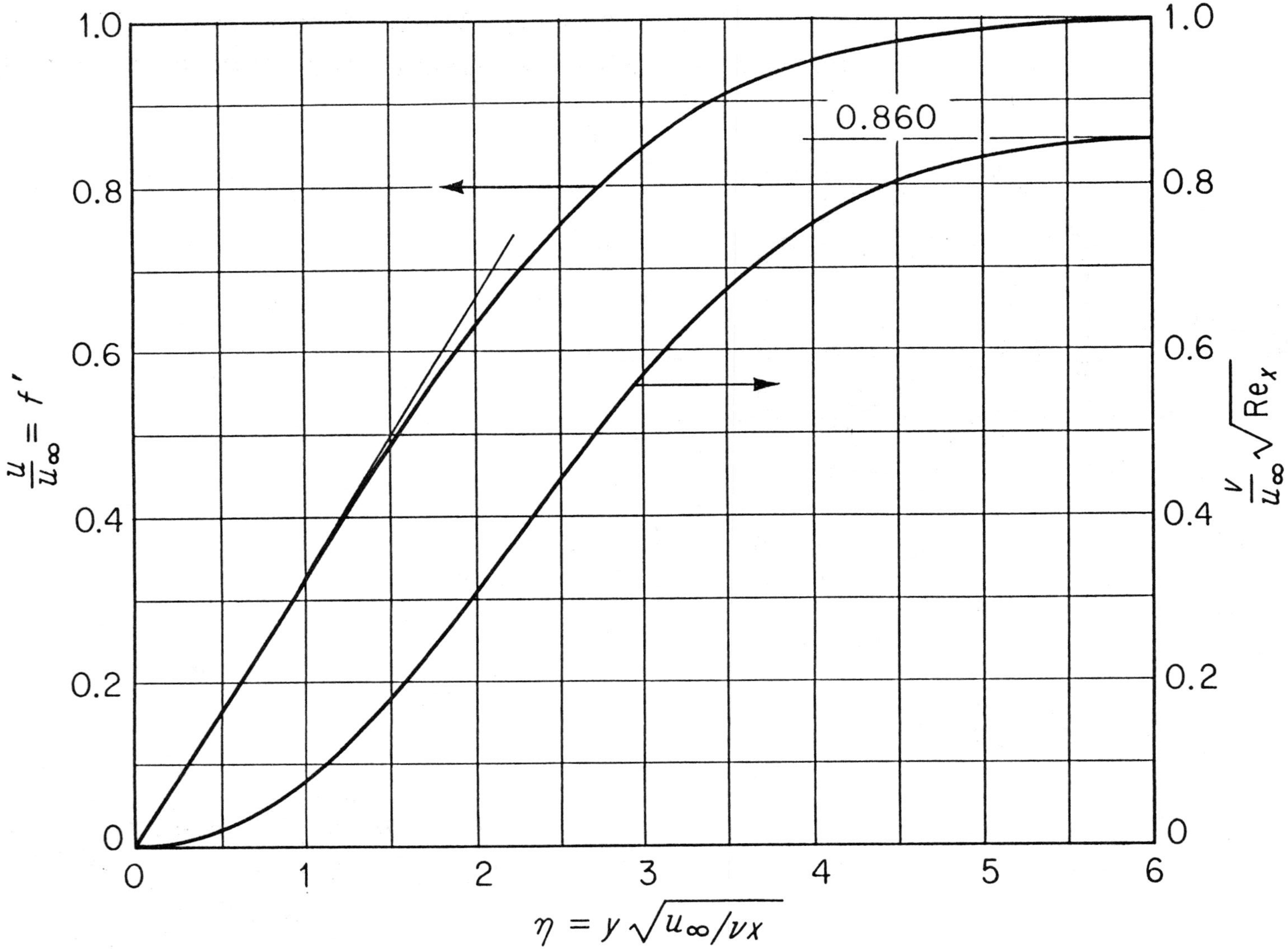

VELOCITY PROFILES IN LAMINAR BOUNDARY LAYER. From Holman, J. P.: HEAT TRANSFER, McGraw-Hill Book Co.

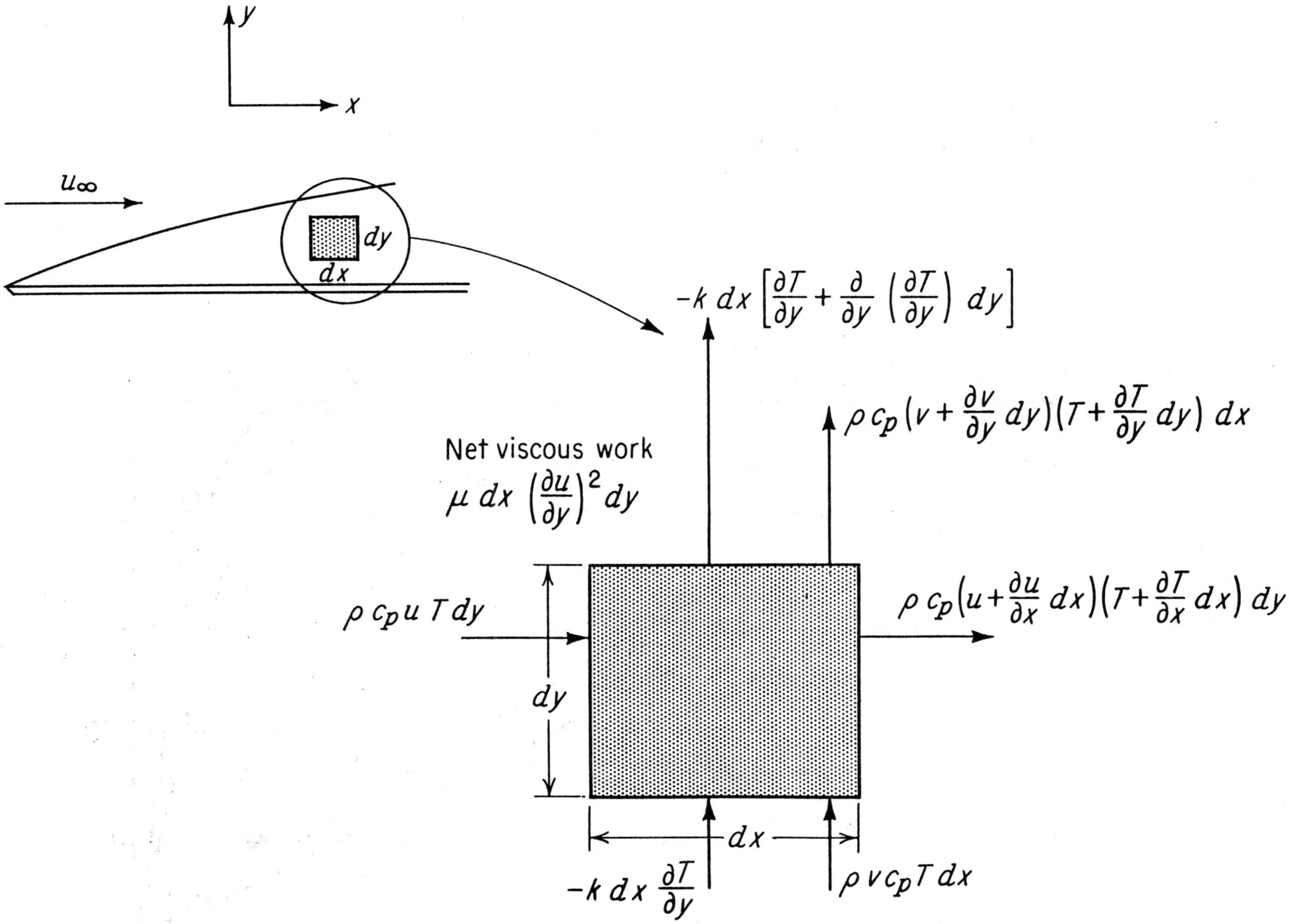

ELEMENTAL CONTROL VOLUME FOR ENERGY ANALYSIS OF LAMINAR BOUNDARY LAYER. From Holman, J. P.: HEAT TRANSFER, McGraw-Hill Book Co.

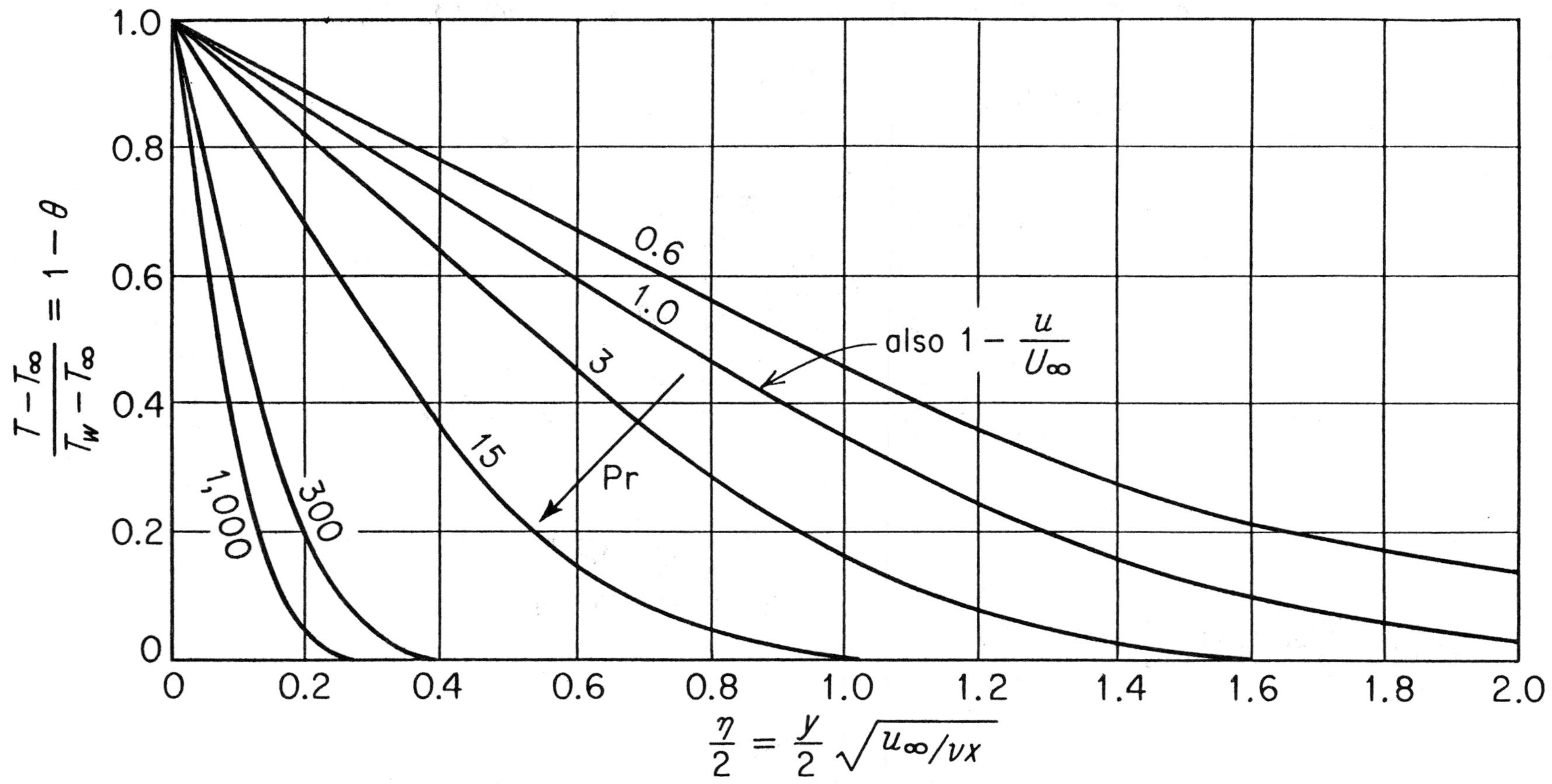

TEMPERATURE PROFILES IN LAMINAR BOUNDARY LAYER WITH ISOTHERMAL WALL.

From Holman, J. P.: HEAT TRANSFER, McGraw-Hill Book Co.

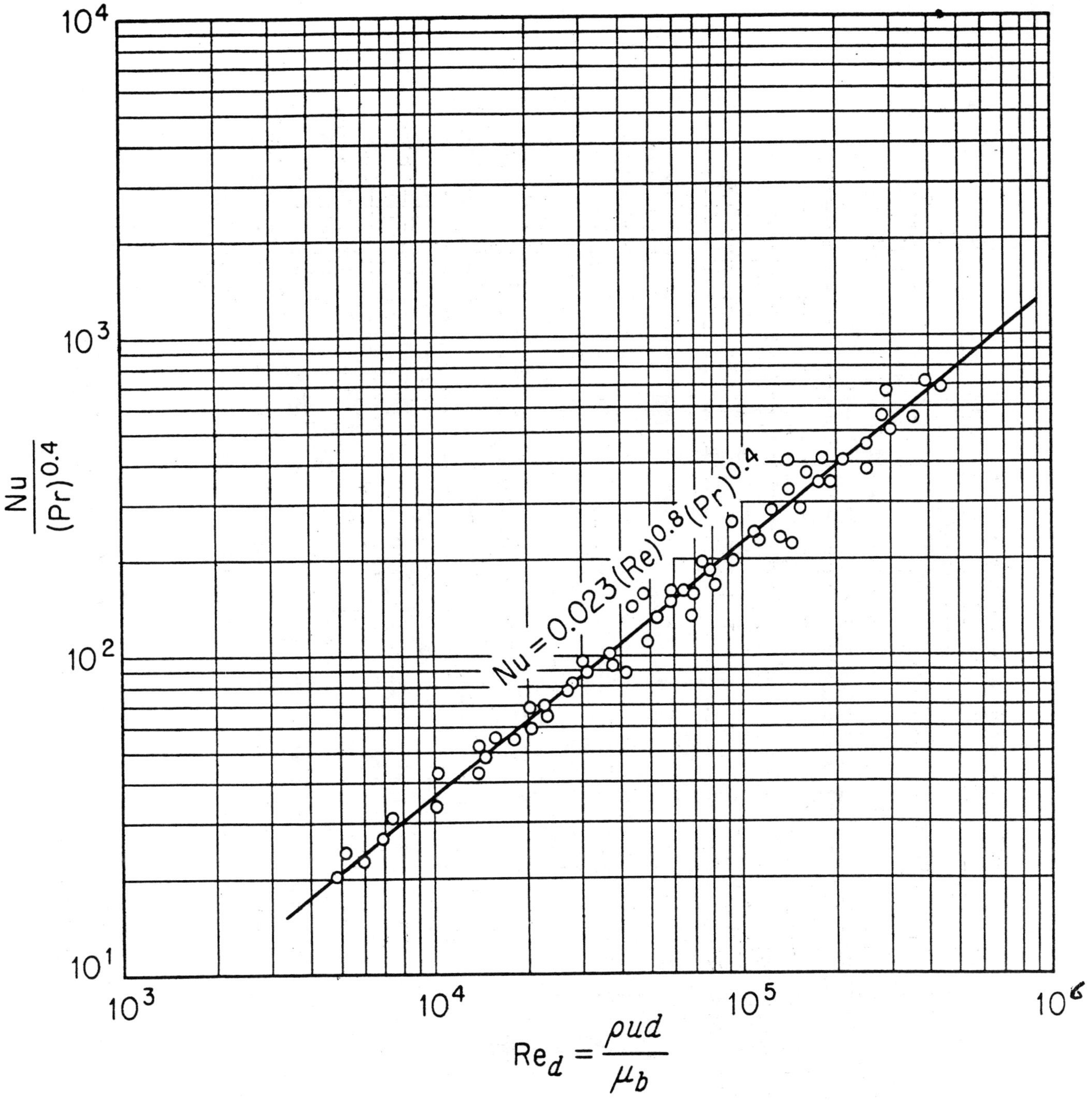

TYPICAL DATA CORRELATION FOR FORCED CONVECTION IN SMOOTH TUBES, TURBULENT FLOW. From Holman, J. P.: HEAT TRANSFER, McGraw-Hill Book Co.

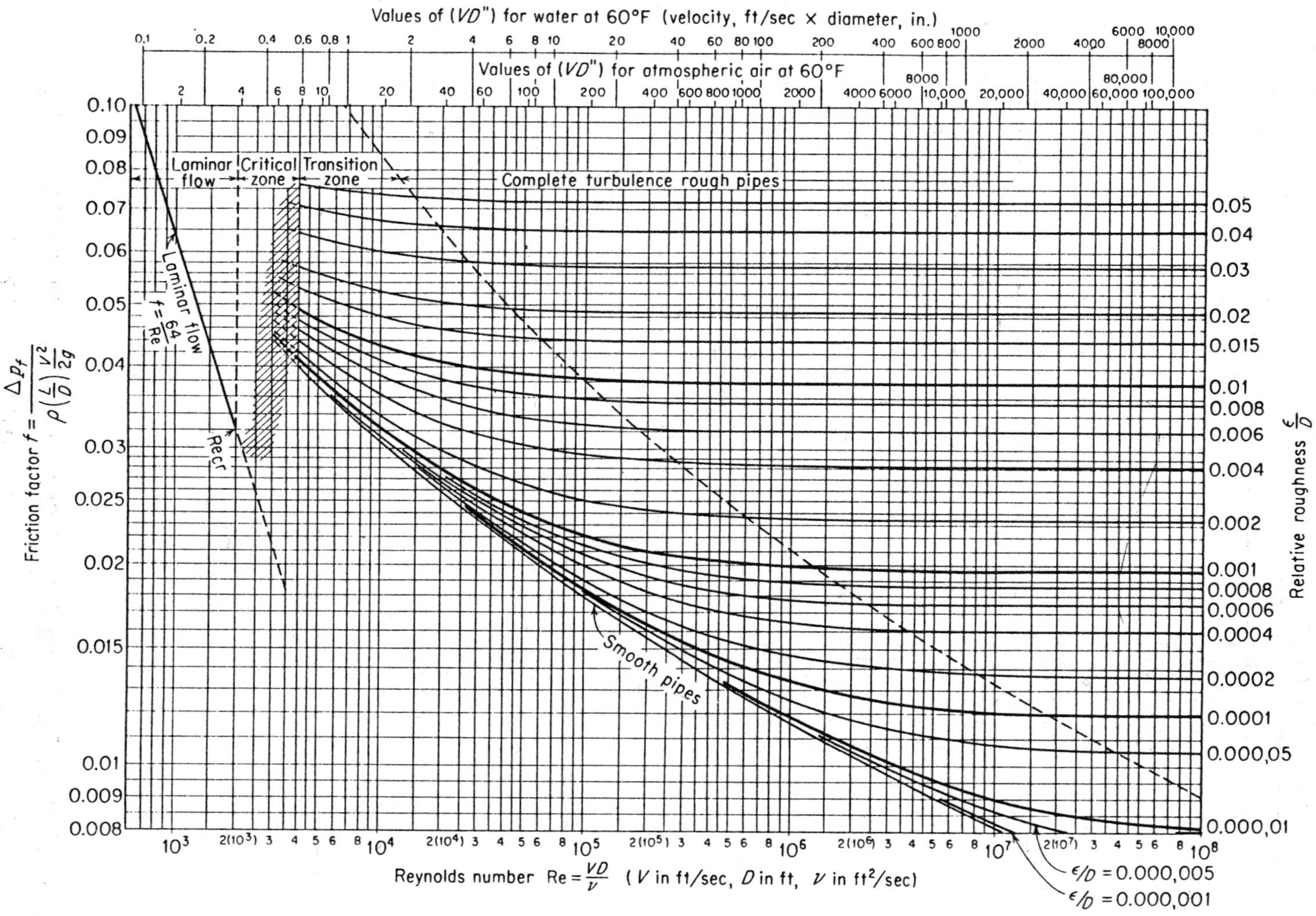

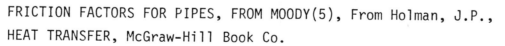

FRICTION FACTORS FOR PIPES, FROM MOODY(5), From Holman, J.P., HEAT TRANSFER, McGraw-Hill Book Co.

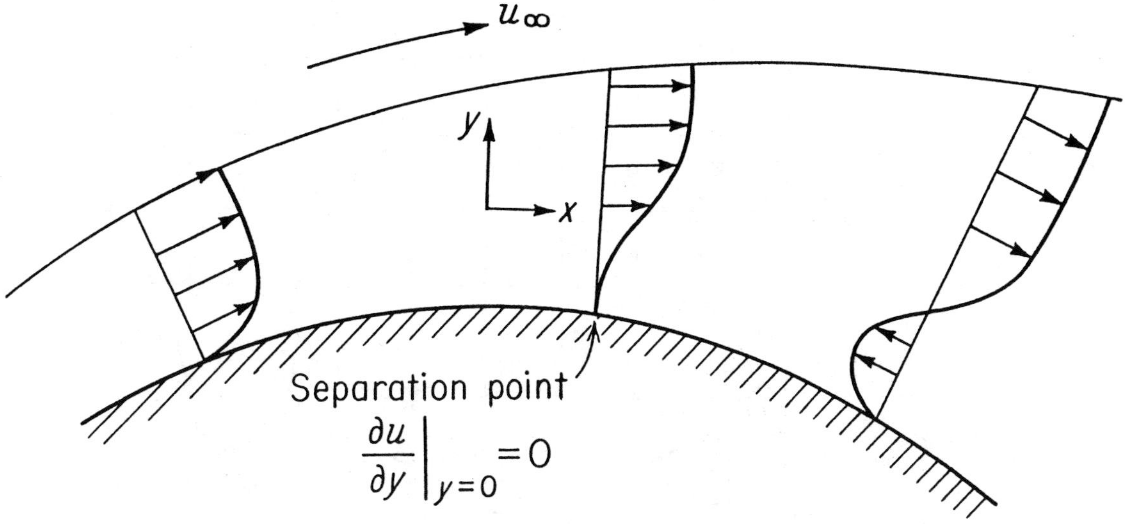

FLOW SEPARATION ON A CYLINDER
IN CROSSFLOW

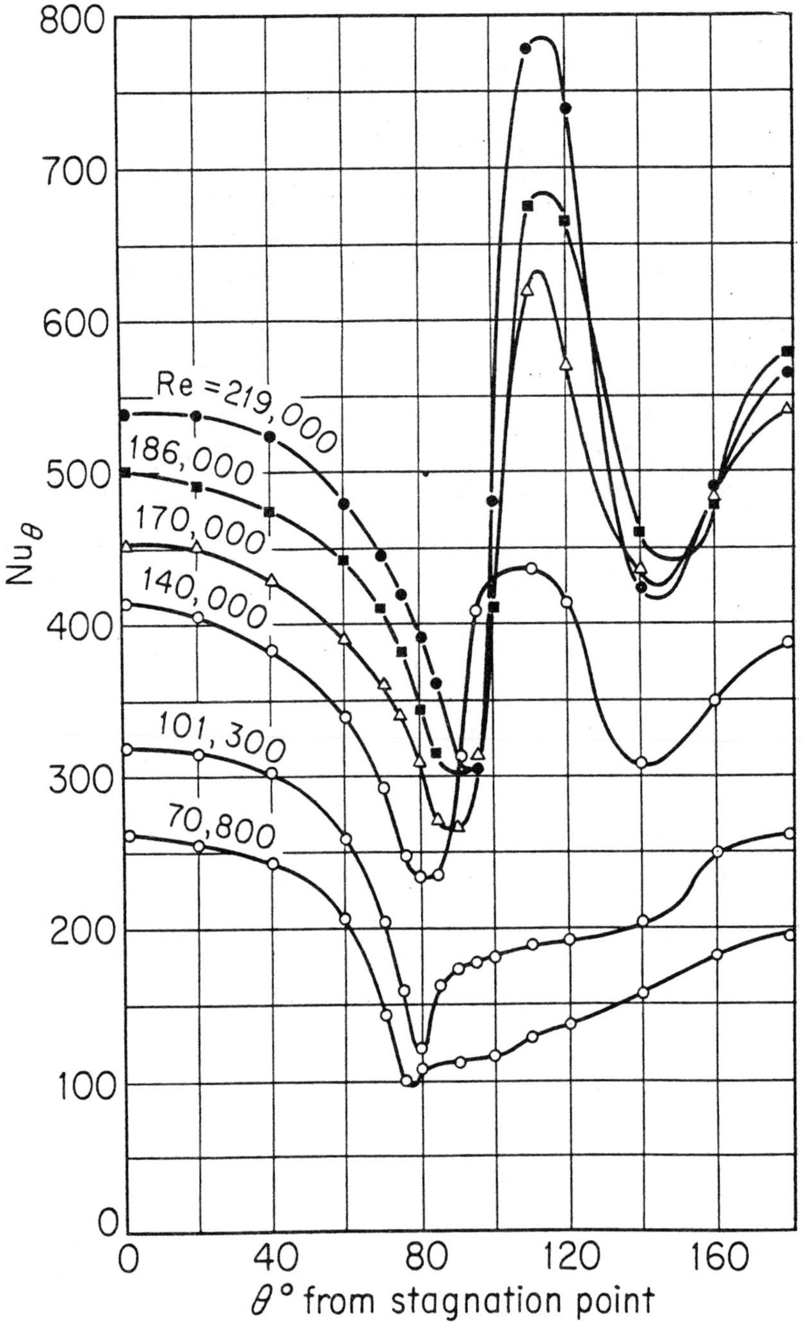

LOCAL NUSSELT NUMBER FOR HEAT TRANSFER FROM A CYLINDER IN CROSS FLOW, FROM GIEDT (7). From Holman, J. P.: HEAT TRANSFER, McGraw-Hill Book Co.

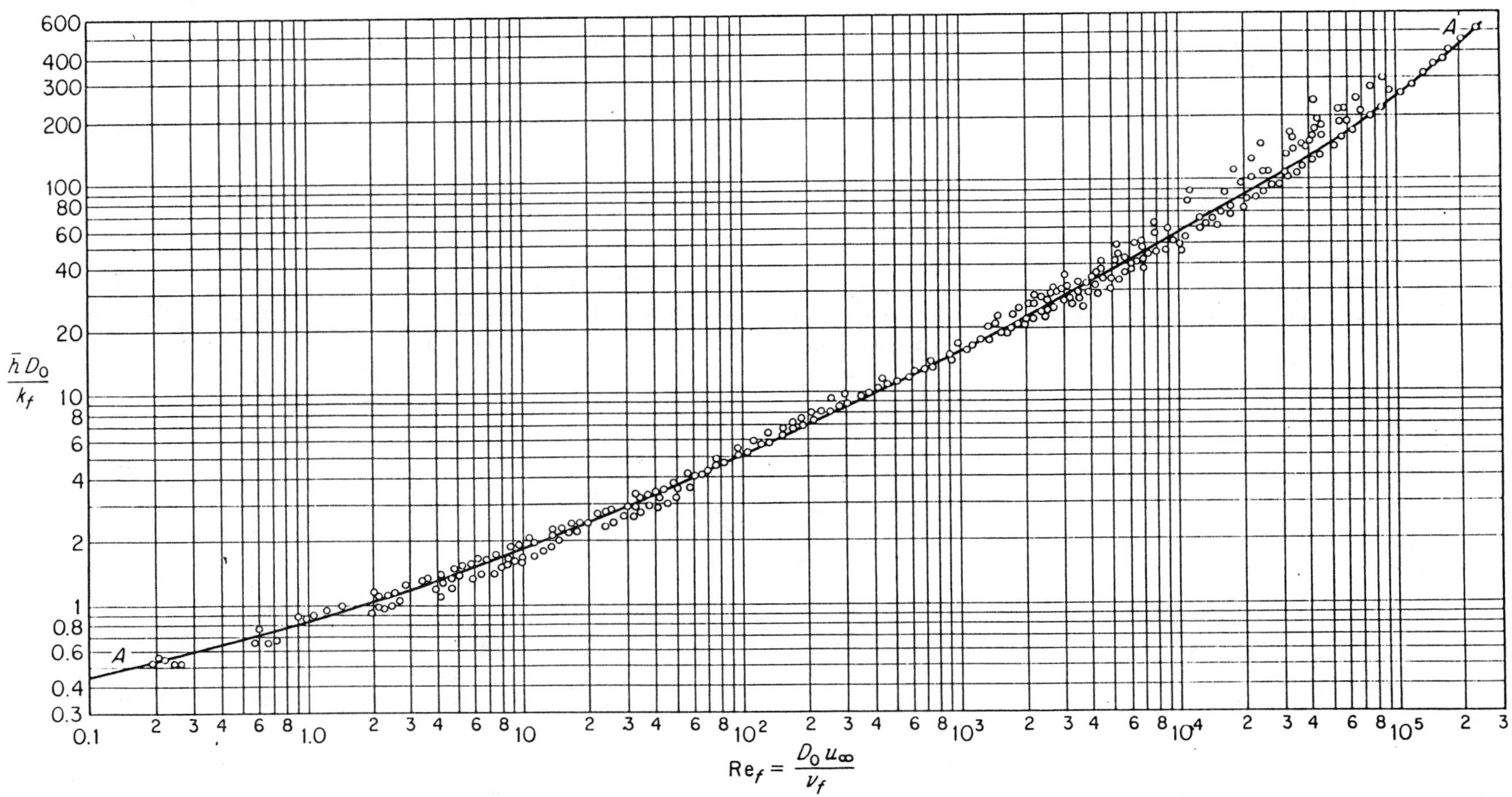

DATA FOR HEATING AND COOLING OF AIR FLOWING NORMAL TO SINGLE CYLINDERS, FROM McADAMS (10). From Holman, J. P.: HEAT TRANSFER, McGraw-Hill Book Co.

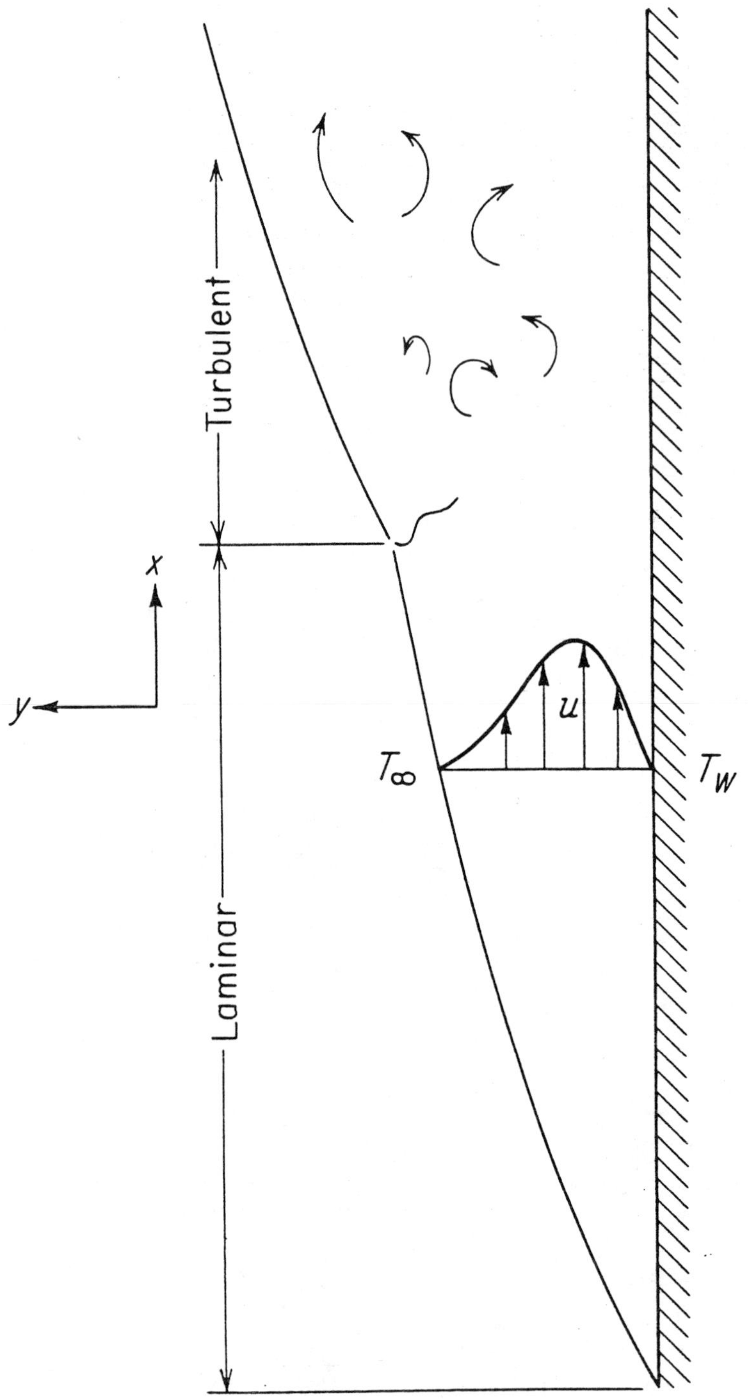

BOUNDARY LAYER ON A VERTICAL FLAT PLATE. From Holman, J. P.: HEAT TRANSFER, McGraw-Hill Book Co.

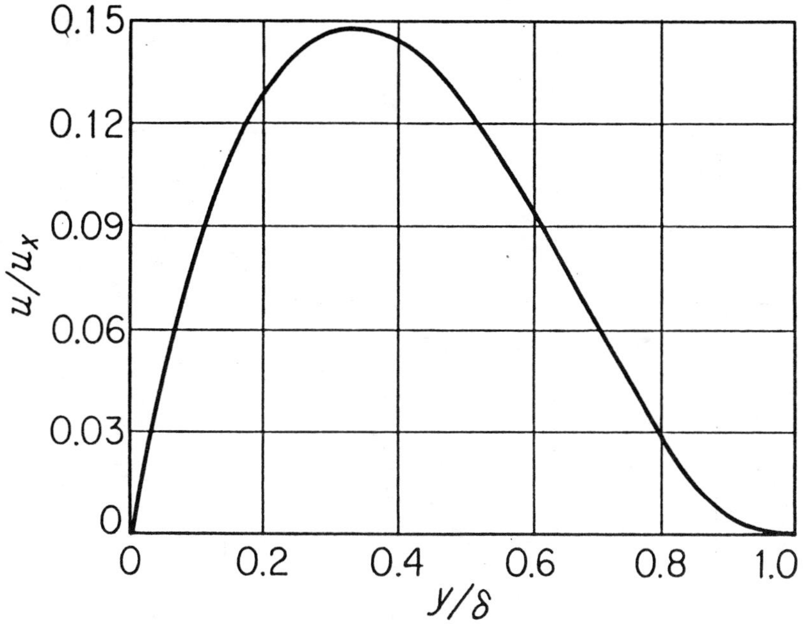

FREE CONVECTION VELOCITY PROFILE from Holman, J.P.:
HEAT TRANSFER, McGraw-Hill Book Co.

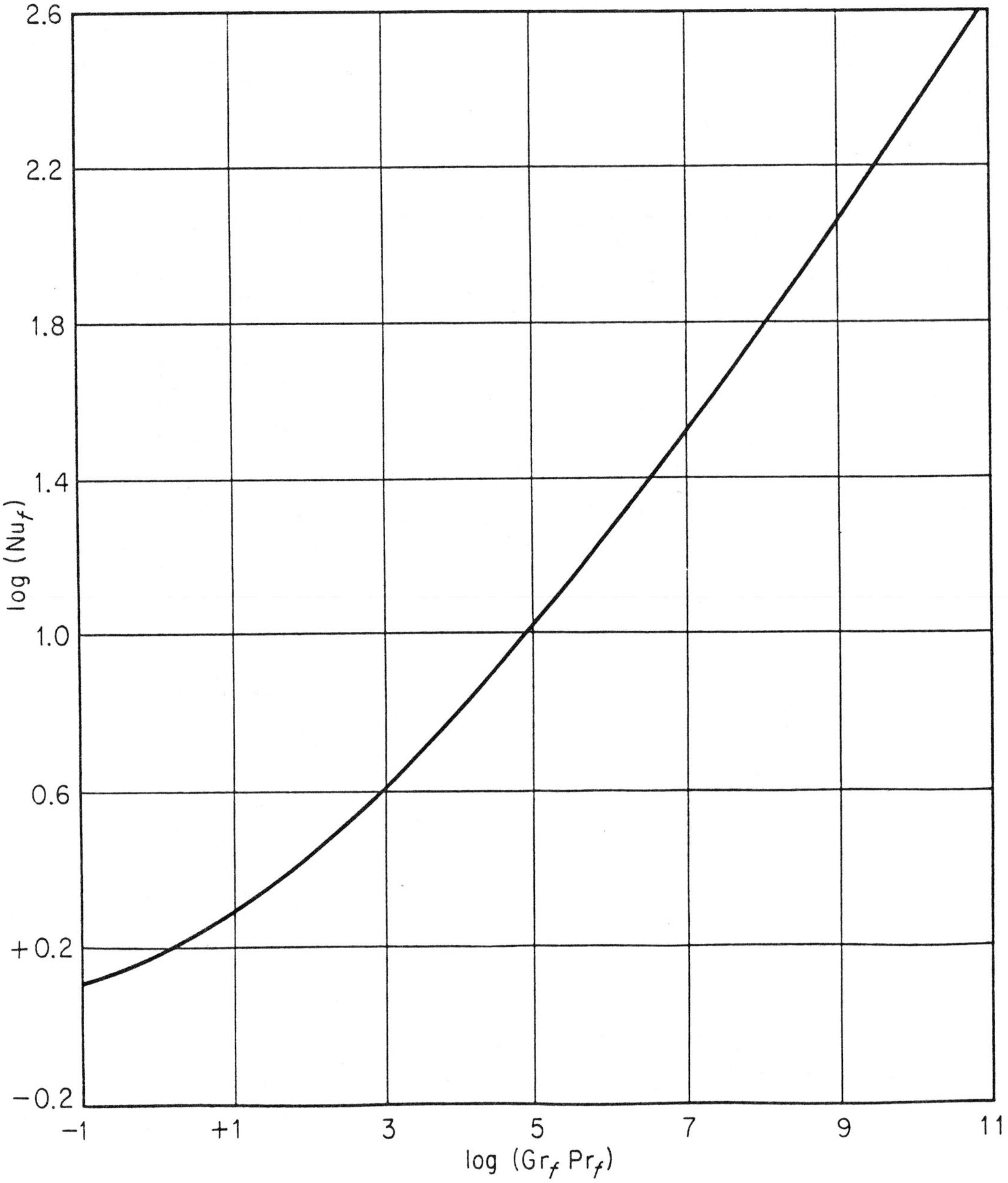

FREE CONVECTION FROM HEATED VERTICAL PLATES, ACCORDING TO McAdams(4). From Holman, J.P. : HEAT TRANSFER, McGraw-Hill Book Co.

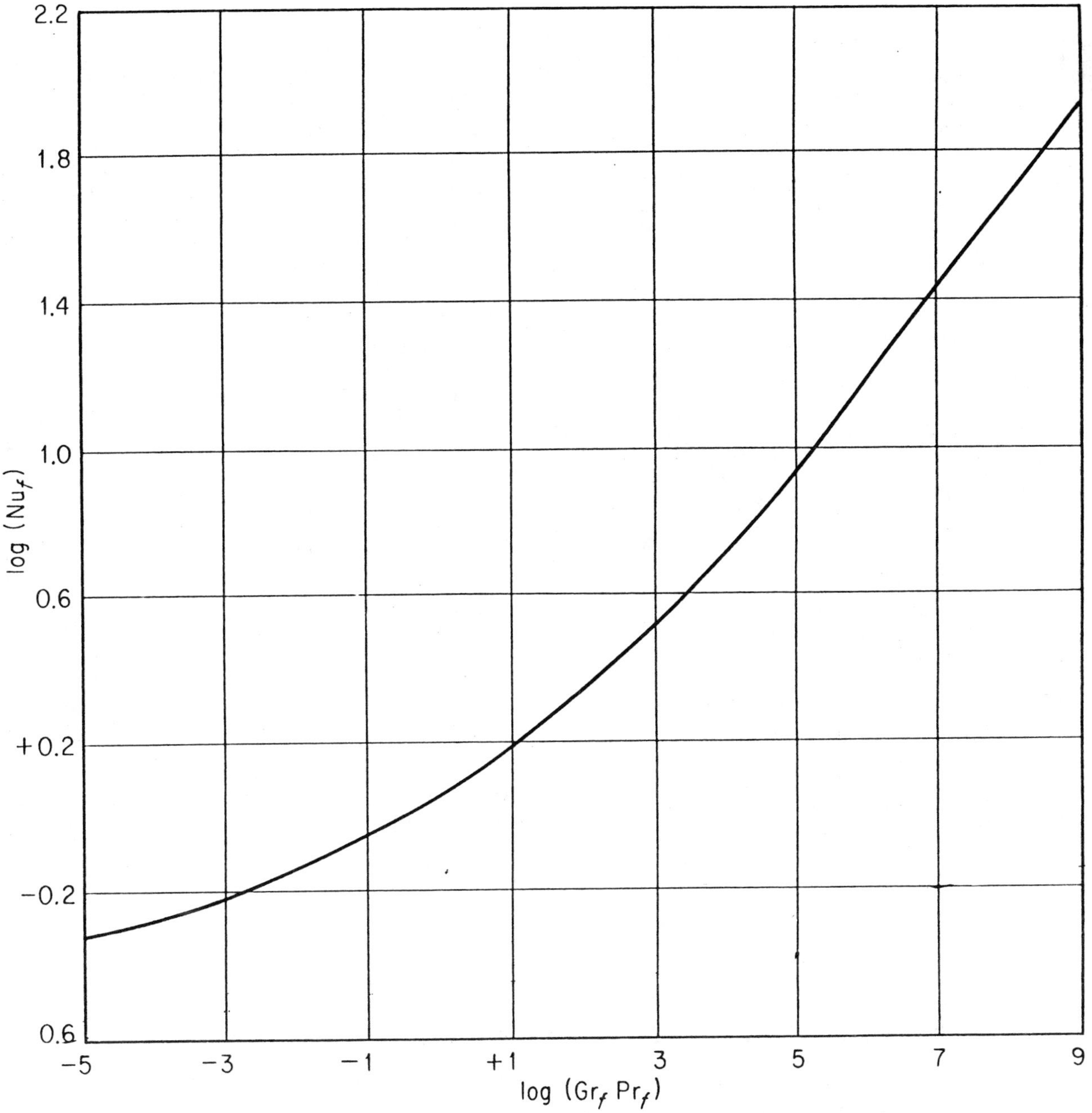

FREE CONVECTION FROM HEATED HORIZONTAL CYLINDERS, ACCORDING TO McAdams (4). From Holman, J.P.: HEAT TRANSFER, McGraw-Hill Book. Co.

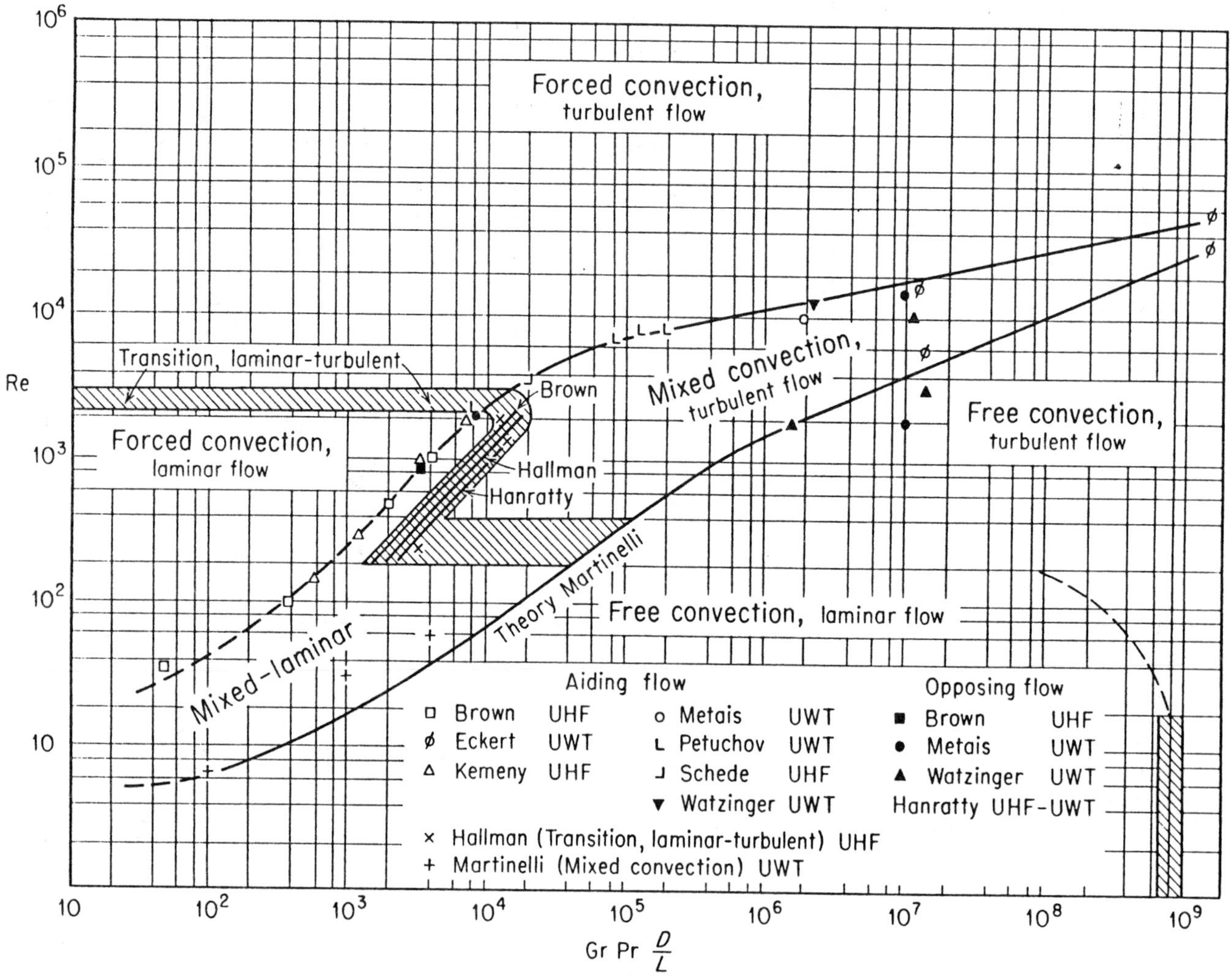

REGIMES OF FREE, FORCED, AND MIXED CONVECTION FOR FLOW THROUGH VERTICAL TUBES, From Holman, J.P. : HEAT TRANSFER McGraw-Hill Book Co.

Forced convection,
turbulent flow

$$Nu = 0.116\left[1 + \left(\frac{D}{L}\right)^{2/3}\right](Re^{2/3} - 125)\,Pr^{1/3}\left(\frac{\eta_F}{\eta_W}\right)^{0.14}$$

Hausen

Transition, laminar turbulent

Mixed convection,
turbulent flow

$$Nu = 4.69\,Re^{0.27}\,Pr^{0.21}\,Gr^{0.07}\left(\frac{D}{L}\right)^{0.36}$$

Metais

Forced convection,
laminar flow

$$Nu = 1.86\,Pe^{1/3}\left(\frac{\eta_F}{\eta_W}\right)^{0.14}$$

Sieder and Tate

Mixed convection,
laminar flow

$$Nu = 1.75\left(\frac{\mu_F}{\mu_W}\right)^{0.14}\left[Gz + 0.0083\,(Gr\,Pr)^{0.75}\right]^{1/3}$$

Oliver

Free convection

□ Holden   △ Petuchov
▽ Kern   × Sherwood
○ Metais

Re

$Gr\,Pr\,\dfrac{D}{L}$

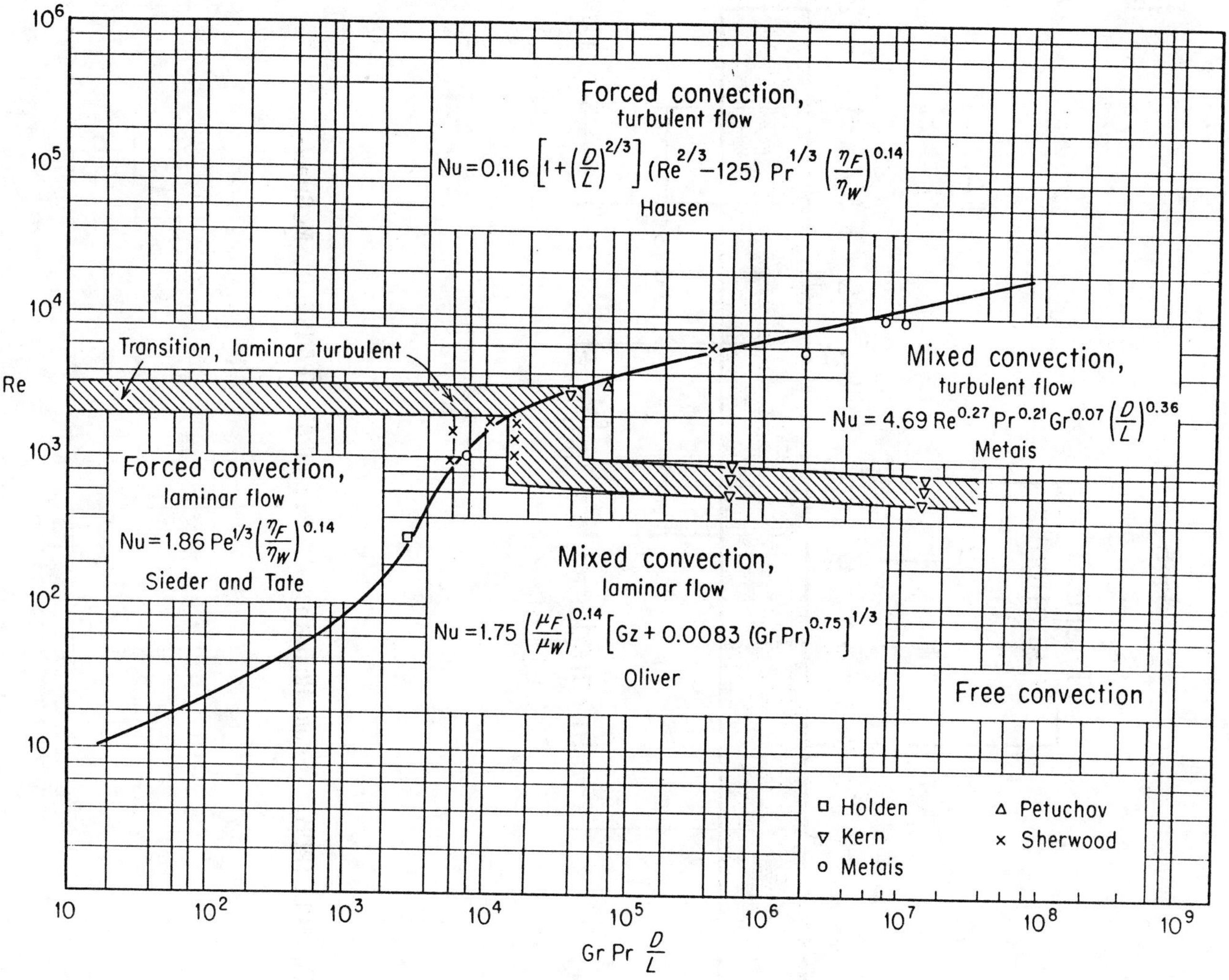

REGIMES OF FREE, FORCED, AND MIXED CONVECTION FOR FLOW
THROUGH HORIZONTAL TUBES, ACCORDING TO METAIS AND ECKERT(10)
From Holman, J.P.: HEAT TRANSFER, McGraw-Hill Book Co.

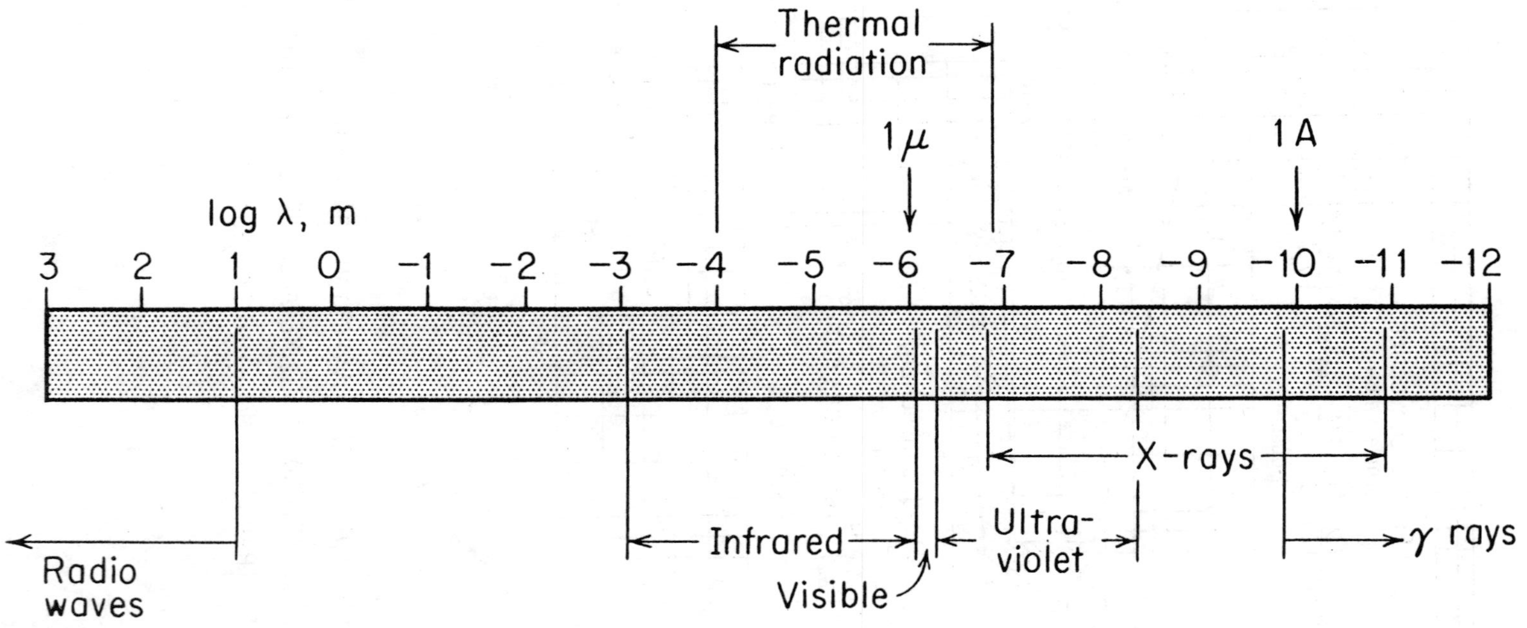

ELECTROMAGNETIC SPECTRUM. From Holman, J. P.: HEAT TRANSFER, McGraw-Hill Book Co.

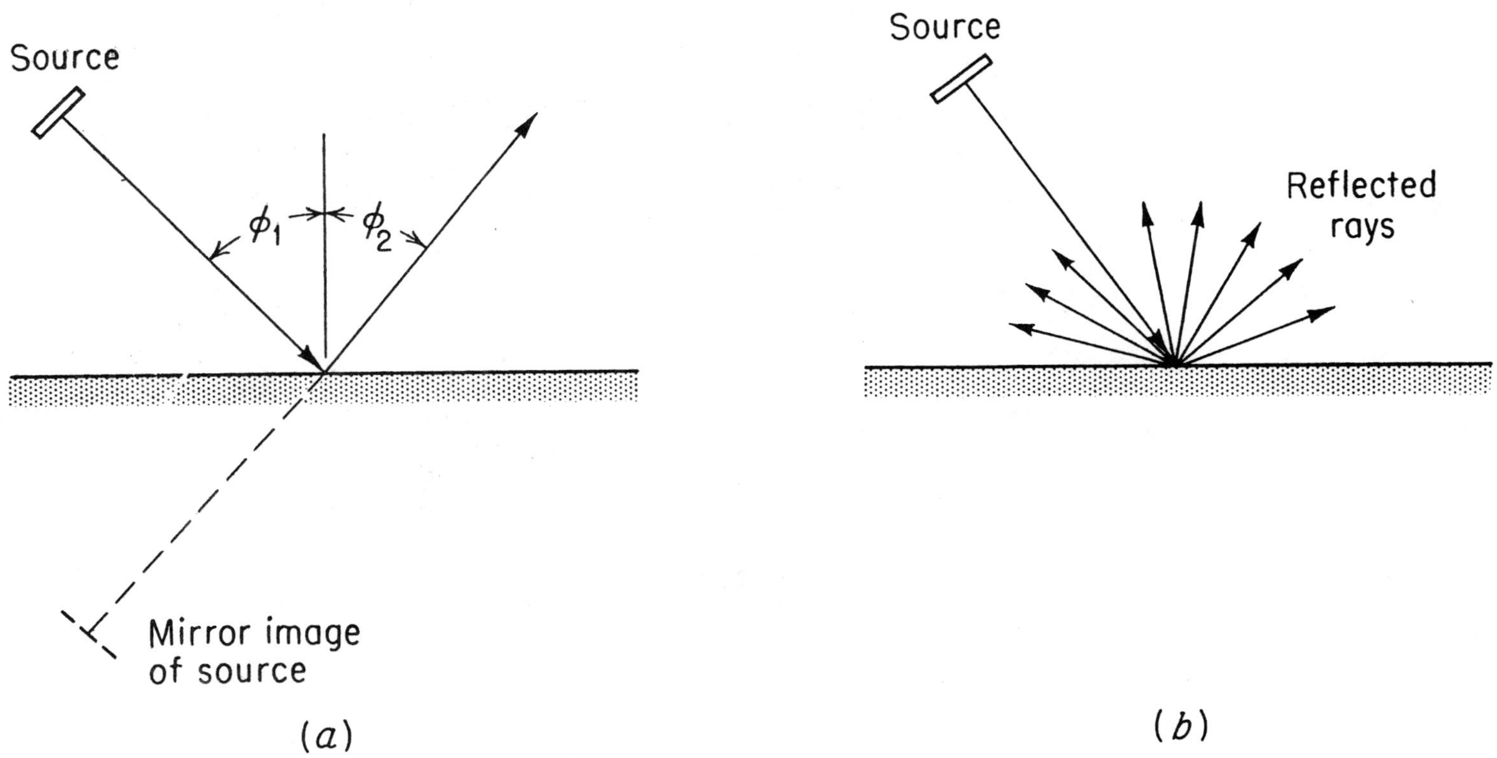

(a) SPECULAR, (b) DIFFUSE REFLECTION

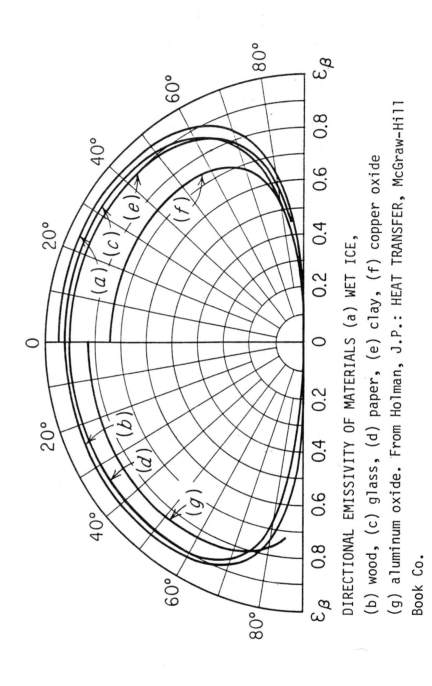

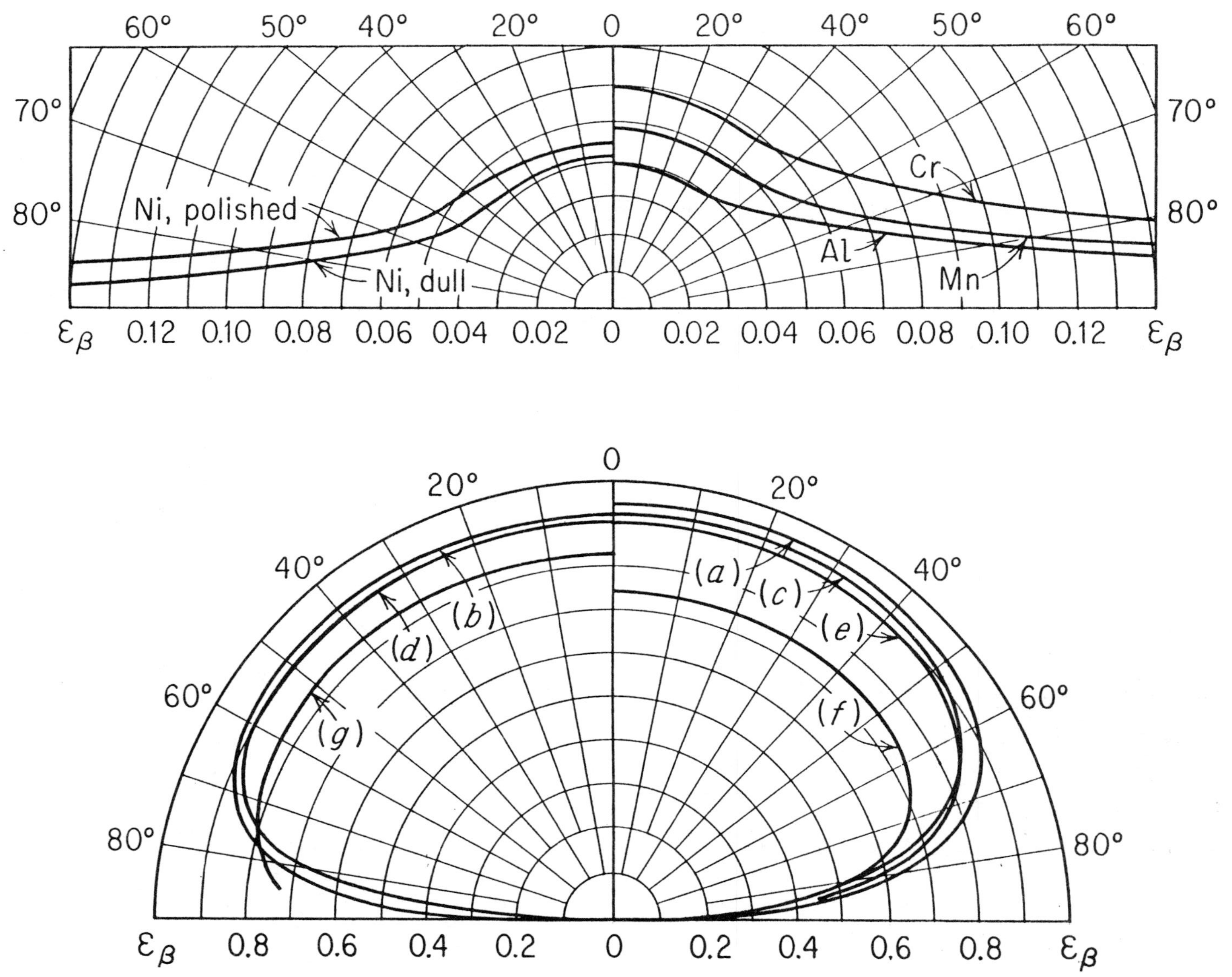

DIRECTIONAL EMISSIVITY OF MATERIALS (a) WET ICE,
(b) wood, (c) glass, (d) paper, (e) clay, (f) copper oxide
(g) aluminum oxide. From Holman, J.P.: HEAT TRANSFER, McGraw-Hill Book Co.

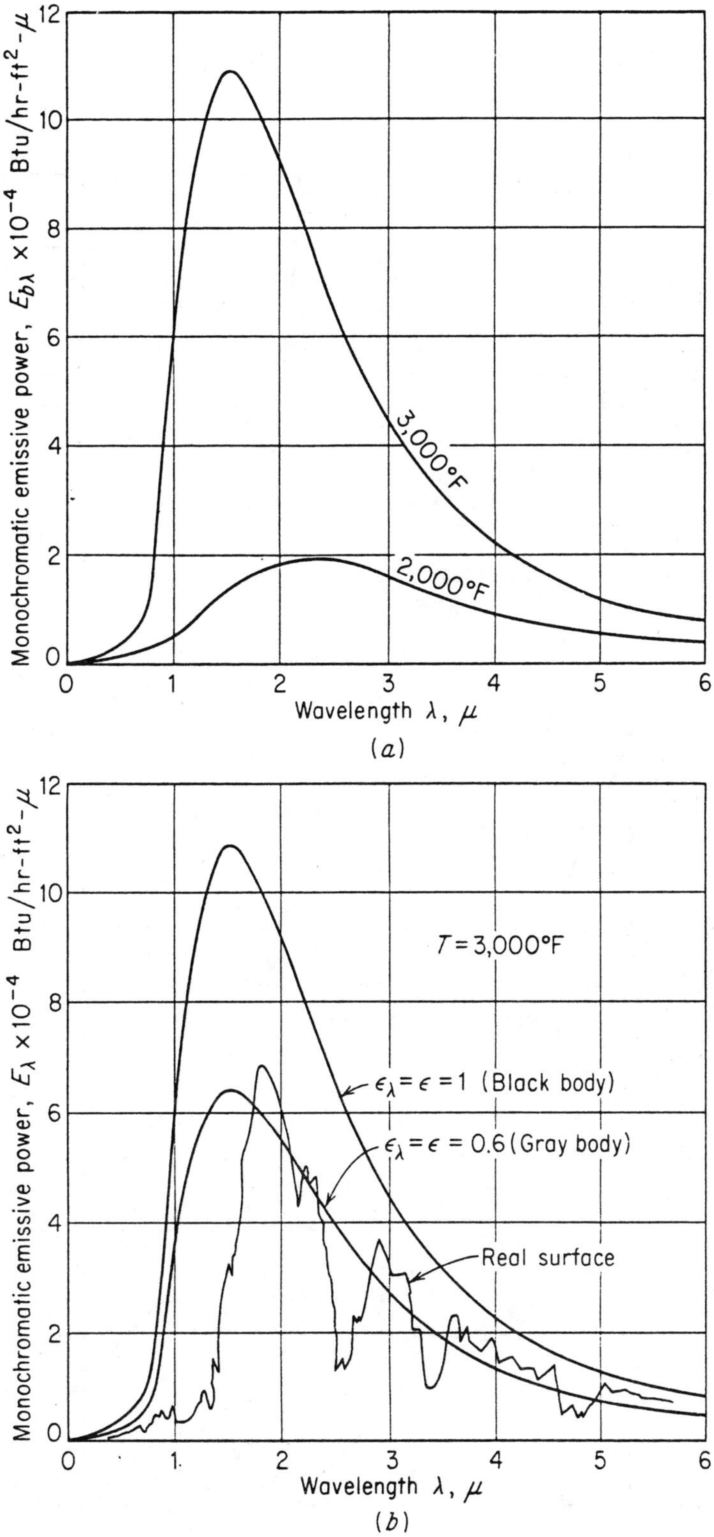

(A) BLACKBODY EMISSIVE POWER AS A FUNCTION OF WAVELENGTH AND TEMPERATURE. (B) COMPARISON OF EMISSIVE POWER OF IDEAL BLACKBODIES AND GRAY BODIES WITH THAT OF A REAL SURFACE. From Holman, J. P.: HEAT TRANSFER, McGraw-Hill Book Co.

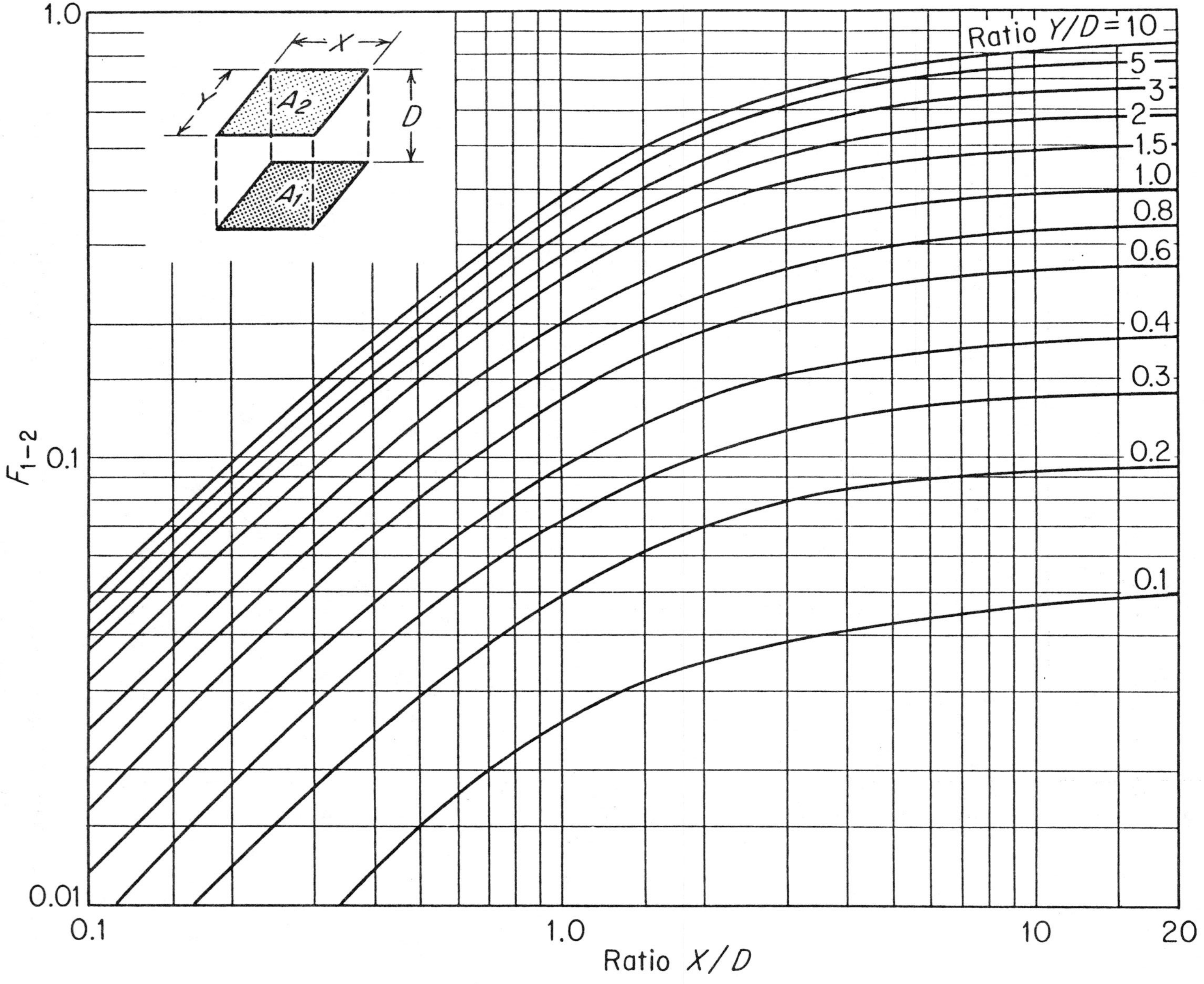

RADIATION SHAPE FACTOR FOR RADIATION BETWEEN PARALLEL RECTANGLES.

From Holman, J. P.: HEAT TRANSFER, McGraw-Hill Book Co.

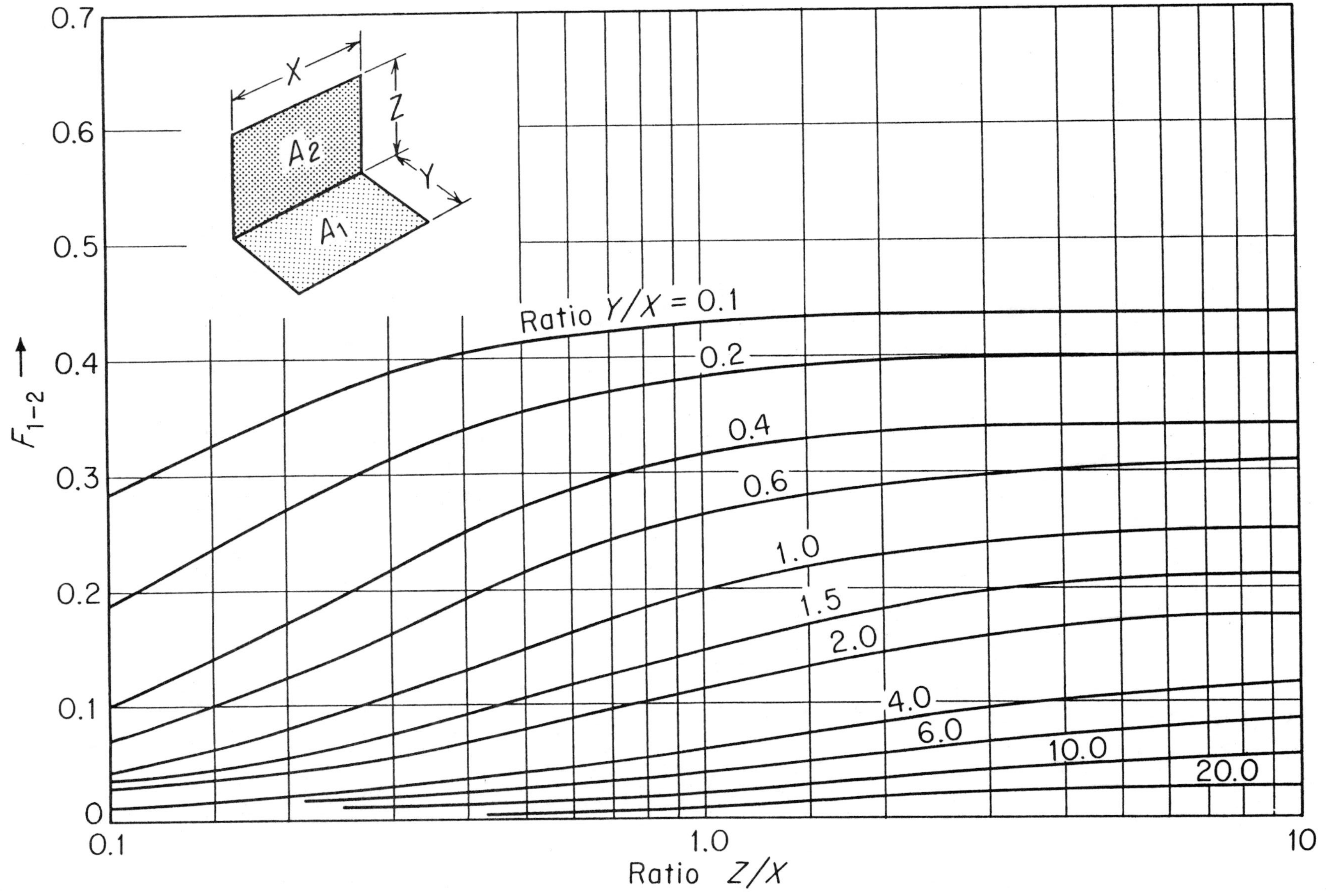

RADIATION SHAPE FACTOR FOR RADIATION BETWEEN PERPENDICULAR RECTANGLES WITH A COMMON EDGE. From Holman, J. P.: HEAT TRANSFER, McGraw-Hill Book Co.

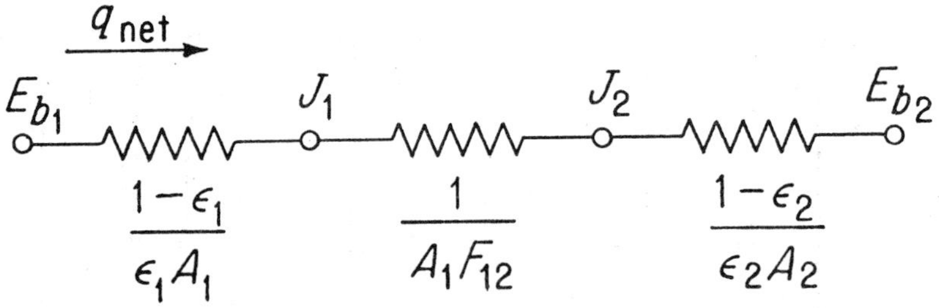

RADIATION NETWORK FOR TWO SURFACES WHICH SEE EACH OTHER AND NOTHING ELSE. From Holman, J. P.: HEAT TRANSFER, McGraw-Hill Book Co.

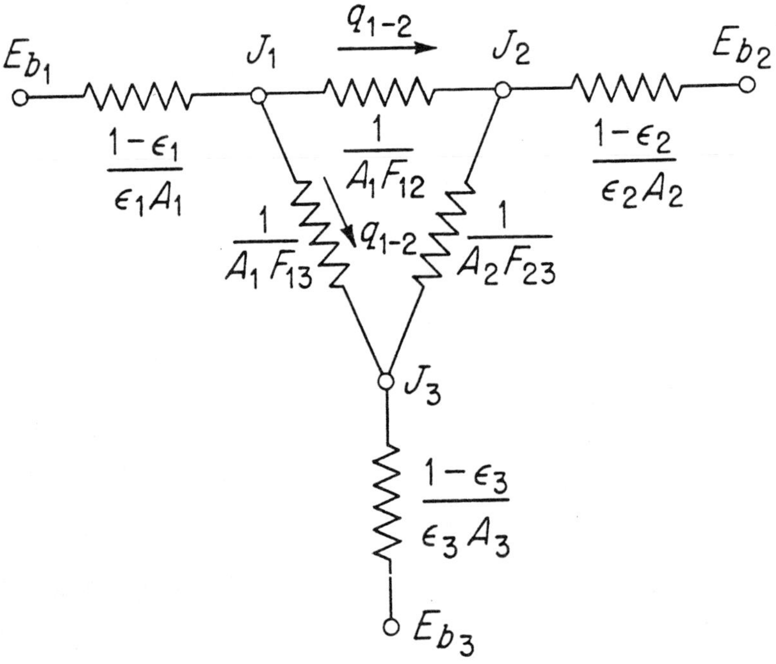

RADIATION NETWORK FOR THREE SURFACES WHICH SEE EACH OTHER AND NOTHING ELSE. From Holman, J. P.: HEAT TRANSFER, McGraw-Hill Book Co.

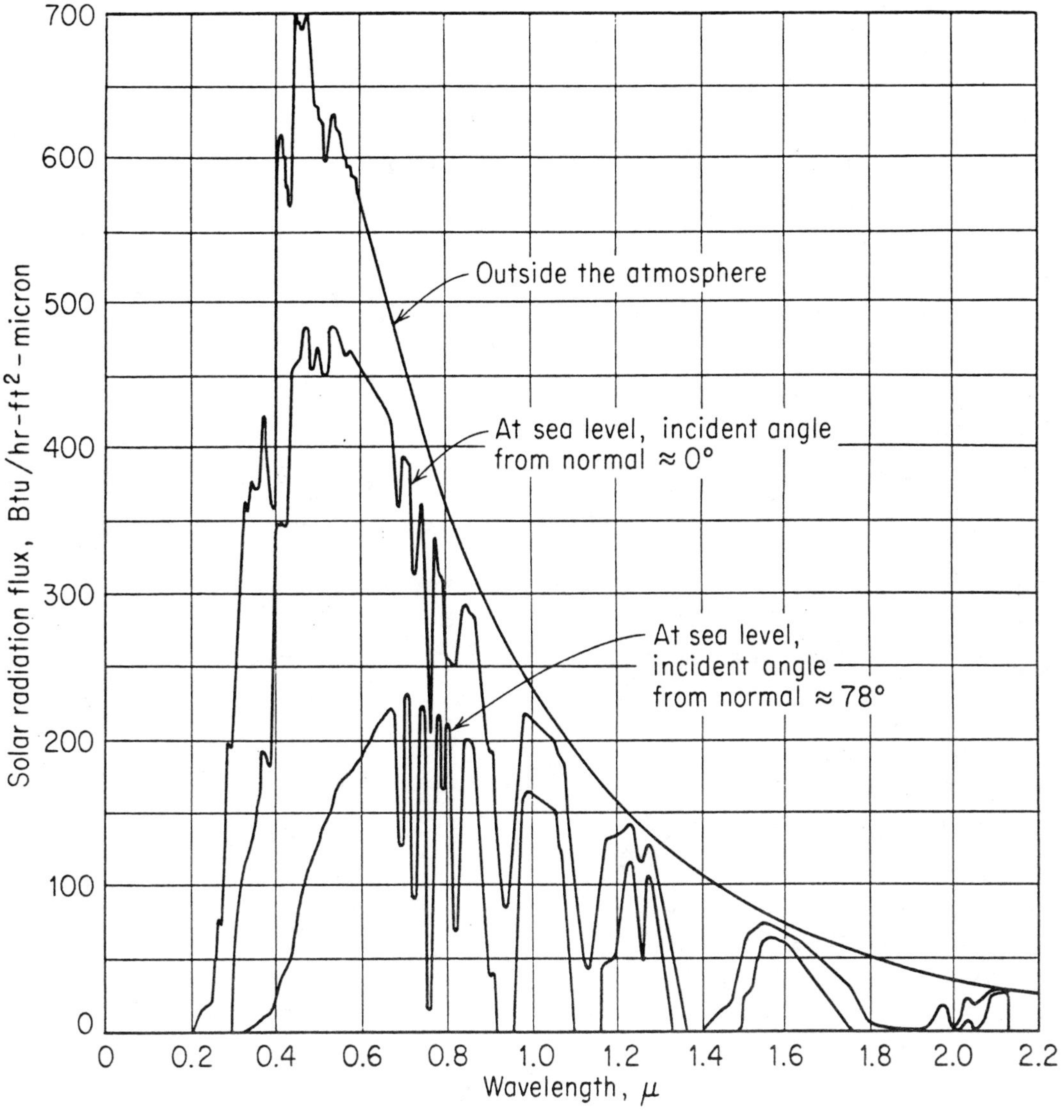

SPECTRAL DISTRIBUTION OF SOLAR RADIATION AS FUNCTIONS OF ATMOSPHERIC CONDITIONS AND ANGLE OF INCIDENCE, ACCORDING TO THRELKELD AND JORDAN (15). From Holman, J. P.: HEAT TRANSFER, McGraw-Hill Book Co.

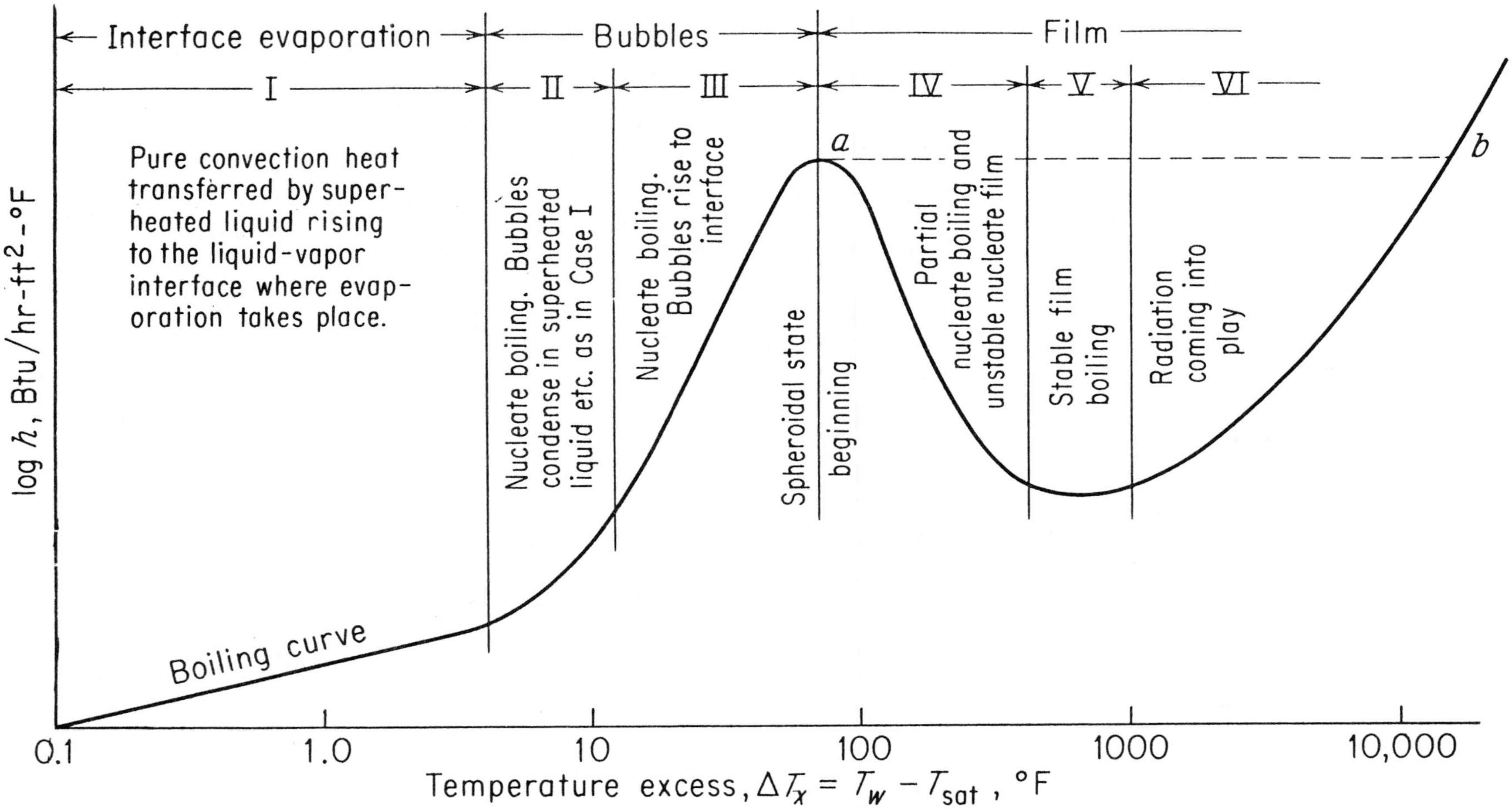

HEAT-FLUX DATA FROM AN ELECTRICALLY HEATED PLATINUM WIRE, FROM FARBER AND SCORAH (9). From Holman, J. P.: HEAT TRANSFER, McGraw-Hill Book Co.

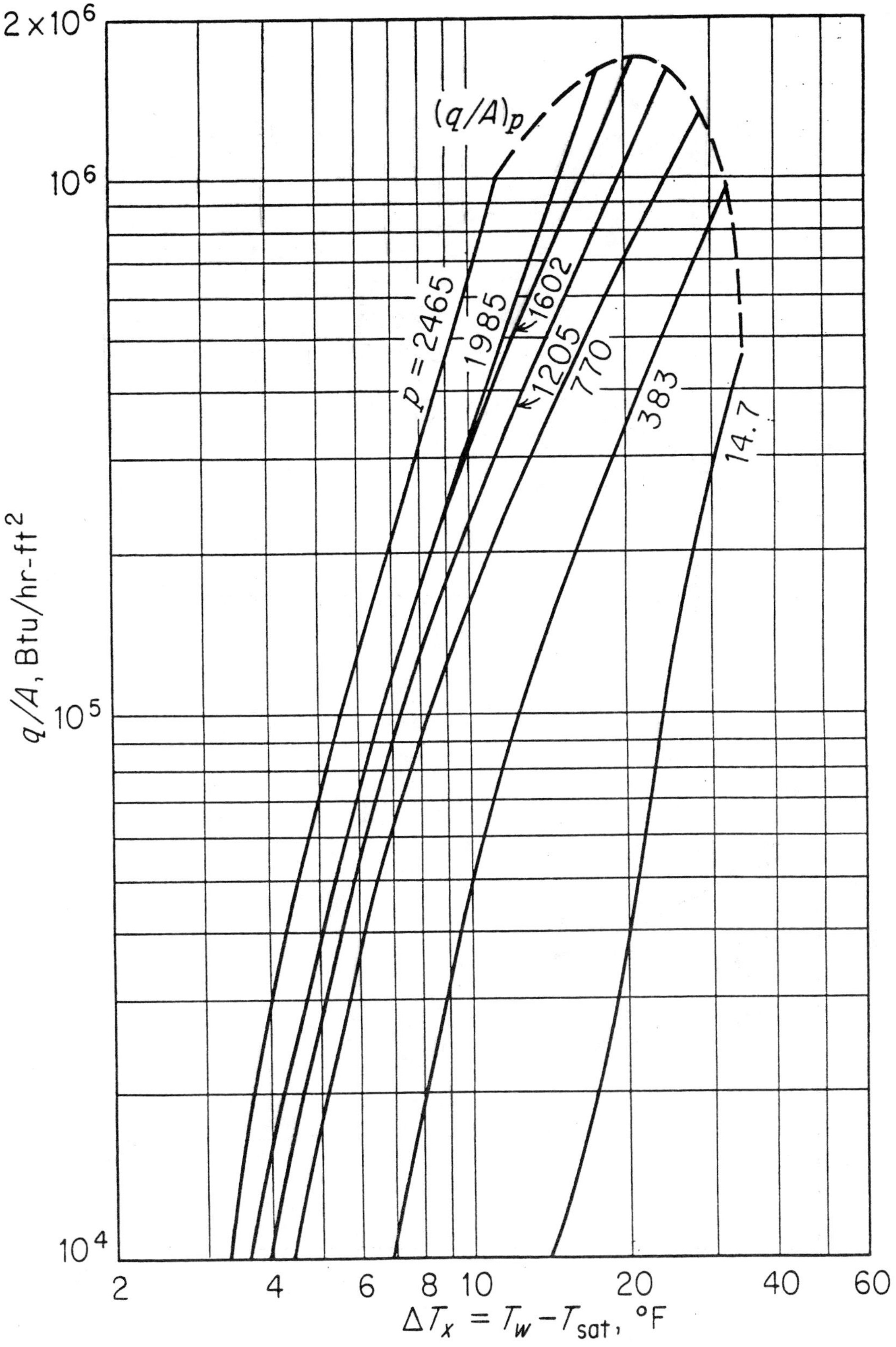

HEAT FLUX DATA FOR WATER BOILING ON A PLATINUM WIRE, FROM McAdams (3), From Holman, J.P.: HEAT TRANSFER, McGraw-Hill Book Co.

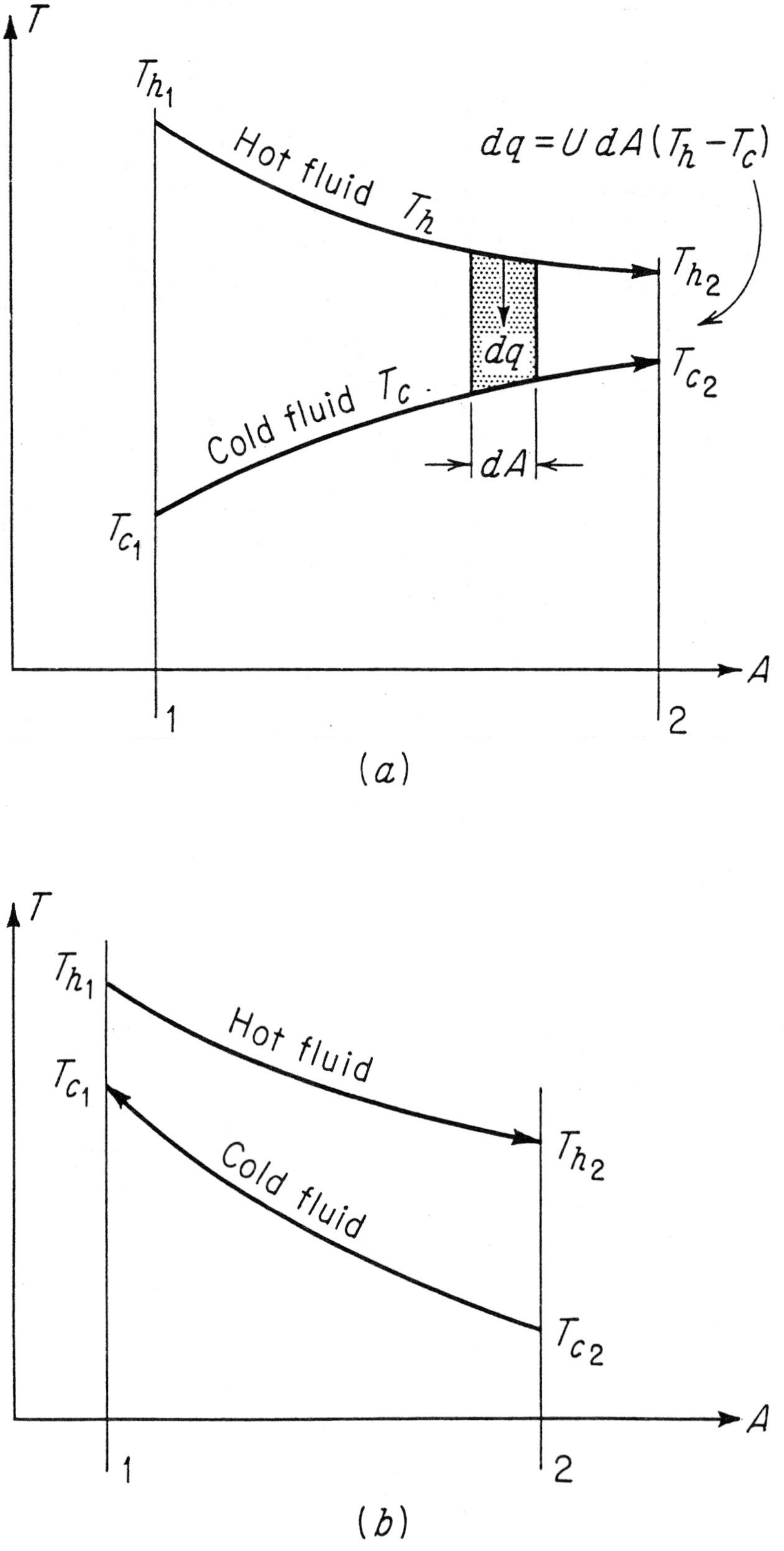

TEMPERATURE PROFILES FOR PARALLEL FLOW AND COUNTERFLOW IN DOUBLE PIPE HEAT EXCHANGER. From Holman, J. P.: HEAT TRANSFER, McGraw-Hill Book Co.

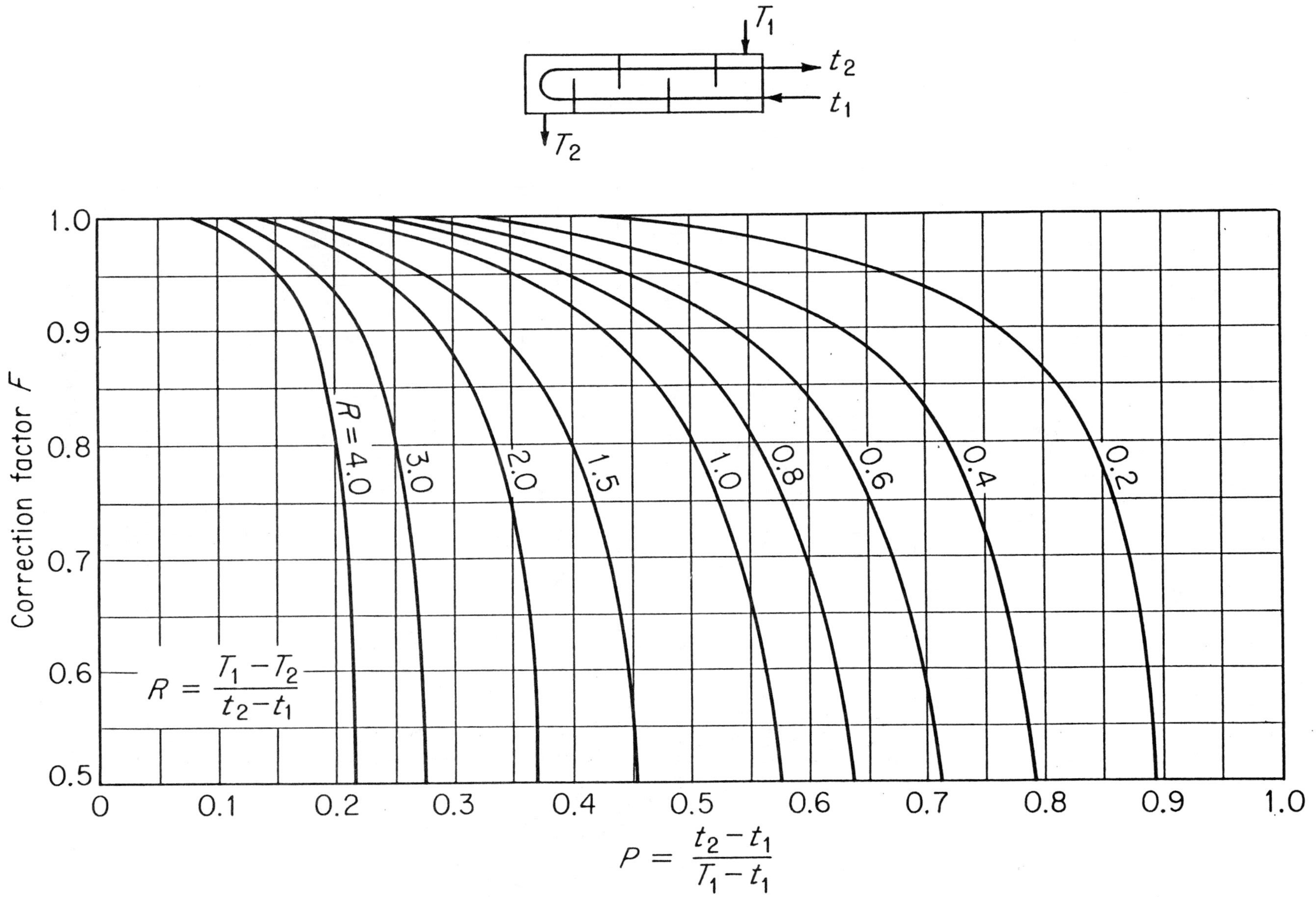

CORRECTION FACTOR PLOT FOR EXCHANGER WITH ONE SHELL PASS AND TWO, FOUR, OR ANY MULTIPLE OF TUBE PASSES. From Holman, J. P.: HEAT TRANSFER, McGraw-Hill Book Co.

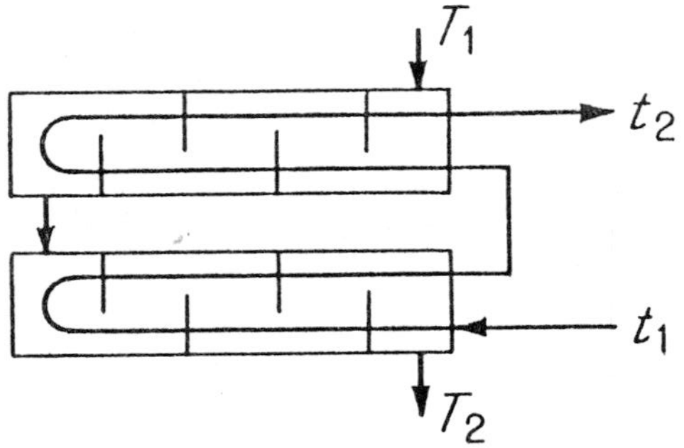

CORRECTION FACTOR PLOT FOR EXCHANGER WITH TWO SHELL PASSES AND FOUR, EIGHT, OR ANY MULTIPLE OF TUBE PASSES. From Holman, J.P. : HEAT TRANSFER, McGraw-Hill Book Co.

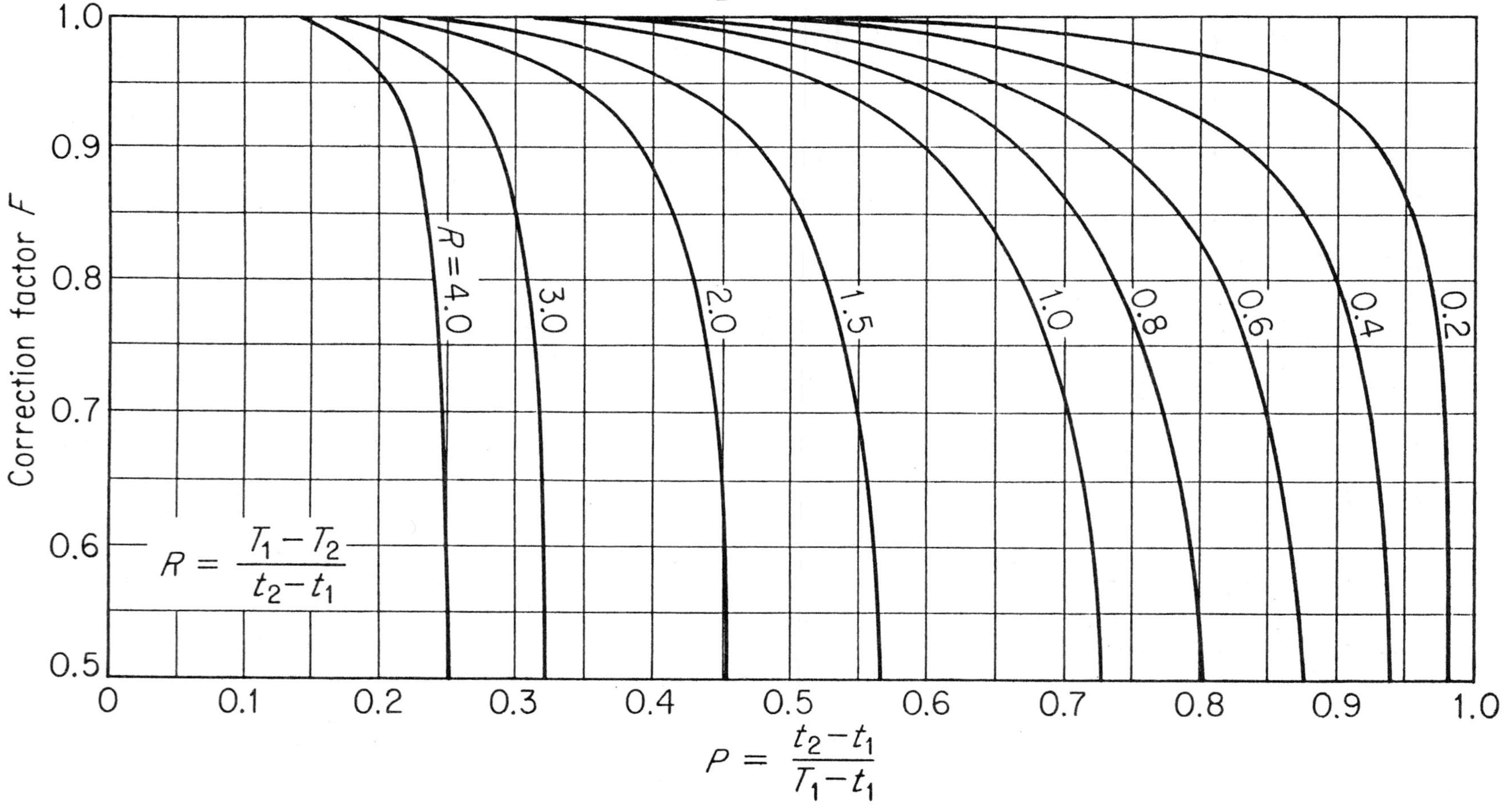

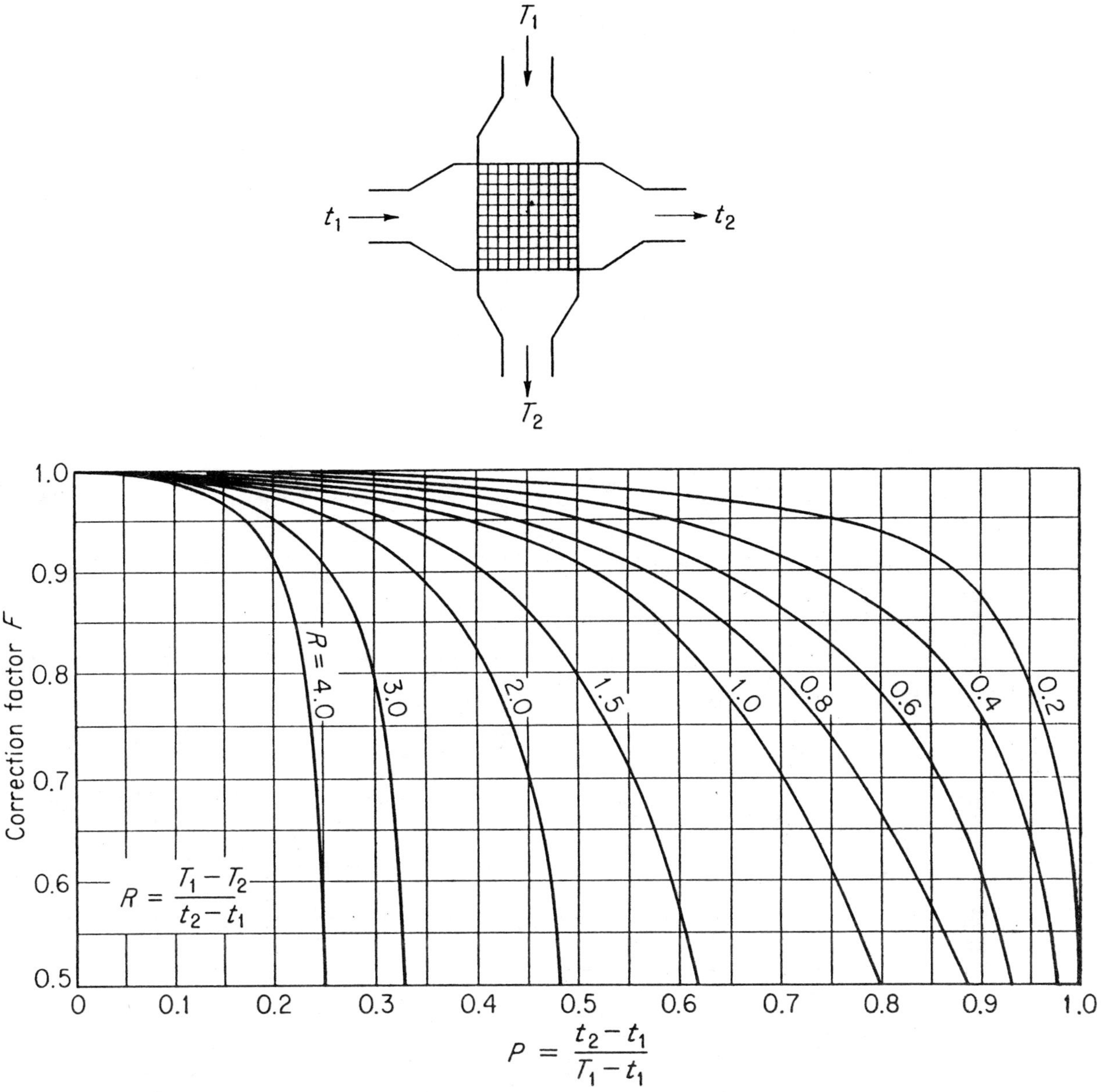

CORRECTION FACTOR PLOT FOR SINGLE-PASS CROSS-FLOW EXCHANGER, BOTH FLUIDS UNMIXED. From Holman, J. P.: HEAT TRANSFER, McGraw-Hill Book Co.

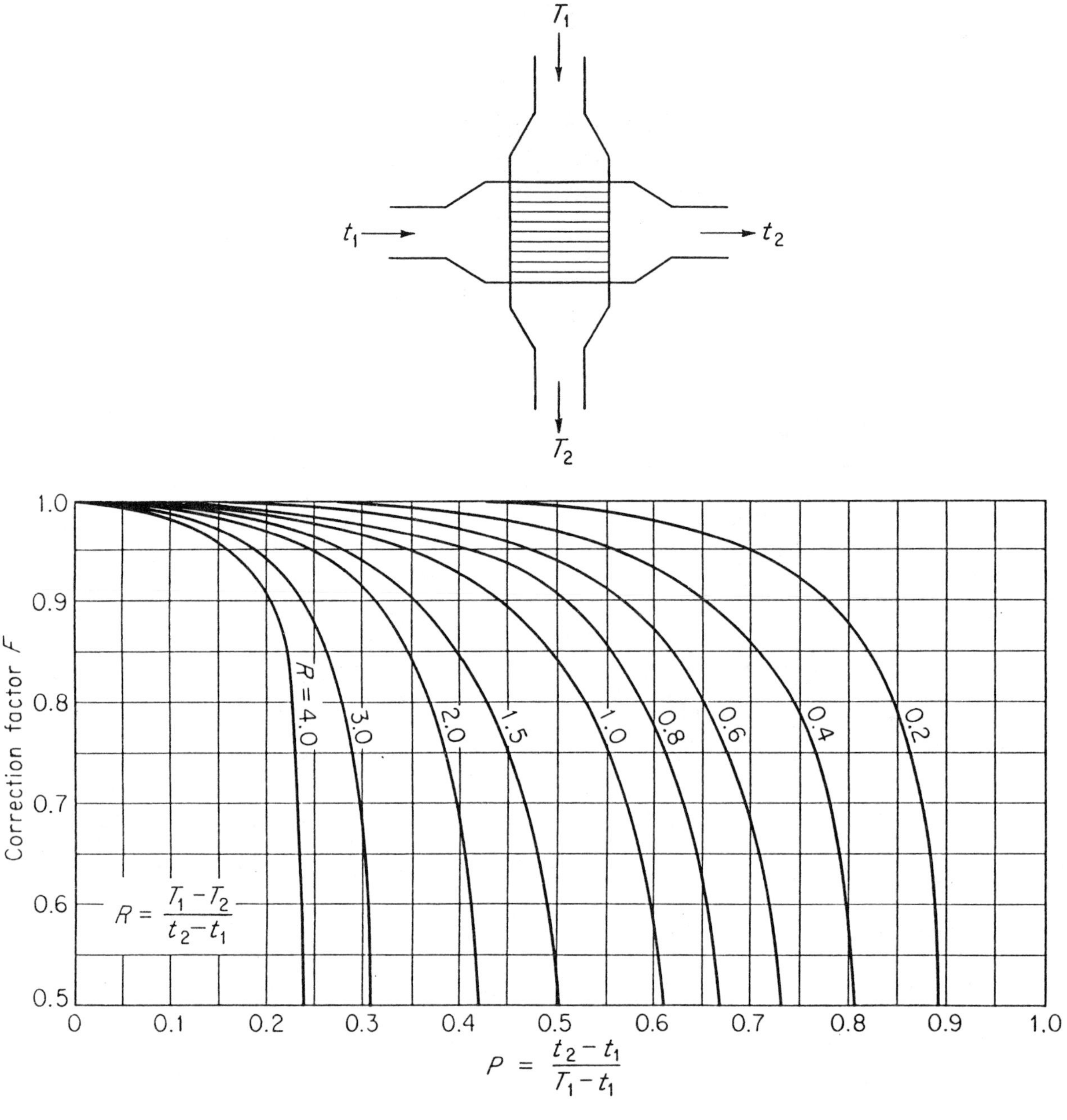

CORRECTION FACTOR PLOT FOR SINGLE-PASS CROSS-FLOW EXCHANGER, ONE FLUID MIXED, THE OTHER UNMIXED. From Holman, J. P.: HEAT TRANSFER, McGraw-Hill Book Co.

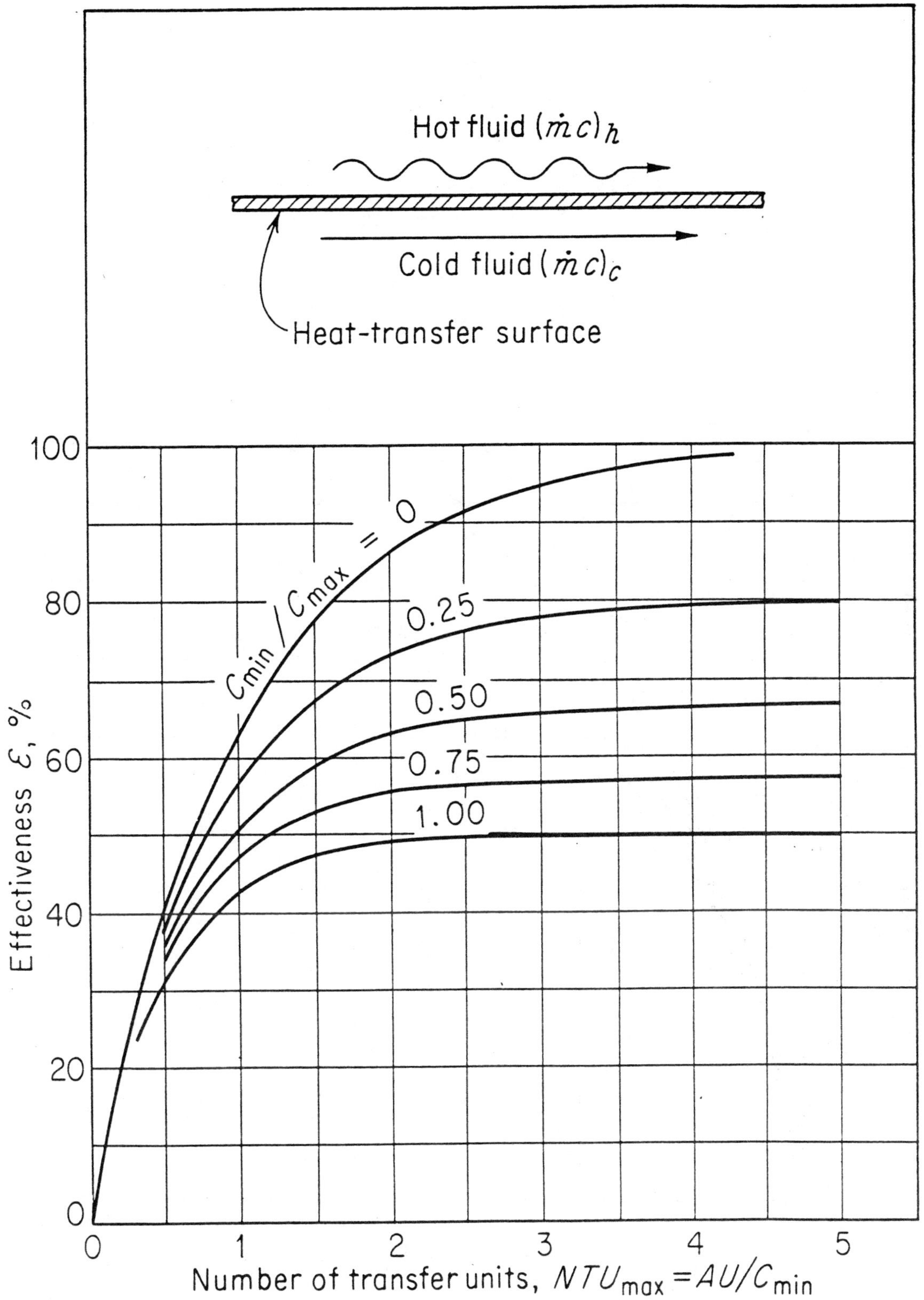

EFFECTIVENESS FOR PARALLEL-FLOW EXCHANGER PERFORMANCE. From Holman, J.P.: HEAT TRANSFER, McGraw-Hill Book Co.

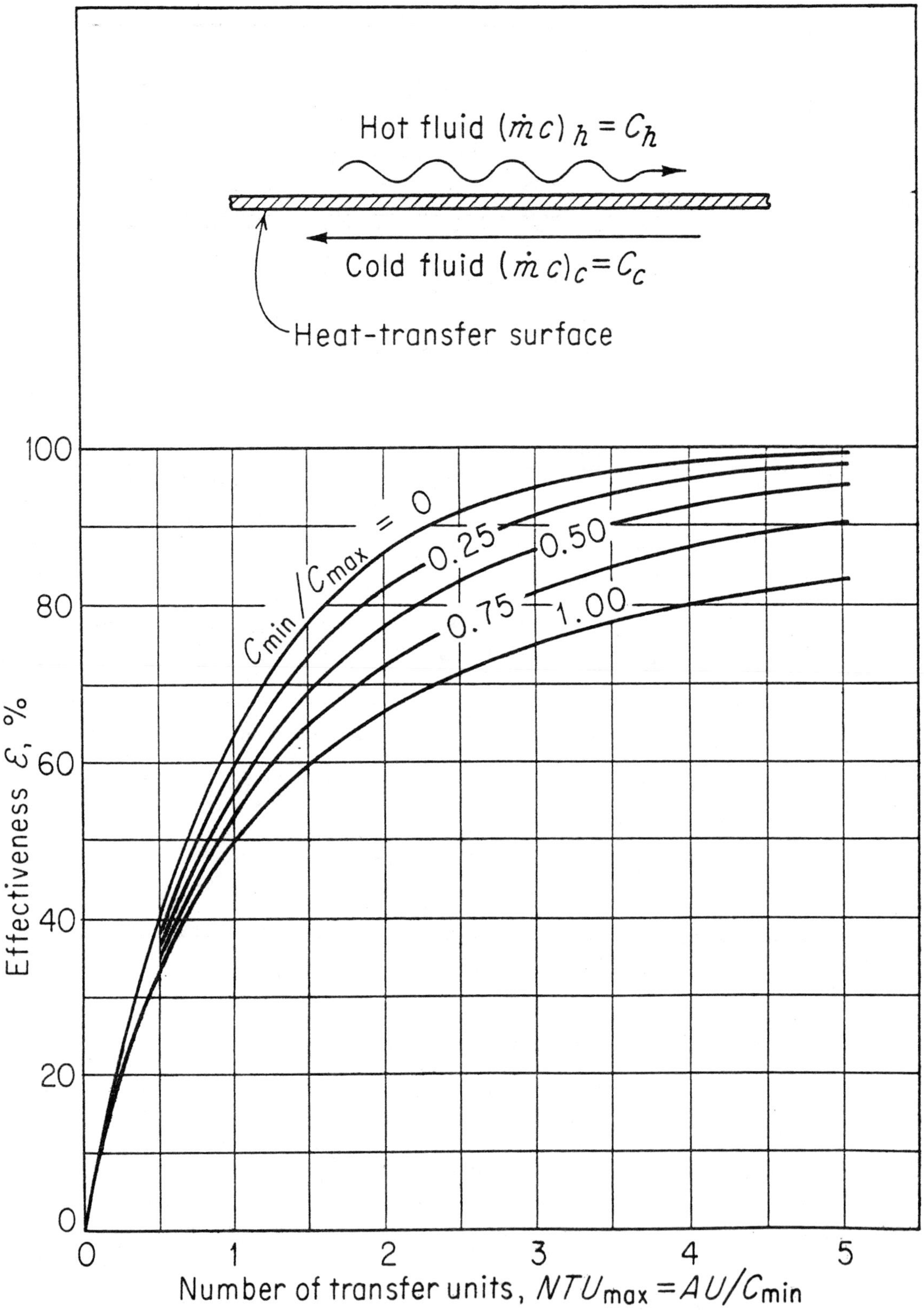

EFFECTIVENESS FOR COUNTERFLOW EXCHANGER PERFORMANCE. From Holman, J.P.: HEAT TRANSFER, McGraw-Hill Book Co.

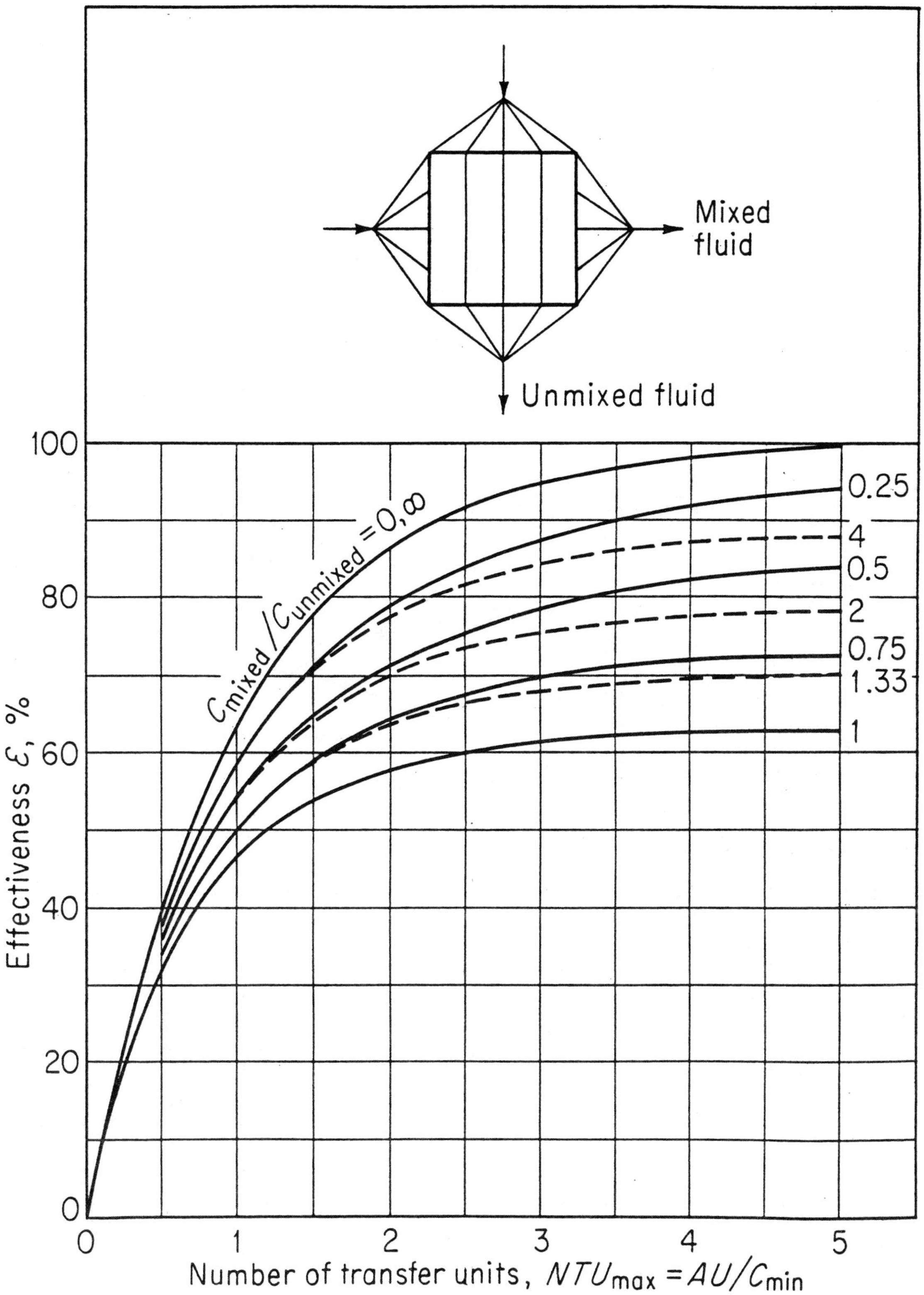

EFFECTIVENESS FOR CROSS-FLOW EXCHANGER WITH ONE FLUID MIXED. From Holman, J.P.: HEAT TRANSFER, McGraw-Hill Book Co.

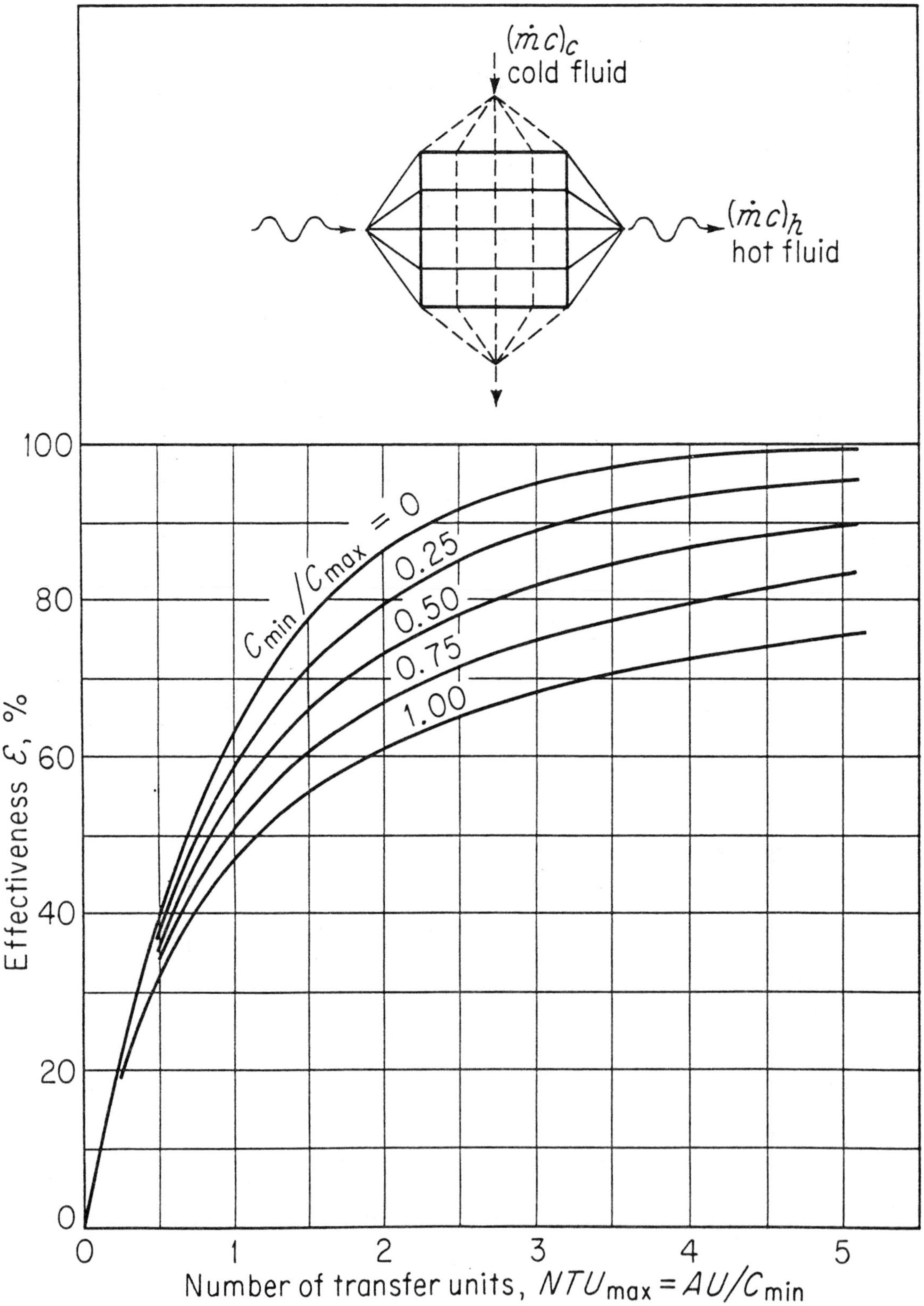

EFFECTIVENESS FOR CROSS-FLOW EXCHANGER WITH FLUIDS UNMIXED. From Holman, J. P.: HEAT TRANSFER, McGraw-Hill Book Co.

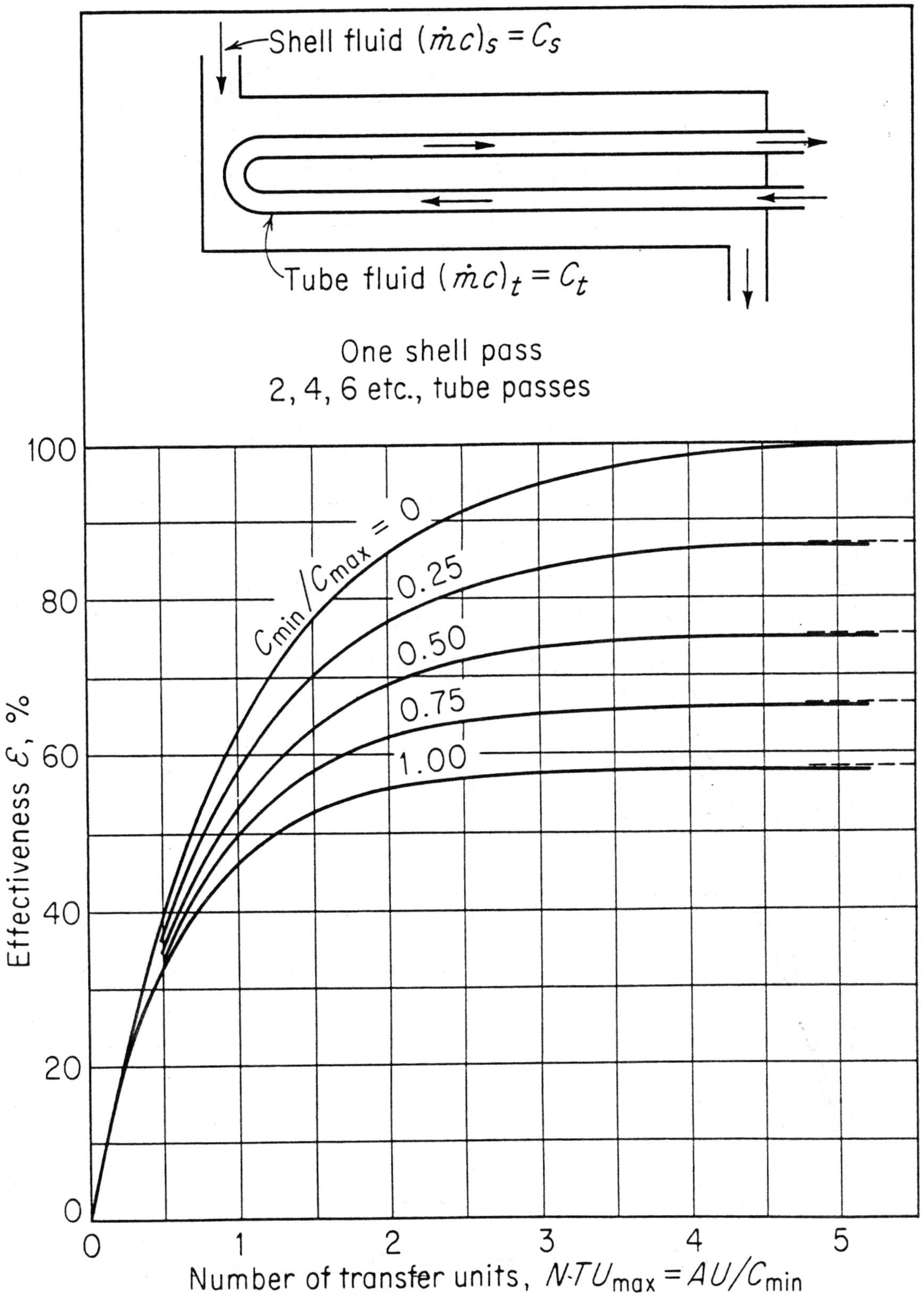

EFFECTIVENESS FOR 1-2 PARALLEL-COUNTERFLOW EXCHANGER PERFORMANCE.
From Holman, J. P.: HEAT TRANSFER, McGraw-Hill Book Co.

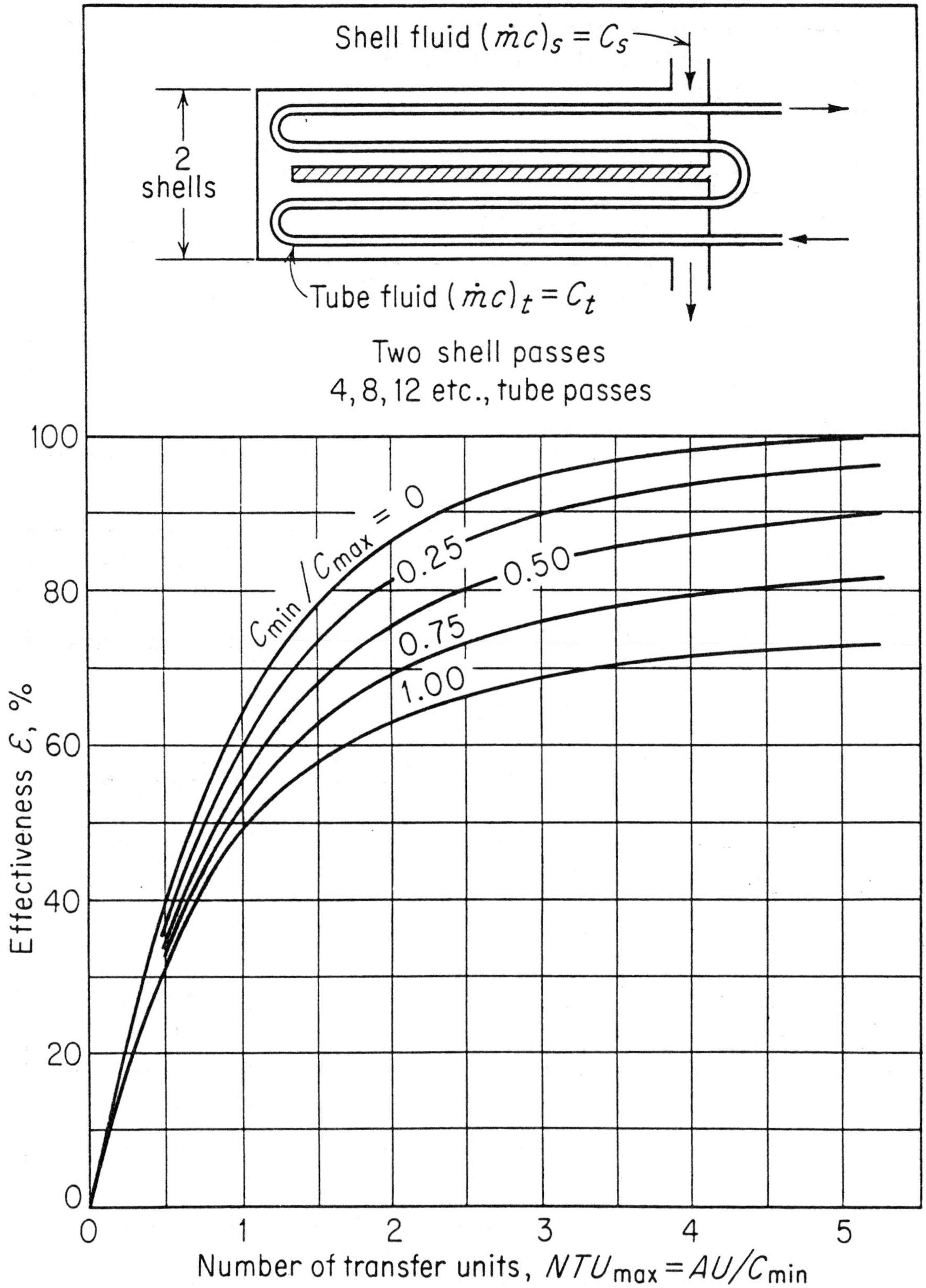

EFFECTIVENESS FOR 2-4 MULTIPASS COUNTERFLOW EXCHANGER PERFORMANCE.
From Holman, J. P.: HEAT TRANSFER, McGraw-Hill Book Co.

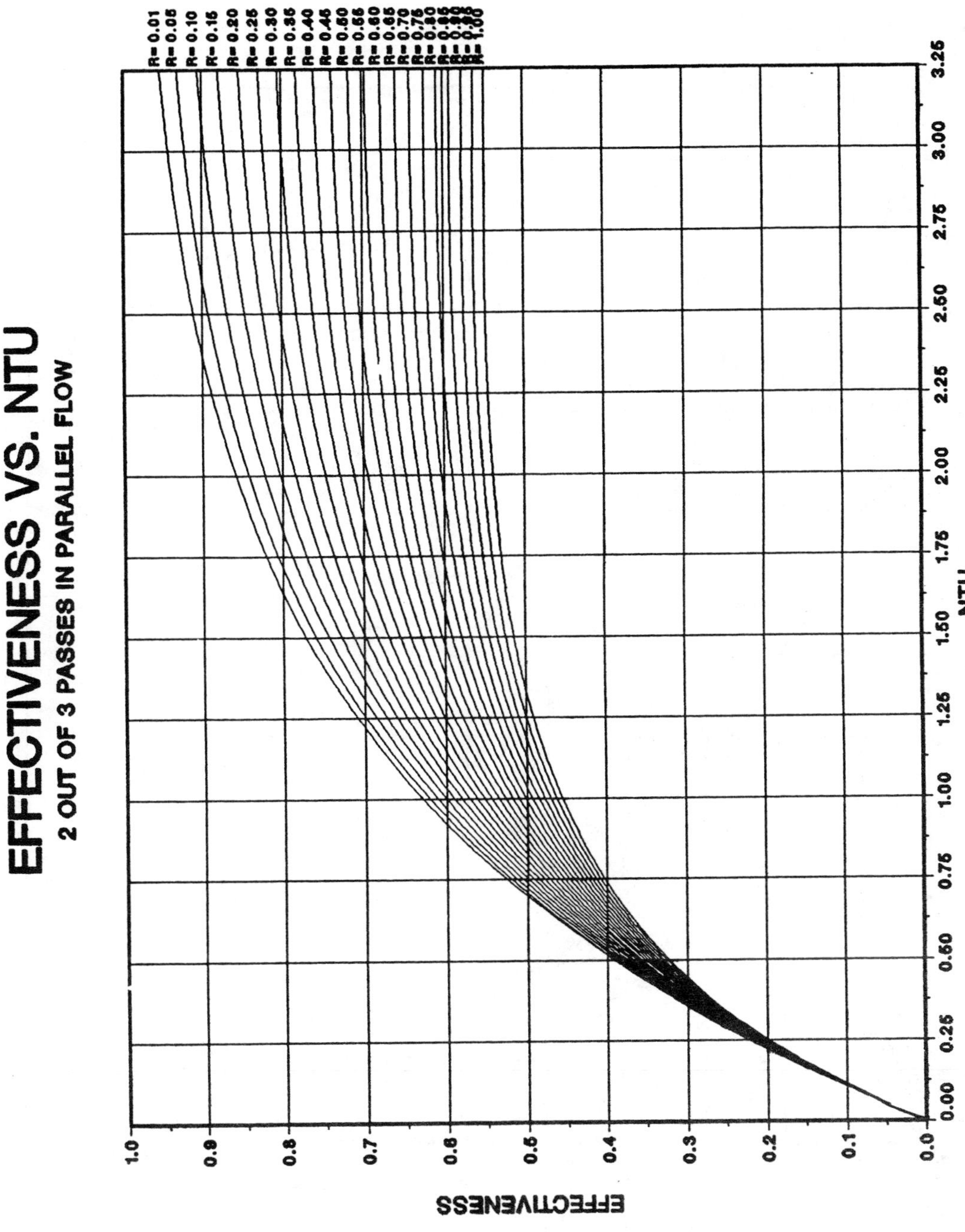

EFFECTIVENESS NTU PLOTS FOR SHELL AND TUBE HEAT EXCHANGER WITH ONE SHELL PASS AND THREE TUBE PASSES. COURTESY MARK S. O'HARE, NAVAL POSTGRADUATE SCHOOL

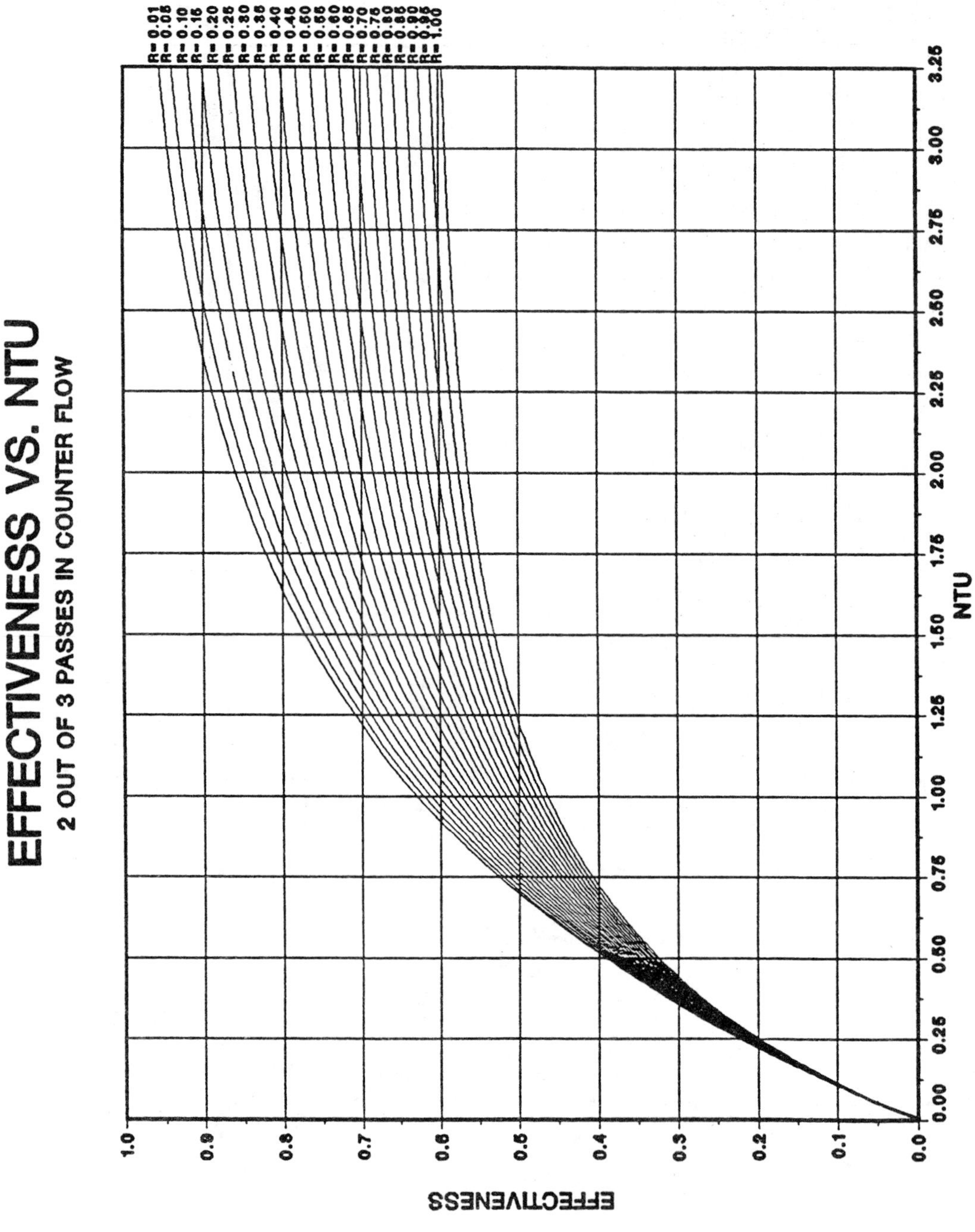

EFFECTIVENESS NTU PLOTS FOR SHELL AND TUBE HEAT EXCHANGER
WITH ONE SHELL PASS AND THREE TUBE PASSES. COURTESY
MARK S. O'HARE, NAVAL POSTGRADUATE SCHOOL

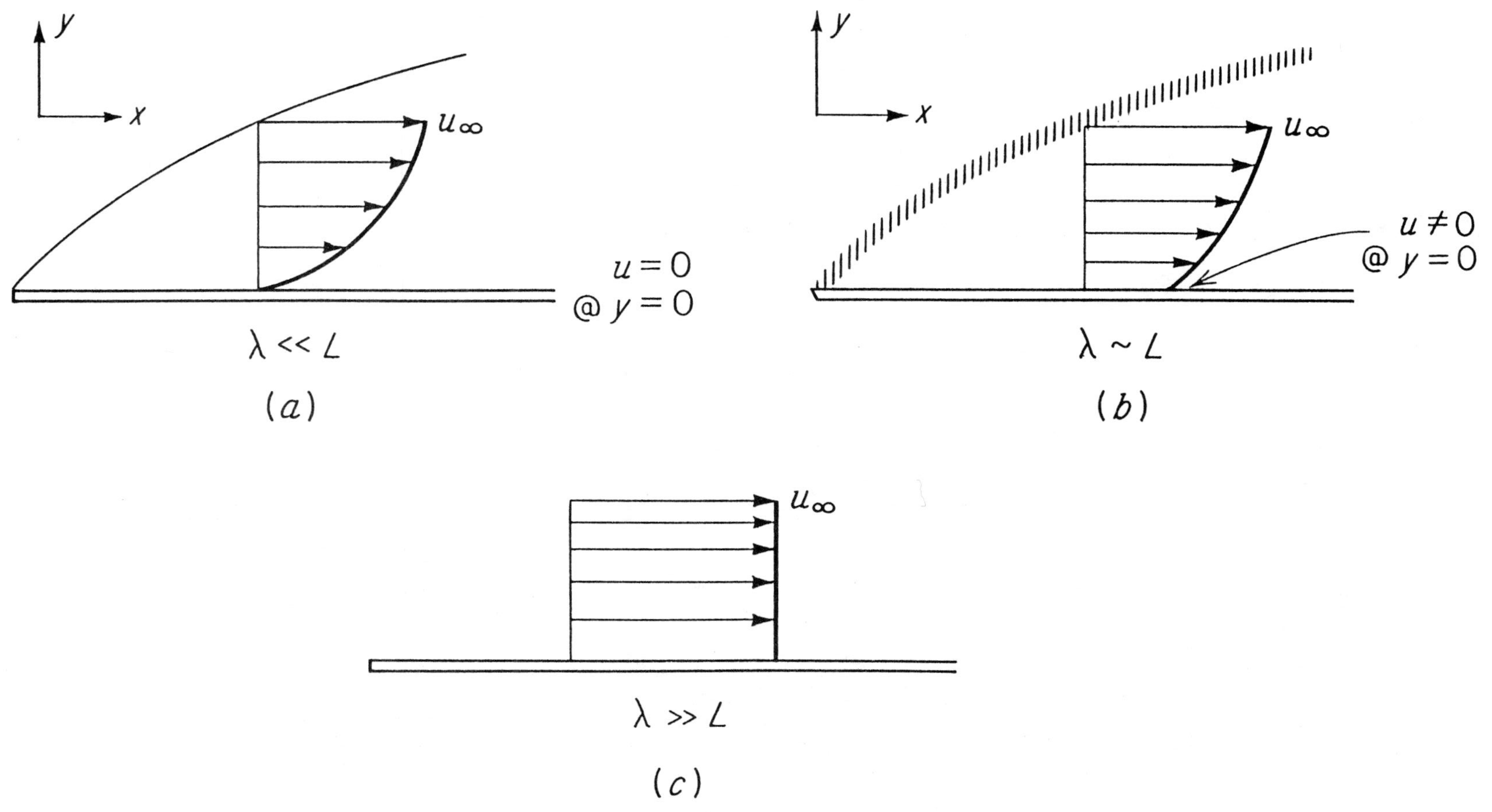

THREE TYPES OF FLOW REGIMES FOR A FLAT PLATE. (A) CONTINUUM FLOW; (B) SLIP FLOW; (C) FREE-MOLECULE FLOW. From Holman, J. P.: HEAT TRANSFER, McGraw-Hill Book Co.

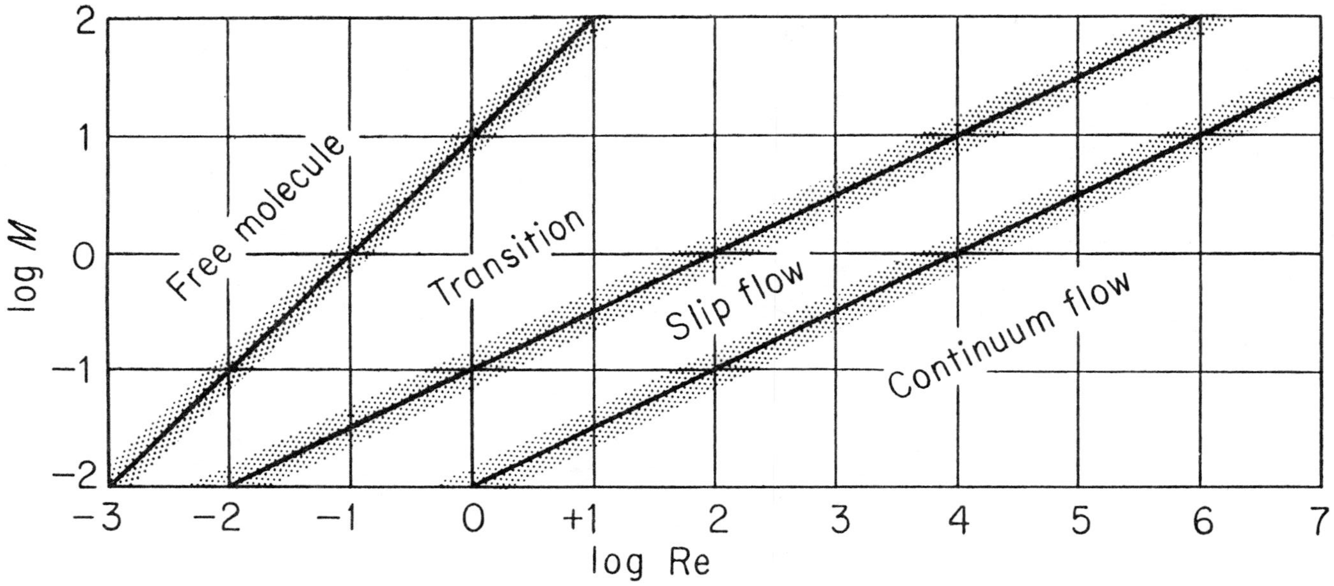

RELATION OF FLOW REGIMES TO MACH AND REYNOLDS NUMBERS,
From Holman, J.P.: HEAT TRANSFER, McGraw-Hill Book Co.

The graphs on the following pages were taken from the paper:

Colakyan, M., R. Turton, and O. Levenspiel: "Unsteady Heat Transfer to Variously Shaped Objects", Heat Transfer Engr., Vol. 5, Nos. 3-4, 1984.

Figures courtesy Richard Turton, Chemical Engineering Department, Oregon State University.

These figures may be used as an alternative to the Heisler charts of Chapter 4 of the text. Nomenclature for the figures is given below.

NOMENCLATURE

b	thickness of semi-infinite plate, m
Bi	Biot number, (–)
C_p	specific heat, J/kg K
Fo	Fourier number, (–)
h	film heat transfer coefficient, W/m^2 K
k	thermal conductivity, W/m K
l	distance, m
L	characteristic length (volume/surface area), m
Q	remaining heat to be taken up by the object, J
Q_{max}	maximum possible heat taken up by the object, J
R	radius of a sphere or cylinder, m
ΔT_{max}	maximum temperature difference between the object and surroundings, K
t	time, s
α	thermal diffusivity, m^2/s
ρ	density, kg/m^3
$\theta = \Delta T/\Delta T_{max}$	unaccomplished temperature ratio, (–)

$$\mathrm{Bi} = \frac{hL}{k} = \frac{\text{interior resistance}}{\text{film or surface resistance}}$$

$$\mathrm{Fo} = \frac{\alpha t}{L^2} = \frac{k}{\rho C_p}\frac{t}{L^2}$$

$$\text{where } L = \frac{\text{volume of object}}{\text{surface area of object}}$$

$= \text{characteristic length of object}$

$= \dfrac{\text{thickness}}{2}$ for infinite flat plate

$= \dfrac{R}{2}$ for infinite cylinder

$= \dfrac{R}{3}$ for sphere

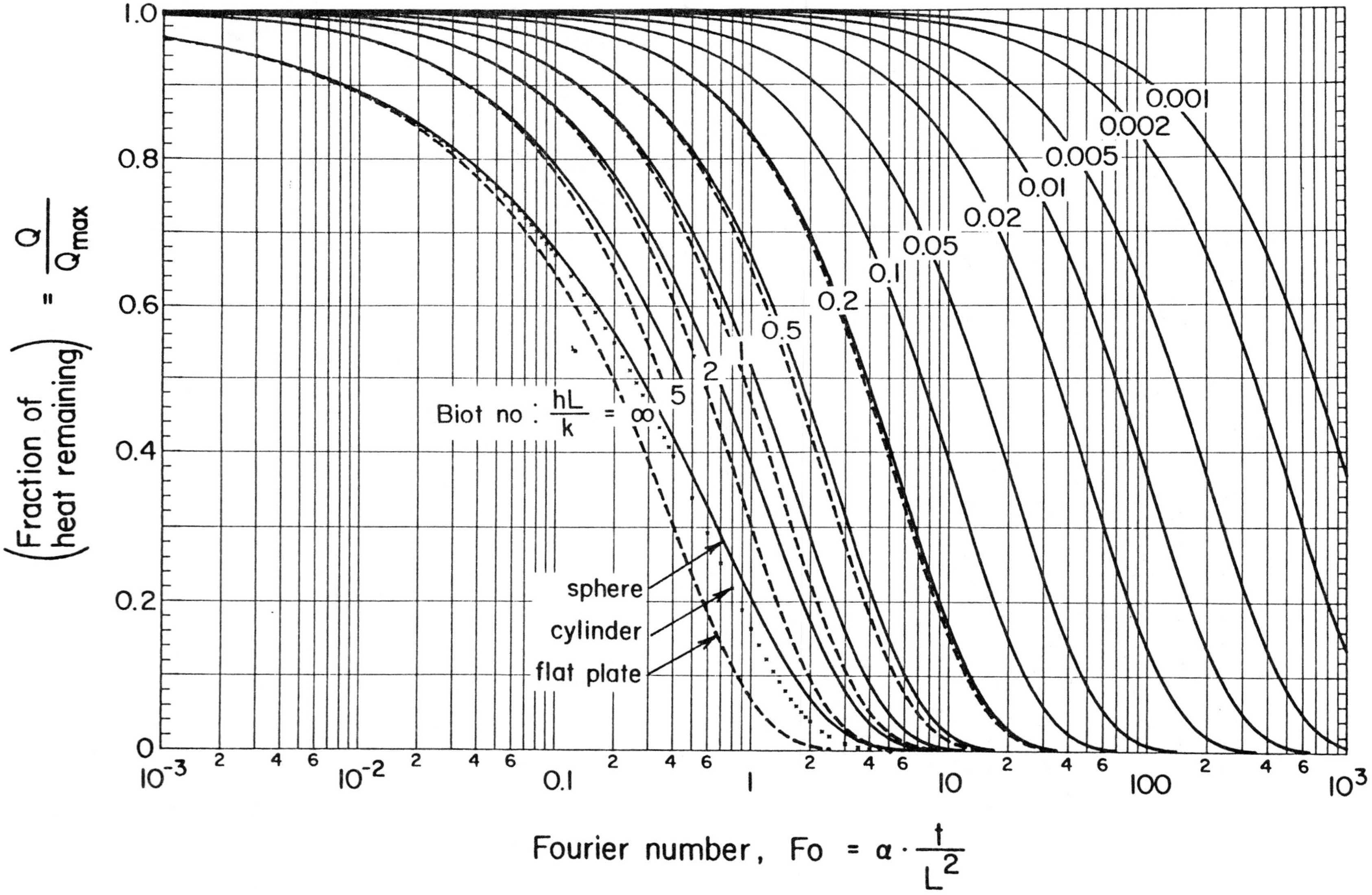

Figure 1 Unsteady state heating or cooling in a constant temperature environment with convective boundary condition for a sphere, infinite cylinder, and infinite flat plate.

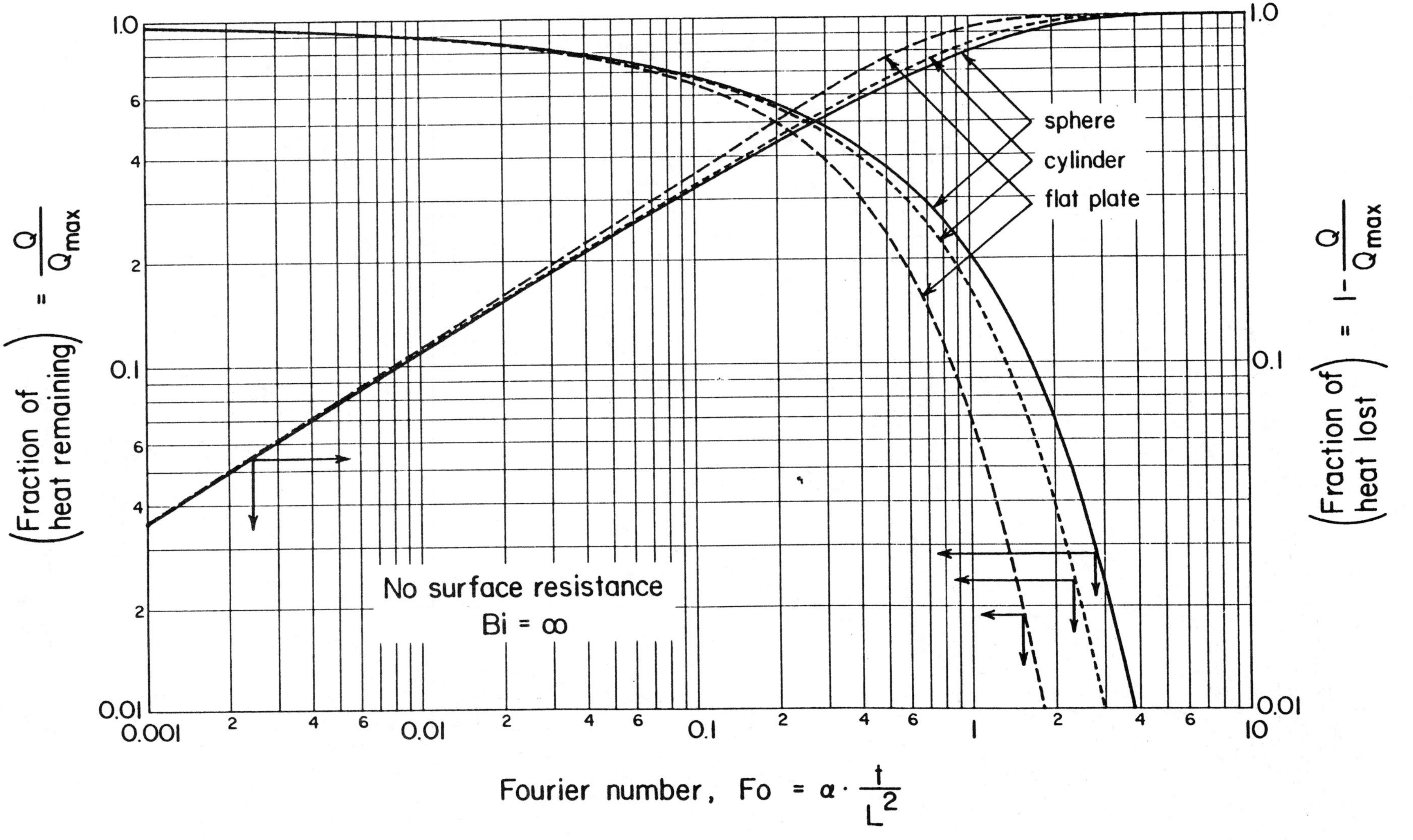

Figure 2 Unsteady state heating or cooling for the case of zero film resistance in a constant temperature environment.

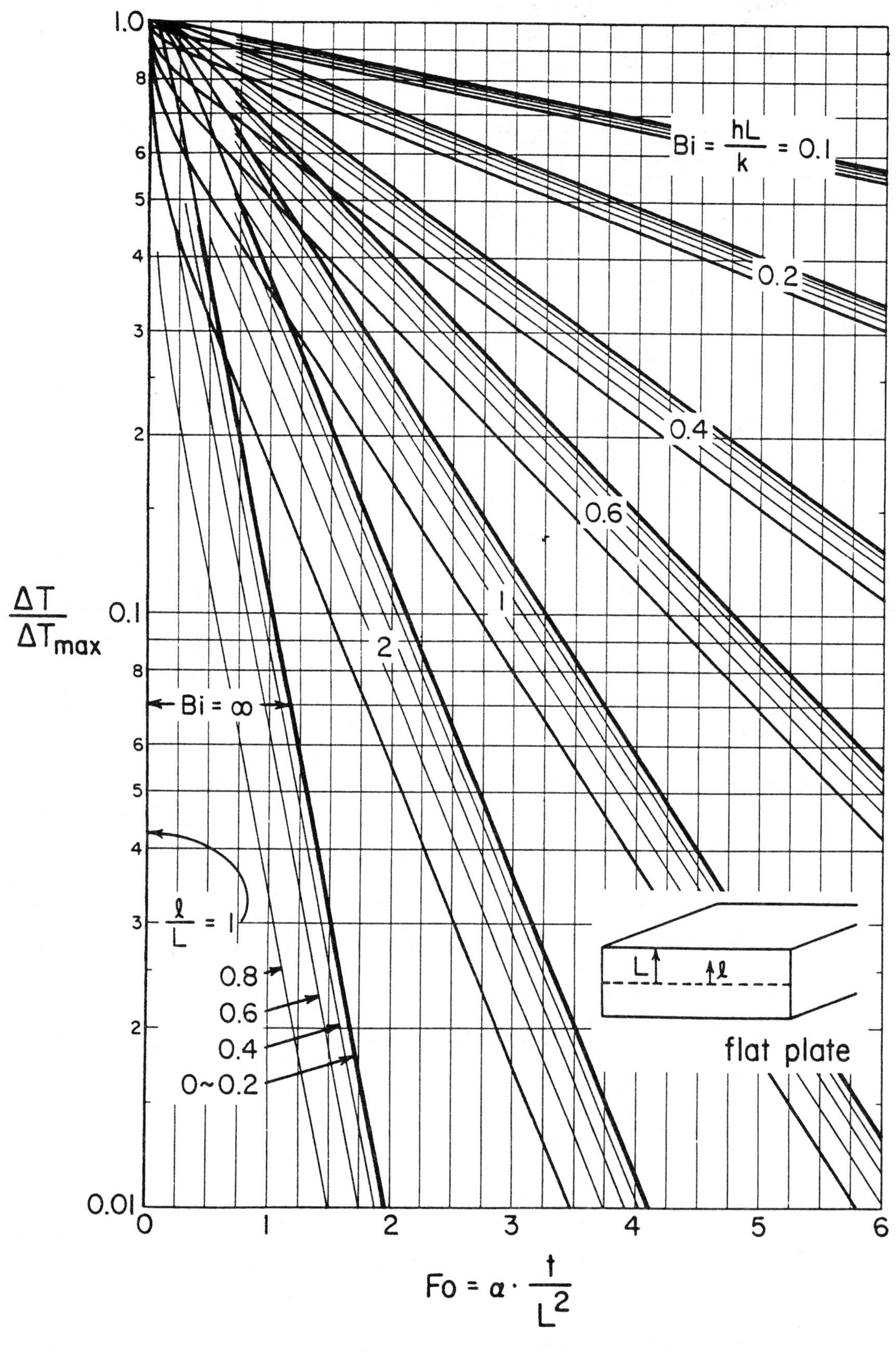

Figure 3 Unsteady state temperature history for an infinite flat plate in a constant temperature environment.

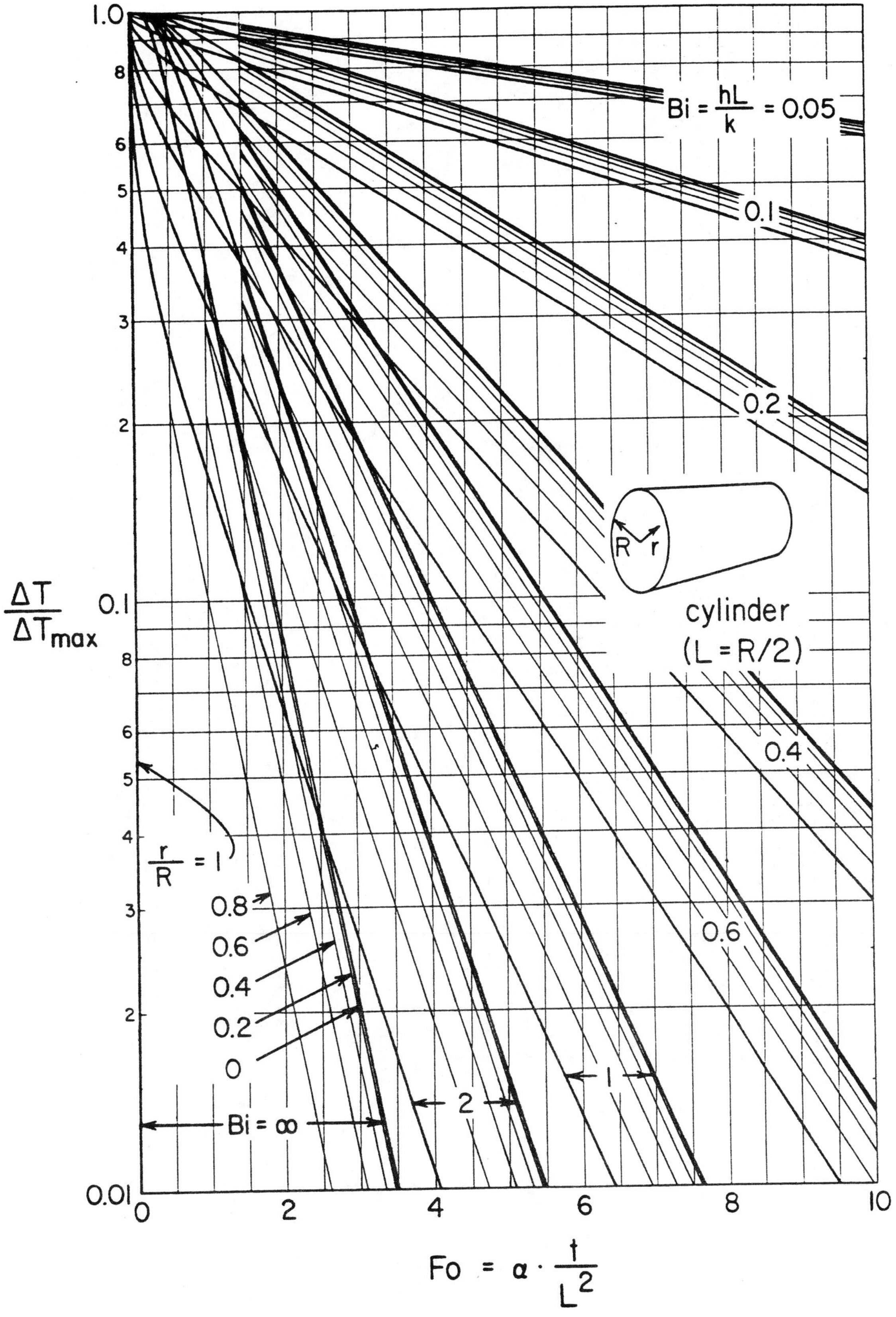

Figure 4 Unsteady state temperature history for an infinite cylinder in a constant temperature environment.

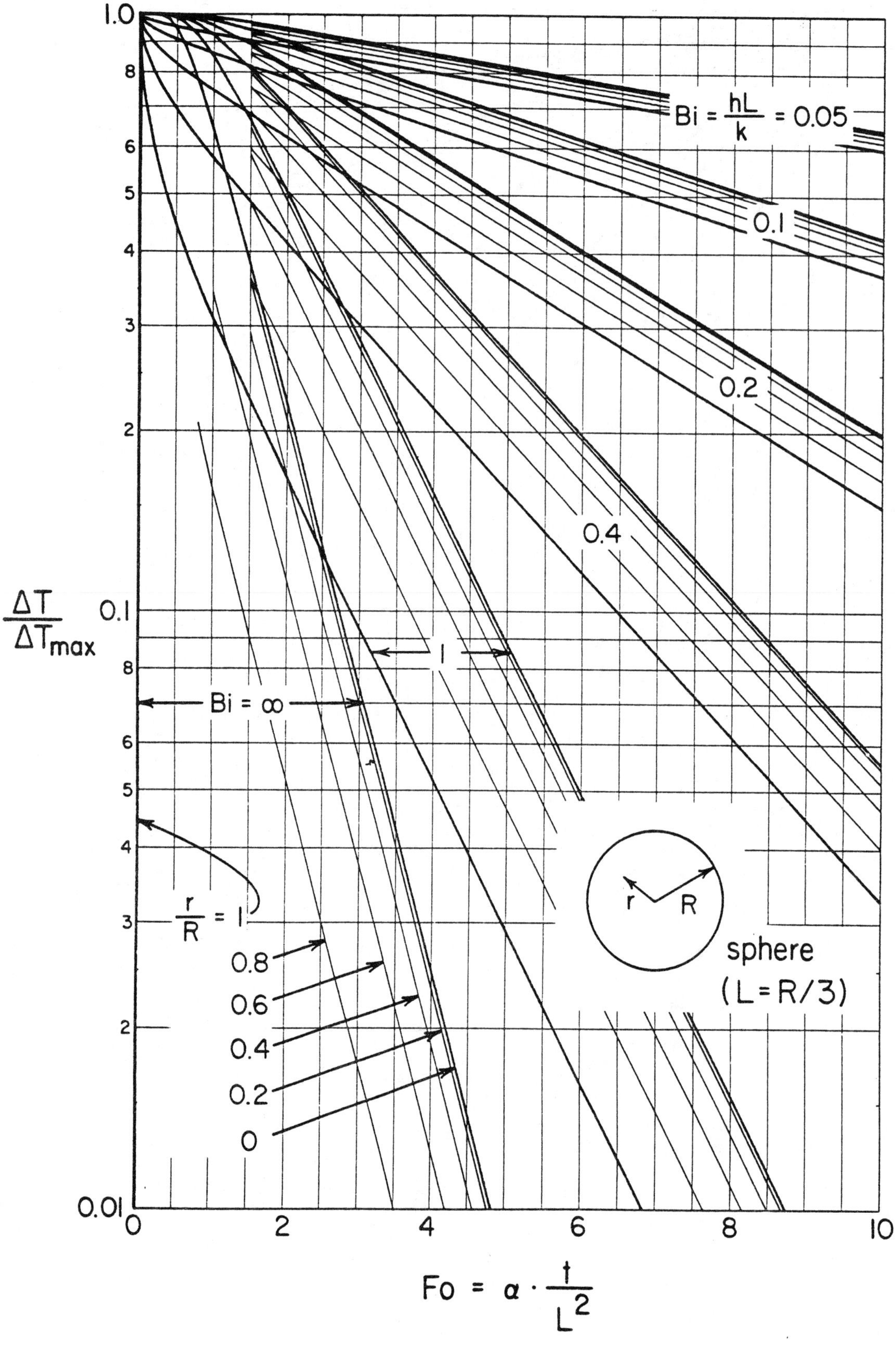

Figure 5 Unsteady state temperature history for a sphere in a constant temperature environment.

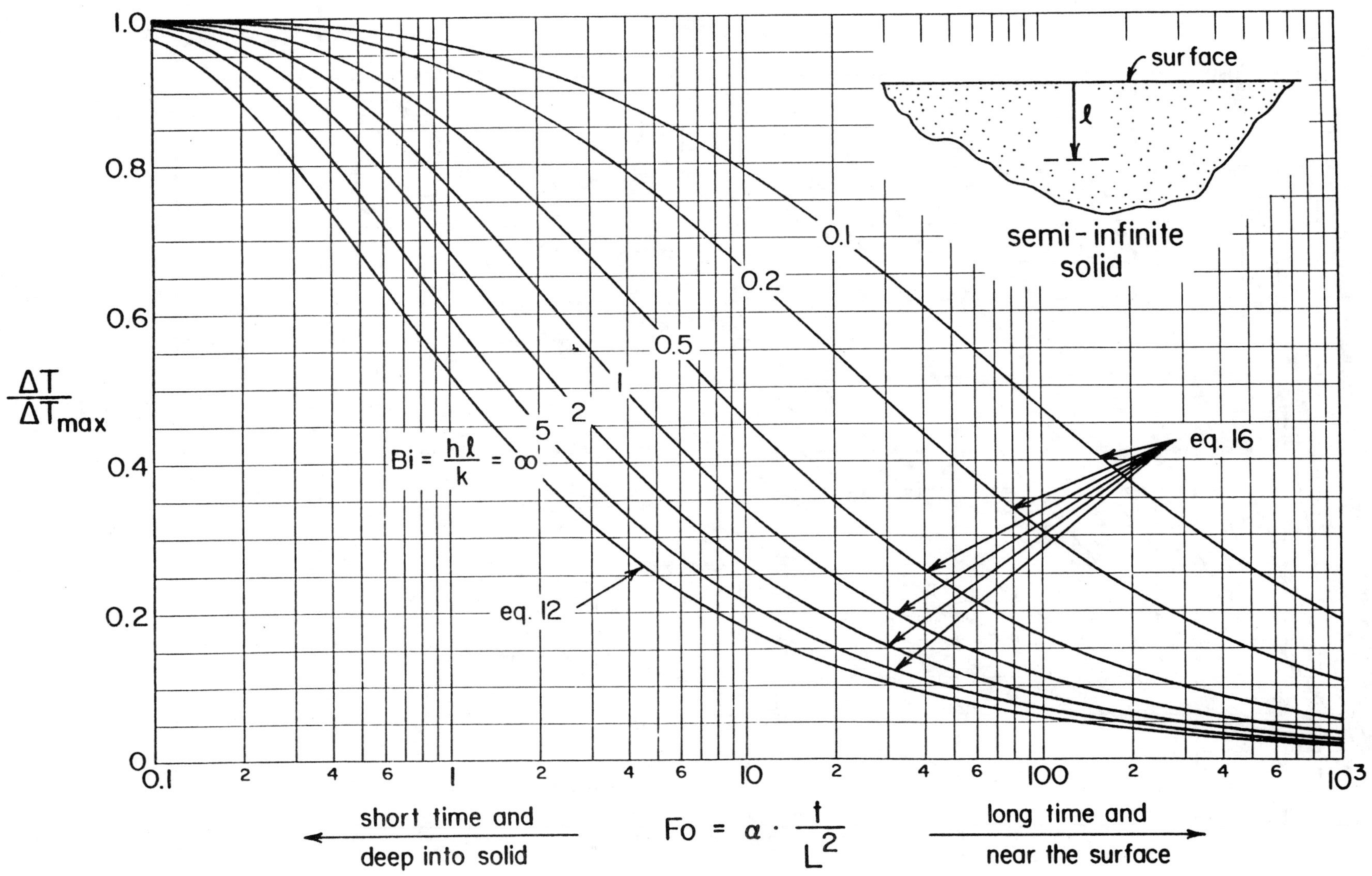

Figure 6 Unsteady state temperature history for a semi-infinite solid exposed to a constant temperature environment.

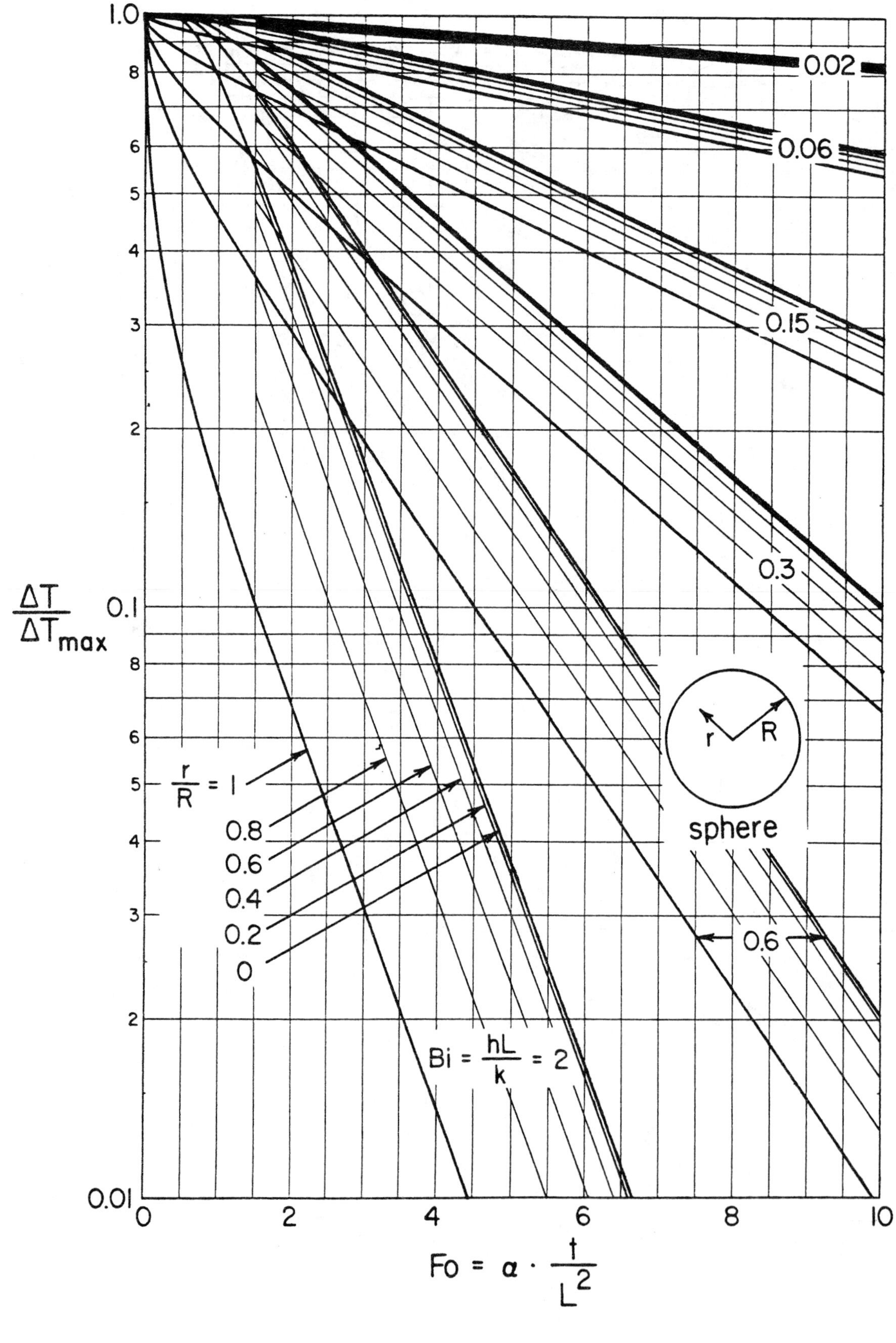
1.0
8
6
5
4
3
2
0.1
0.01
ΔT / ΔT_max
0.02
0.06
0.15
0.3
0.6
r/R = 1
0.8
0.6
0.4
0.2
0
Bi = hL/k = 2
r
R
sphere
0
2
4
6
8
10
Fo = α · t/L²

TWO SIMPLE PROGRAMS FOR USE WITH THE IBM PC FOR NUMERICAL SOLUTIONS OF STEADY AND UNSTEADY STATE HEAT TRANSFER PROBLEMS

NOTE: THE FOLLOWING COMPUTER PROGRAM FOR THE I.B.M.-PC HAS BEEN USED FOR SEVERAL OF THE CHAPTER 3 PROBLEMS. IT USES THE GAUSS-SEIDEL ITERATION METHOD FOR THE SOLUTION OF THE NODAL EQUATIONS.

```
610 REM SSHT Gauss Seidel solution of steady state heat transfer"
620 REM "X = number of temperature nodes"
630 REM "Z = number of iterations"
640 REM "The user is to enter the equations in the form of Eq.(3-32)"
650 REM "of Holman: HEAT TRANSFER starting at line 800."
652 REM "If resistor values are to be calculated enter at line 800"
653 REM " before listing equations."
660 INPUT "No. of equations ="; X
661 DIM T(X)
670 FOR N = 1 TO X:T(N) = 0: NEXT N
680 Y = 0
690 INPUT "No. of iterations = ";Z: IF Z = 0 THEN 1050
1000 Y = Y + 1:IF Y < Z THEN 800
1010 PRINT "NUMBER OF ITERATIONS: ";Y
1020 FOR N = 1 TO X: PRINT "T(";N;") = "; T(N):NEXT N
1030 PRINT "If you want more iterations enter them when asked"
1040 GOTO 690
1050 END
```

NOTE: THE FOLLOWING COMPUTER PROGRAM FOR THE I.B.M-PC HAS BEEN USED FOR SEVERAL OF THE CHAPTER 4 PROBLEMS. IT USES EXPLICIT FORMULATION OF EQUATION 4-47 OF THE TEXT.

```
10 REM "TSHT Transient solution of heat transfer problems. "
11 REM "Enter equations starting at line 130 in the form of Eq.(4-47)"
12 REM "of Holman: HEAT TRANSFER. TP(N) = fcn of T(N)'s"
13 REM "X = number of equations"
14 REM" Z = number of time increments"
15 REM "If resistor or capacitance variables are to be calculated"
16 REM "enter at line 130 before listing equations. Time increment"
17 REM "magnitude may also be listed at this point if desired"
20 REM "Enter the initial values of T(N) for N = 1 to X"
21 REM "starting at line 40."
100 P = 0
110 INPUT "Number of equations = ";X
111 DIM T(X+1),TP(X+1)
120 INPUT "Number of time increments = ";Z
121 IF Z = 0 THEN END
500 P = P+1
505 FOR N = 1 TO X: T(N) = TP(N): NEXT N
510 IF P<Z GOTO 130
520 PRINT "Number of time increments:";P
530 FOR N = 1 TO X: PRINT "T(";N;") = ";T(N):NEXT N
540 PRINT "For more time increments, enter value when asked."
550 GOTO 120
```

SOLUTIONS TO PROBLEMS IN THE TEXT

1-1

$$\Delta T = \frac{(3000)(0.025)}{(0.2)(1.0)} = 375°C$$

1-2

$$\frac{q}{A} = \frac{(0.035)(85)}{0.13} = 82385 \text{ J/hr-m}^2$$

1-3

$$q = -kA\frac{dT}{dx} \qquad \frac{dx}{\pi r^2} = -k\frac{dT}{q}$$

$r = ax + b$; $x = 0$; $r = 0.0375$

$x = 0.3$, $r = 0.0625$

$r = 0.0833x + 0.0375$

$$\int \frac{dx}{\pi(0.0833x + 0.0375)^2} = -\frac{(204)(93-540)}{q}$$

$$\frac{-1}{\pi(0.0833)}\left(\frac{1}{0.0833x + 0.0375}\right)^{-1}\Bigg|_{x=0}^{x=0.3} = -\frac{(204)(-447)}{q}$$

$$q = 2238 \text{ W}$$

1-4

$$\frac{q}{A} = \frac{(0.78)(370-93)}{0.15} = 1440 \ W/m^2$$

1-5

$A = \pi r^2 \qquad q = -k\,4\pi r^2 \frac{dT}{dr} \qquad q = \dfrac{-4\pi k(T_o - T_i)}{\dfrac{1}{r_i} - \dfrac{1}{r_o}}$

$$q = \frac{-4\pi\,(2\times10^{-4})(21.1 + 195.6)}{\dfrac{1}{0.3048} - \dfrac{12}{(13)(0.3048)}} = 2.158\,\text{W}$$

$$\text{mass evaporated} = \frac{2.158}{(85.8)(1055)} = 2.384\times10^{-5}\ \frac{\text{lbm}}{\text{sec}}$$

$$= 2.06\ \frac{\text{lbm}}{\text{day}}$$

1-7

$$\frac{q}{L} = \frac{T_i - T_\infty}{\dfrac{\ln(r_o/r_i)}{2\pi k} + \dfrac{1}{h\pi d_o}} = \frac{30+20}{\dfrac{\ln\left(\frac{30}{25}\right)}{2\pi(7)(10^{-3})} + \dfrac{1}{12(\pi)(0.6)}}$$

$$= 12.05\ \text{W/m}$$

1-8

$$\frac{q}{A} = \frac{(0.161)(200-100)}{0.05} = 322 \ W/m^2$$

1-9

$$\Delta x = \frac{kA\Delta T}{q}$$

$$= \frac{\left(10 \times 10^{-3} \frac{W}{m \cdot {}^\circ C}\right)(500)}{400 \frac{W}{m^2}}$$

$$= 0.0125 \ m = 1.25 \ cm$$

1-10

$$2.158 = (2.7)(4\pi)\left(\frac{13}{12}\right)(0.3048)^2 \Delta T$$

$$\Delta T = 0.632 \ {}^\circ C$$

1-11

$$\frac{q}{A} = (5.669 \times 10^{-8})\left((1073)^4 - (523)^4\right)$$

$$= 70.9 \ \frac{kW}{m^2}$$

1-12

$$\frac{q}{A} = (5.669 \times 10^{-8})((1373)^4 - (698)^4)$$

$$= 188 \ \frac{kW}{m^2}$$

1-13

$$q = (5.669 \times 10^{-8})(4\pi)(0.3048)^2((294.1)^4 - (77.4)^4)$$

$$= 492.8 \ W$$

1-14

(a) $$q = 5.669 \times 10^{-8} [(773)^4 - (373)^4]$$

$$= 1.914 \times 10^4 \ W/m^2$$

(b) $$q = 5.669 \times 10^{-8} [(773)^4 - (T_p)^4]$$

$$= 5.669 \times 10^{-8} [(T_p)^4 - (373)^4]$$

$$T_p = 659 \ K$$

$$q = 9549 \ W/m^2$$

Reduced by 50%

1 - 15

$$q = h A (T_w - T_{fluid})$$

From Table 1-2 $\quad h = 3500 \ \frac{W}{m^2 \cdot °C}$

$$q = (3500)\, \pi d L \,(40)$$

$$= (3500)\, \pi\, (0.025)(3)(40)$$

$$= 32987 \text{ W}$$

$$q = \dot{m}\, c_p\, \Delta T_{fluid}$$

$$32987 \text{ W} = \left(0.5 \ \text{kg/s}\right)\left(4180 \ \text{J/kg °C}\right) \Delta T$$

$$\Delta T = 15.78 \ °C$$

1-16 $\quad h_{fg} = 2257 \ \text{kJ/kg}$

$$q = \dot{m}\, h_{fg} = (3.78 \ \text{kg/hr})(2257 \ \text{kJ/kg})$$

$$= 8531 \ \frac{kJ}{hr} = 2.37 \ \frac{kJ}{s} = 2.37 \text{ kW}$$

From Table 1-2 $\quad h \sim 7500 \ \frac{W}{m^2 \cdot °C}$

$$q = h A (T_w - T_{fluid})$$

$$2370 \text{ W} = (7500)(0.3)^2 (T_w - 100)$$

$$T_w \approx 96.5 \ °C$$

1-17

$$q = hA\Delta T$$

$$3 \times 10^4 \ \frac{Btu}{hr \cdot ft^2} = h\,(232 - 212)\,°F$$

$$h = 1500 \ \frac{Btu}{hr \cdot ft^2} = 8517 \ \frac{W}{m^2 \cdot °C}$$

1-18

$$q = \sigma \epsilon A\left((T_1)^4 - (T_2)^4\right)$$

$$1600 \ W = (5.669 \times 10^{-8})(0.85)(0.006)(3)\left((T_1)^4 - (298)^4\right)$$

$$T_1 = 1167 \ K$$

1-19

$$\frac{q}{A} = \sigma T^4 = (5.669 \times 10^{-8})(1000 + 273)^4$$

$$= 1.489 \times 10^5 \ \frac{W}{m^2}$$

1-20

$$\frac{q}{A} = \sigma T^4$$

$$54 \times 10^6 = (5.669 \times 10^{-8})\,T^4$$

$$T = 5556 \ K$$

1 − 21

$$q = \sigma \epsilon A_1 (T_1^4 - T_2^4)$$

$$= (5.669 \times 10^{-8})(0.65)(4\pi)(0.04)^2(423^4 - 293^4)$$

$$= 18.26 \text{ W}$$

1 − 22

$$q = kA\frac{\Delta T}{\Delta x} = hA(T_0 - T_\infty)$$

$$\frac{(1.4)(315 - 41)}{0.025} = h(41 - 38)$$

$$h = 5114 \ \frac{\text{W}}{\text{m}^2 \cdot {}^\circ\text{C}}$$

1 − 24

$$q = (1.6)\frac{100 - T_{w_2}}{0.4} = 10(T_{w_2} - 10)$$

$$T_{w_2} = 35.7\ ^\circ\text{C}$$

$$q = 10(35.7 - 10) = 257 \ \frac{\text{W}}{\text{M}^2}$$

1-25

From Table 1-2

$$h = 4.5 \ \frac{W}{m^2 \cdot {}^\circ C} \qquad \text{for } \Delta T = 30\,{}^\circ C$$

$$q = hA\Delta T = (4.5)(0.3)^2(30)$$

$$= 12.15 \text{ W}$$

Conduction

$$q = kA\frac{\Delta T}{\Delta x}$$

$$k \text{ for air} = 0.03 \ \frac{W}{m \cdot {}^\circ C}$$

$$q = \frac{(0.03)(0.3)^2(30)}{0.025} = 3.24 \text{ W}$$

1-26

$$\Delta T = \frac{(1500)\left(\frac{0.25}{12}\right)}{25} = 1.25\ {}^\circ F$$

$$T = 100 - 1.25 = 98.75\ {}^\circ F$$

1-27

$$700 = (11)(T_w - 30)$$

$$T_w = 93.6\ {}^\circ C$$

1-28

$$q = q_{conv} + q_{rad}$$

$$q_{conv} = hA(T_w - T_\infty)$$

From Table 1-2 $\quad h = 180 \ \frac{W}{m^2 \cdot {}^\circ C}$

$$q_{conv} = (180)\pi(0.05)(1)(200-30)$$

$$= 4807 \ \frac{W}{m} \text{ length}$$

$$q_{rad} = \sigma \epsilon A_1 (T_1^4 - T_2^4)$$

$$= (5.669 \times 10^{-8})(0.7)\pi(0.05)(1)(473^4 - 283^4)$$

$$= 272 \ \frac{W}{m} \text{ length}$$

$$q_{total} = 4807 + 272 = 5079 \ \frac{W}{m}$$

Most heat transfer is by convection

1-29

$$q = q_{conv} + q_{rad}$$

$$q_{conv} = h A (T_w - T_\infty)$$

From Table 1-2 $h = 4.5 \ \frac{W}{m^2 \cdot °C}$

$$q_{conv} = (4.5)(0.3)^2 (50 - 20)(2 \text{ sides})$$

$$= 24.3 \text{ W}$$

$$q_{rad} = \sigma \epsilon A_1 (T_1^4 - T_2^4)$$

$$= (5.669 \times 10^{-8})(0.8)(0.3)^2 (323^4 - 293^4)(2 \text{ sides})$$

$$= 28.7 \text{ W}$$

$$q_{total} = 24.3 \text{ W} + 28.7 \text{ W} = 53 \text{ W}$$

Convection and radiation are about the same magnitude

1-30

$$q = q_{conv} + q_{rad} = 0 \quad \text{(insulated)}$$

$$q_{conv} = hA(T_w - T_\infty)$$

From Table 1-2 $h = 12 \ \frac{W}{m^2 \cdot °C}$

$$q_{rad} = \sigma \epsilon A_1 (T_1^4 - T_2^4), \quad \epsilon = 1.0 \qquad T_2 = 35°C = 308\ K$$

$$0 = hA_1(T_1 - T_\infty) + \sigma \epsilon A_1 (T_1^4 - T_2^4)$$

$$0 = (12)(T_1 - 273) + (5.669 \times 10^{-8})(1.0)(T_1^4 - 308^4)$$

Solution by iteration:

$$T_1 = T_W = 285\ K = 12°C$$

1-31

$$(100)(353 - T_{w_1}) = (5.669 \times 10^{-8})(T_{w_1}^4 - T_{w_2}^4)$$

$$= 15(T_{w_2} - 293)$$

$$15(T_{w_2} - 293) - (5.669 \times 10^{-8})\left[(397 - 0.15\,T_{w_2})^4 - (T_{w_2})^4\right] = 0 = f(T_{w2})$$

T_{w_2}	$f(T_{w_2})$
320	158.41
350	907.22
310	−77.03
313.3	0.058

$$T_{w_1} = 397 - (0.15)(313.3) = 350\ K$$

1-38

$$q = q_{conv} + q_{rad}$$

$$q_{conv} = hA(T_w - T_\infty)$$

$$= (2)\pi(1)(6)(78-68)$$

$$= 377 \text{ Btu/hr}$$

For $T_2 = 45°F = 505°R$

$$q_{rad} = \sigma \epsilon A_1 (T_1^4 - T_2^4)$$

$$= (0.1714 \times 10^{-8})(0.9)\pi(1)(6)(538^4 - 505^4)$$

$$= 544 \text{ Btu/hr}$$

$$q_{total} = 377 + 544 = 921 \text{ Btu/hr}$$

For $T_2 = 80°F = 540°R$

$$q_{rad} = (0.1714 \times 10^{-8})(0.9)\pi(1)(6)(538^4 - 540^4)$$

$$= -36.4 \text{ Btu/hr}$$

$$q_{total} = 377 - 36.4 = 340.6 \text{ Btu/hr}$$

Conclusion: Radiation plays a very important role in "thermal comfort".

2-1

$$q = \frac{T_i - T_o}{\left.\frac{\Delta x}{kA}\right)_w + \left.\frac{\Delta x}{kA}\right)_{ins}} \qquad 1830 = \frac{1300 - 30}{\frac{0.02}{1.3} + \frac{\Delta x}{0.35}}$$

$$\Delta x = 0.238 \text{ m}$$

2-2

Assume Linear variation : $k = k_o(1 + \beta T)$

$$q = -\frac{k_o A}{\Delta x}\left[T_3 - T_1 + \frac{\beta}{2}(T_3^2 - T_1^2)\right]$$

$$-\frac{k_o A}{\Delta x/2}\left[T_2 - T_1 + \frac{\beta}{2}(T_2^2 - T_1^2)\right]$$

$T_3 = 95°C \quad T_2 = 62°C \quad T_1 = 35°C \quad \Delta x = 0.025$

$$2\left[62 - 35 + \frac{\beta}{2}(62^2 - 35^2)\right] = \left[95 - 35 + \frac{\beta}{2}(95^2 - 35^2)\right]$$

$$\beta = -4.68 \times 10^{-3}$$

$$1000 = q = \frac{k_o(0.1)}{0.025}\left[95 - 35 - \frac{4.68 \times 10^{-3}}{2}(95^2 - 35^2)\right]$$

$$k_o = 5.988$$

$$k = 5.988(1 - 4.68 \times 10^{-3}\, T) \ \frac{\text{W}}{\text{m} \cdot °\text{C}}$$

2-3

$$\frac{q}{A} = \frac{560}{\frac{0.025}{386} + \frac{3.2\times 10^{-3}}{0.16} + \frac{0.05}{0.038}} = 419 \ \frac{W}{m^2}$$

2-4

$$R = \frac{\Delta x}{kA}$$

$$R_A = \frac{0.025}{(150)(0.1)} = 1.667 \times 10^{-3}$$

$$R_B = \frac{0.075}{(30)(0.05)} = 0.05$$

$$R_c = \frac{0.05}{(50)(0.1)} = 0.01$$

$$R_D = \frac{0.075}{(70)(0.05)} = 0.02143$$

$$R = R_A + R_c + \frac{1}{\frac{1}{R_B} + \frac{1}{R_D}} = 2.667 \times 10^{-2}$$

$$q = \frac{\Delta T}{R} = \frac{370 - 66}{2.667 \times 10^{-2}} = 11\,400 \ W$$

2-5

$$\frac{44000}{A} = \frac{260 - 38}{\frac{0.05}{386} + \frac{0.025}{0.038}} \qquad A = 130.4 \ m^2$$

2-6

$$\frac{q}{A} = \frac{45}{\frac{0.10}{0.69} + \frac{0.025}{0.05}} = 69.78 \text{ W/m}^2$$

2-7

$$\frac{q}{A} = \frac{\Delta T}{R}$$

$$\frac{300}{A} = \frac{175 - 80}{\frac{0.04}{386} + \frac{0.015}{0.038}}$$

$$A = 1.247 \text{ m}^2$$

2-8

Assume one directional No Heat sources

$$q = -kA\frac{dT}{dx} = -k_0 A\left[1 + \beta T^2\right]\frac{dT}{dx}$$

$$= -k_0 A\frac{dT}{dx} - k_0 \beta A T^2 \frac{dT}{dx}$$

Integrating: $q\,\Delta x = -k_0 A \int_{T_1}^{T_2} dT - A k_0 \beta \int_{T_1}^{T_2} T^2\,dT$

$$q = -\frac{k_0 A}{\Delta x}\left[(T_2 - T_1) + \frac{\beta}{3}(T_2^3 - T_1^3)\right]$$

2-9

$$\frac{\partial T}{\partial x} = 300x - 30$$

$$\frac{\partial^2 T}{\partial x^2} = 300 = \frac{1}{\alpha}\frac{\partial T}{\partial \tau} \quad \text{heating up}$$

$$\frac{\partial T}{\partial x} = -30 \quad \text{at } x = 0$$

$$= 60 \quad \text{at } x = 0.3$$

$$\frac{q}{A} = -(0.04)(-30) = +1.2 \ \frac{W}{m^2} \quad @ \ x = 0$$

$$= -(0.04)(60) = -2.4 \ \frac{W}{m^2} \quad @ \ x = 0.3$$

2-10

$$R_{Cu} = \frac{0.02}{374} = 5.35 \times 10^{-5}$$

$$R_{As} = \frac{0.003}{0.166} = 0.0181$$

$$R_{Fi} = \frac{0.06}{0.038} = 1.579$$

$$\frac{q}{A} = \frac{500}{\Sigma R} = 313 \ \frac{W}{m^2}$$

2-1

$$R_c = (6/12)/(1.2)(0.5778) = 0.721$$

$$R_f = (2/12)/(0.038)(0.5778) = 7.59$$

$$R_g = (0.375/12)/(0.05)(0.5778) = 1.082$$

$$R_i = \frac{1}{2.0} = 0.5 \qquad R_o = \frac{1}{7} = 0.143$$

$$\frac{q}{A} = \frac{72-20}{\Sigma R} = 5.18 \ \frac{Btu}{hr \cdot ft^2}$$

$$U = \frac{1}{\Sigma R} = 0.0996$$

2-12

$$R_{ss} = \frac{\Delta x}{k} = \frac{0.004}{16} = 0.00025$$

$$R_{overall} = \frac{1}{U} = \frac{1}{120} = 0.00833$$

$$\frac{\Delta T_{ss}}{\Delta T_{overall}} = \frac{R_{ss}}{R_{overall}} = \frac{0.0025}{0.00833} = 0.03$$

$$\Delta T_{ss} = (0.03)(60) = 1.8\ °C$$

2-13

Ice at 0°C $\rho = 999.8$ kg/m3

$V = (0.25)(0.4)(1.0) = 0.1 \ m^3$

$m = 100$ kg

$q = (100)(330 \times 10^3) = 3.3 \times 10^7$ J

$A_i = (2)(0.25)(0.4) + (2)(0.4)(1.0) + (2)(0.25)(1.0)$

$= 1.5 \ m^2$

$A_o = (2)(0.35)(0.5) + (2)(0.5)(1.1) + (2)(0.35)(1.1)$

$= 2.22 \ m^2$

$A_m = 1.86 \ m^2$

$$R_s = \frac{\Delta x}{k A} = \frac{0.05}{(0.033)(1.86)} = 0.8146$$

$$R_o = \frac{1}{h A_o} = 0.045$$

$R = 0.8596$

$$\frac{Q}{\Delta T} = \frac{3.3 \times 10^7}{\Delta T} = \frac{25 - 0}{0.8596}$$

$\Delta T = 1.135 \times 10^6$ sec

$= 315$ hr.

$= 13$ days

2-14

$$q \text{ (no ins.)} = hA(T_w - T_\infty)$$

$$= (25)(4\pi)(0.5)^2(120-15)$$

$$= 8247 \text{ W}$$

$$k_{foam} = \frac{18 \text{ mW}}{\text{m} \cdot ^\circ\text{C}}$$

$$q = \frac{4\pi k (T_i - T_o)}{\frac{1}{r_i} - \frac{1}{r_o}} = h 4\pi r_o^2 (T_o - T_\infty)$$

$$\frac{(0.018)(120-40)}{\frac{1}{0.5} - \frac{1}{r_o}} = (25) r_o^2 (40-15)$$

$$r_o = 0.5023 \text{ m}$$

$$\text{thk} = r_o - r_i = 0.023 \text{ m}$$

$$q \text{ (w/ins)} = (25)(4\pi)(0.5023)^2(40-15)$$

$$= 1982 \text{ W}$$

2-15

$$q = \frac{4\pi k (T_i - T_o)}{\frac{1}{r_i} - \frac{1}{r_o}} \qquad k = 204 \frac{\text{W}}{\text{m} \cdot ^\circ\text{C}}$$

$$= \frac{(4)\pi(204)(100-50)}{\frac{1}{0.02} - \frac{1}{0.04}} = 5127 \text{ W}$$

2-16 $q = \dfrac{\Delta T}{\Sigma R}$

$$R_{alum} = \frac{\frac{1}{0.02} - \frac{1}{0.04}}{4\pi\,(204)} = 9.752 \times 10^{-3}$$

$$R_{ins} = \frac{\frac{1}{0.04} - \frac{1}{0.05}}{4\pi\,(0.05)} = 7.958$$

$$R_{conv} = \frac{1}{hA} = \frac{1}{(20)(4\pi)(0.05)^2}$$

$$= 1.592$$

$$q = \frac{100 - 10}{0.00975 + 7.958 + 1.592} = 9.41 \text{ W}$$

2-17 $d_i = 2.90$ in $\quad d_o = 3.50$ in $\quad k = 43 \frac{W}{m \cdot °C}$

$$R_{steel} = \frac{\ln(3.5/2.9)}{(2\pi)(43)(1)} = 6.96 \times 10^{-4}$$

$$R_{ins} = \frac{\ln(5.5/3.5)}{(2\pi)(0.06)(1)} = 1.1999$$

$$R_{conv} = 1/hA_o = \frac{1}{(10)\,\pi\,(5.5)(0.0254)}$$

$$= 0.2278$$

$$R_{tot} = 1.427$$

$$q = \frac{\Delta T}{R} = \frac{250 - 20}{1.427} = 161.1 \ \frac{W}{m}$$

2-18

$k_A = 0.166 \quad k_S = 0.0485 \ \frac{W}{m \cdot °C}$

$$\frac{315 - T_i}{\frac{\ln\left(\frac{31.4}{25}\right)}{0.166}} = \frac{T_i - 38}{\frac{\ln\left(\frac{56.4}{31.4}\right)}{0.0485}}$$

$$0.7283(315 - T_i) = 0.0828(T_i - 38)$$

$$T_i = 286.7 °C$$

2-19

$$q_r = -k 4\pi r^2 \frac{dT}{dr}$$

$$q_r \int_{r_i}^{r_o} 1/r^2 \, dr = -k 4\pi \int_{T_i}^{T_o} dT$$

$$q_r \left(\frac{1}{r_o} - \frac{1}{r_i}\right) = -4\pi k (T_o - T_i)$$

$$q = -\frac{4\pi k (T_o - T_i)}{\left(\frac{1}{r_o} - \frac{1}{r_i}\right)}$$

$$R_{th} = \frac{\left(\frac{1}{r_i} - \frac{1}{r_o}\right)}{4\pi k}$$

2-20

$r_i = 0.5\ mm = 5 \times 10^{-4}\ m$

$r_o = \frac{k}{h} = 5 \times 10^{-4} + 2 \times 10^{-4} = 7 \times 10^{-4}$

$k = (7 \times 10^{-4})(120) = 0.084\ W/m \cdot °C$

$q\ (\text{bare wire}) = \pi(0.001)(120)(400 - 40) = 135.7\ W/m$

$q\ (\text{insulated}) = (135.7)(0.25) = 33.93\ W/m$

$$q = \frac{400 - 40}{\dfrac{\ln(r_o / 5 \times 10^{-4})}{2\pi\,(0.084)} + \dfrac{1}{\pi(2)(120)\,r_o}} = 33.93$$

By iteration: $r_o = 135\ mm$

thickness = 134.5 mm

2-21

$R_i = (1)(12) / (30)\pi(2.067) = 6.16 \times 10^{-2}$

$R_p = \ln(2.375/2.067) / 2\pi(27) = 8.188 \times 10^{-4}$

$R_i = \ln(3.375/2.375) / 2\pi(0.023) = 2.432$

$R_o = (1)(12) / 2\pi(3.375) = 5.659 \times 10^{-1}$

$$q/L = \frac{320 - 70}{\Sigma R} = 81.7\ \frac{Btu}{hr \cdot ft}$$

2-22

$$q = -k\,4\pi r^2 \frac{dT}{dr}$$

$$= \frac{k\,4\pi\,(T_o - T_i)}{\frac{1}{r_o} - \frac{1}{r_i}} = h\,4\pi r_o^2\,(T_o - T_\infty)$$

$$q = \frac{T_i - T_\infty}{\frac{1}{4\pi k}\left(\frac{1}{r_i} - \frac{1}{r_o}\right) + \frac{1}{4\pi r_o^2 h}}$$

Take $\frac{dq}{dr_o} = 0$

Result is: $r_o = 2\frac{k}{h}$

2-23

M_w @ 90% full $= (0.9)(970)\,\pi\,(0.8)^2\,(2) = 3511$ Kg

@ 2 °C/hr $q = (3511)(4191)(2)/3600 = 8174$ W

$A = 2\pi\,(0.8)^2 + \pi\,(0.8)(2) = 9.048\ m^2$

Fiberglass boards with $k = 40$ mW/m·°C

$$\Delta x = \frac{(40\times10^{-3})(9.048)(80-20)}{8174} = 2.66\times10^{-3}\ m$$

2-24

For 1 m Length

$$R\ (\text{pipe}) = \frac{\ln(9.1/8)}{2\pi(47)} = 4.363 \times 10^{-4}$$

$$R\ (\text{ins } ①) = \frac{\ln(27.1/9.1)}{2\pi(0.5)} = 0.3474$$

$$R\ (\text{ins } ②) = \frac{\ln(35.1/27.1)}{2\pi(0.25)} = \sout{0.8246} \quad .1646$$

$R\ (\text{tot}) = 1.172$

$$q = \frac{\Delta T}{R} = \frac{250 - 20}{1.172} = 196.2\ \text{W/m} \qquad 448\ \frac{\text{W}}{\text{m}}$$

2-25

Fiberglass $k = 0.038$ $\Delta x = 1.2$ cm $\times 2$

Asbestos $k = 0.154$ $\Delta x = 8.0$ cm

brick $k = 0.69$ $\Delta x = 10$ cm $h = 15\ \frac{\text{W}}{\text{m}^2 \cdot {}^\circ\text{C}} \times 2$

$$U = \frac{1}{\frac{(2)}{15} + \frac{(2)(0.012)}{0.038} + \frac{0.08}{0.154} + \frac{0.1}{0.69}} = 0.70\ \frac{\text{W}}{\text{m}^2 \cdot {}^\circ\text{C}}$$

2-26 $R = 1/k$

	k	R
Fiberglass	0.046	21.74
Urethane	0.018	55.6
Mineral Wool	0.091	11.0
Calcium Silicate	0.058	17.2

2-27

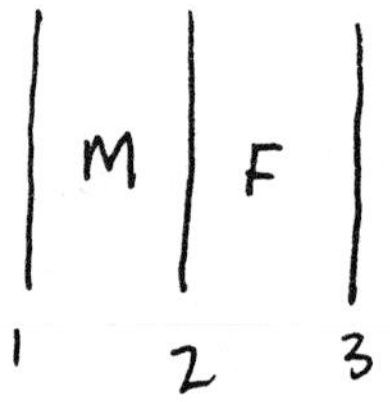

$T_1 = 1000\ °C$

$T_2 = 400\ °C$

$T_3 = 55\ °C$

$k_m = 90\ mW/m \cdot °C$ $\quad h = 15\ W/m^2 \cdot °C$

$k_F = 42\ mW/m \cdot °C$ $\quad T_\infty = 40\ °C$

$$q/A = h(T_3 - T_\infty) = (15)(55-40) = 225\ W/m^2$$

$$q/A = k_m \frac{(1000-400)}{\Delta x_m} \qquad \Delta x_m = 0.24\ m$$

$$q/A = k_F \frac{(400-55)}{\Delta x_F} \qquad \Delta x_F = 0.0644\ m$$

2-28

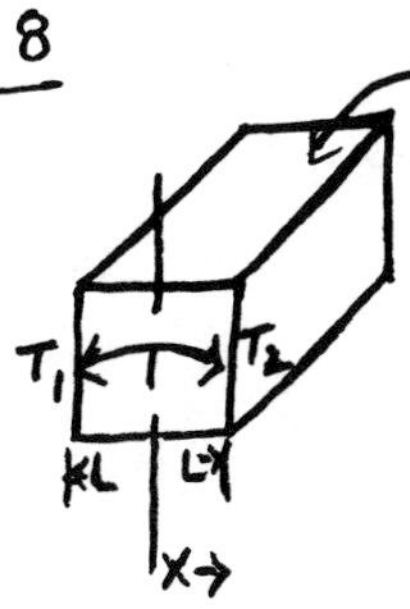

uniformly distributed volume sources.

$$\frac{d^2T}{dx^2} + \frac{\dot{q}}{k} = 0$$

Boundary conditions:

① at $x = -L$ $T = T_1$ ② at $x = L$ $T_1 = T_2$

The general solution is $T = -\frac{\dot{q}x^2}{2k} + c_1 x + c_2$

Substituting Boundary conditions yields

$$T = \frac{\dot{q}}{2k}(L^2 - x^2) + \frac{T_2 - T_1}{2L}(x) + \frac{T_1 + T_2}{2}$$ for Temp. distribution on wall.

2-29

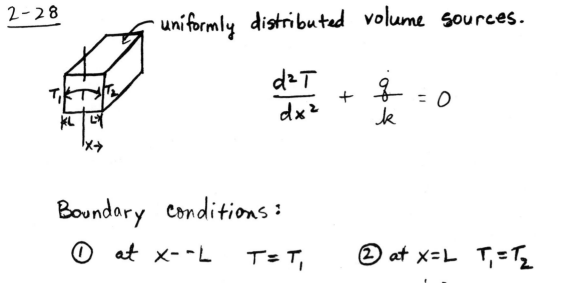

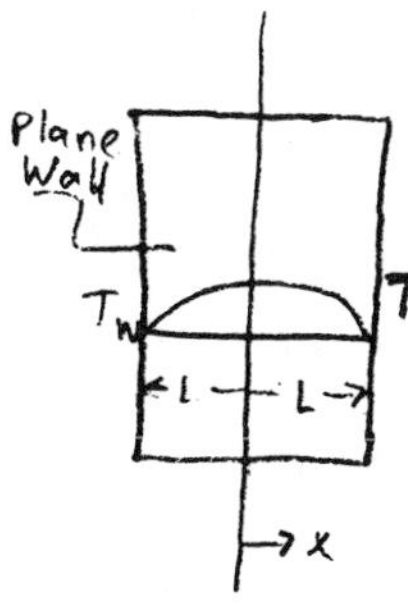

$$\dot{q} = \dot{q}_w\left[1 + \beta(T_r T_w)\right] \quad \frac{d^2T}{dx^2} + \frac{\dot{q}}{k} = 0$$

Boundary conditions:

$T = T_w$ at $x = \pm L$

General Soln.:

$$T - T_w = c_1\left[\cos\left(\sqrt{q_N \beta/k}\right]x + c_2\left[\sin\left(\sqrt{q_N \beta/k}\right)\right]x - 1/\beta$$

From Boundary Conditions: $c_2 = 0$, $C_1 = \frac{1}{\beta \cos\sqrt{q_N \beta/k}\, L}$

2-30

$\dot{q} = 0.30\ MW/m^3$ Same as half of wall 15 cm thick with convection on each side.

$$T_0 - T_w = \frac{\dot{q}L^2}{2k} = \frac{(0.30 \times 10^6)(0.060)^2}{(2)(21)} = 25.7\ °C$$

$$\dot{q}LA = hA(T_w - T_\infty) \qquad T_w - T_\infty = \frac{(0.30 \times 10^6)(0.060)}{570} = 31.6\ °C$$

$$T_0 = T_{max} = 93 + 25.7 + 31.6 = 150.3\ °C$$

2-31

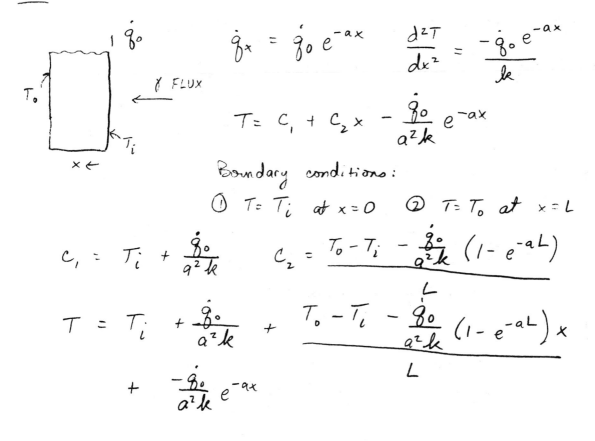

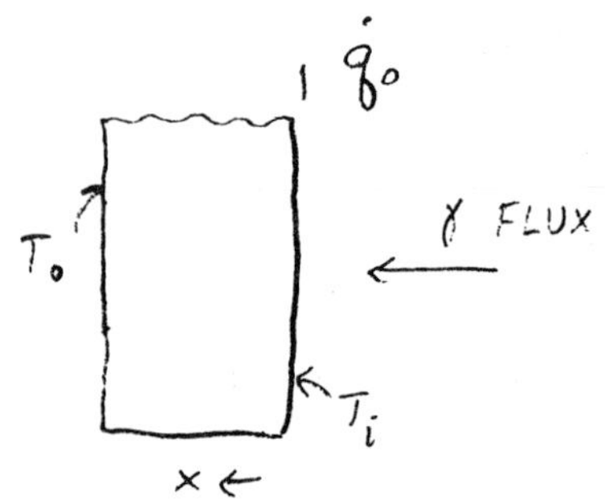

$$\dot{q}_x = \dot{q}_0 e^{-ax} \qquad \frac{d^2T}{dx^2} = \frac{-\dot{q}_0 e^{-ax}}{k}$$

$$T = C_1 + C_2 x - \frac{\dot{q}_0}{a^2 k} e^{-ax}$$

Boundary conditions:

① $T = T_i$ at $x = 0$ ② $T = T_0$ at $x = L$

$$C_1 = T_i + \frac{\dot{q}_0}{a^2 k} \qquad C_2 = \frac{T_0 - T_i - \frac{\dot{q}_0}{a^2 k}(1 - e^{-aL})}{L}$$

$$T = T_i + \frac{\dot{q}_0}{a^2 k} + \frac{T_0 - T_i - \frac{\dot{q}_0}{a^2 k}(1 - e^{-aL})}{L} x$$

$$+ \frac{-\dot{q}_0}{a^2 k} e^{-ax}$$

2-32

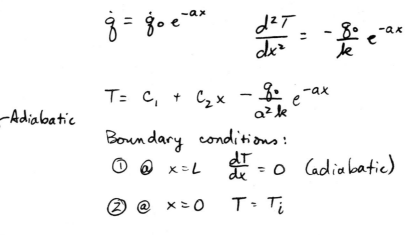

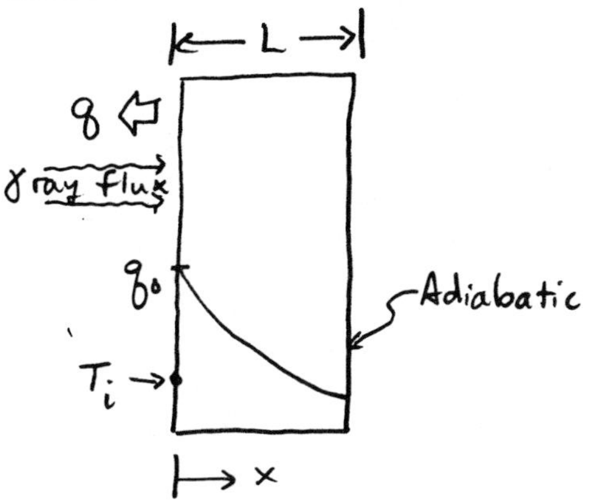

$$\dot{q} = \dot{q}_0 e^{-ax} \qquad \frac{d^2T}{dx^2} = -\frac{q_0}{k} e^{-ax}$$

$$T = C_1 + C_2 x - \frac{q_0}{a^2 k} e^{-ax}$$

Boundary conditions:

① @ $x = L$ $\quad \frac{dT}{dx} = 0$ (adiabatic)

② @ $x = 0$ $\quad T = T_i$

$$C_1 = T_i + \frac{q_0}{a^2 k} \qquad C_2 = -\frac{q_0}{a k} e^{-aL}$$

$$T = T_i + \frac{q_0}{a^2 k} - \frac{q_0 e^{-aL}}{a k} x - \frac{q_0}{a^2 k} e^{-ax}$$

2-33

$$T - T_w = C_1 \cos\sqrt{q_w \beta/k}\, x + C_2 \sin\sqrt{q_w \beta/k}\, x - \frac{1}{\beta}$$

$T = T_1$ @ $x = \pm L$ $\quad C_2 = 0$

$$-kA \left.\frac{\partial T}{\partial x}\right|_{x=L} = hA(T_1 - T_\infty) \qquad C_1 = \frac{T_1 - T_w + 1/\beta}{\cos\sqrt{q_w \beta/k}\, L}$$

$$T = T_w + \frac{T_1 - T_w + 1/\beta}{\cos\sqrt{q_w \beta/k}\, L} \cos\sqrt{q_w \beta/k}\, x - \frac{1}{\beta}$$

Solving for $\left.\frac{\partial T}{\partial x}\right|_{x=L}$ and substituting in above equation:

$$T_1 = \frac{1}{(1-h)}\left[T_w - \frac{1}{\beta} - \frac{h T_\infty}{k\sqrt{q_w \beta/k}\, \tan\sqrt{q_w \beta/k}\, L}\right]$$

2-34

$$\dot{q} AL = h P L (T_w - T_\infty)$$

$$(35.3 \times 10^6)(0.025)^2 = (4000)(4)(0.025)(T_w - 20)$$

$$T_w = 75.16\ ^\circ C$$

2-35

$$\frac{d^2T}{dx^2} + \frac{\dot{q}_0 \cos(ax)}{k} = 0 \qquad \frac{dT}{dx} = \frac{-\dot{q}_0}{ak} \sin(ax) + C_1$$

$T = T_w$ at $x = \pm L$ $\therefore C_1 = 0$ $\qquad T = \frac{\dot{q}_0}{a^2 k} \cos(ax) + C_1 x + C_2$

$$T_w = \frac{\dot{q}_0}{a^2 k} \cos(aL) + C_2 \qquad T - T_w = \frac{\dot{q}_0}{a^2 k} \left(\cos(ax) - \cos(aL)\right)$$

$$\frac{q}{A} = -2k \left.\frac{dT}{dx}\right|_{x=L} = -2k \frac{\dot{q}_0}{ak}\left(-\sin(aL)\right) = \frac{2k\dot{q}_0}{ak} \sin(aL)$$

2-36 $k = 0.0124 \text{ W/cm}\cdot°C = 1.24 \text{ W/m}\cdot°C$

$\rho = 1.5 \times 10^{-3} \ \Omega\text{-cm}$

$R = (1.5 \times 10^{-3})(3/1) = 4.5 \times 10^{-3}$

$q = I^2 R = (50)^2 (4.5 \times 10^{-3}) = 11.25 \text{ W}$

$\dot{q} = q/V = \frac{11.25}{3 \times 10^{-6}} = 3.75 \text{ MW/m}^3 \qquad T = -\frac{\dot{q}}{2k} x^2 + c_1 x + C_2$

$L = 1.5 \text{ cm} = 0.015 \text{ m} \qquad T = 300 \text{ at } x = -0.015$

$T = 100 \text{ at } x = +0.015$

$300 - 100 = c_1(-0.015 - 0.015) \quad c_1 = -6667$

$300 = \frac{(-3.75 \times 10^6)(0.015)^2}{(2)(1.24)} - (6667)(-0.015) + C_2 \quad C_2 = 540.2$

$T \text{ at } x = 0 = C_2 = 540.2°C$

2-37

k = constant $\dot{q} = \dot{q}_0$ at $x = 0$ Assume one directional with no heat storage.

$$\frac{d^2T}{dx^2} + \frac{\dot{q}}{k} = 0 \qquad T - T_1 = (T_2 - T_1)(C_1 + C_2 x^2 + C_3 x^3)$$

$$T = T_1 + (T_2 - T_1)(C_1 + C_2 x^2 + C_3 x^3)$$

$$\frac{d^2T}{dx^2} = (T_2 - T_1)(2C_2 + 3C_3 x)$$

$T = T_1$ @ $x = 0$

$T = T_2$ @ $x = L$

$\dot{q} = \dot{q}_0$ @ $x = 0$

$$q = -k(T_2 - T_1)(2C_2 + 3C_3 x)$$

so: $C_1 = 0 \qquad C_2 = -\dfrac{q_0}{2k(T_2 - T_1)} \qquad C_3 = \dfrac{1}{L^3} + \dfrac{q_0}{2kL(T_2 - T_1)}$

$$\dot{q}_x = \dot{q}_0 - \left[\frac{3k}{L^3}(T_2 - T_1) + \frac{3q_0}{2L}\right]x$$

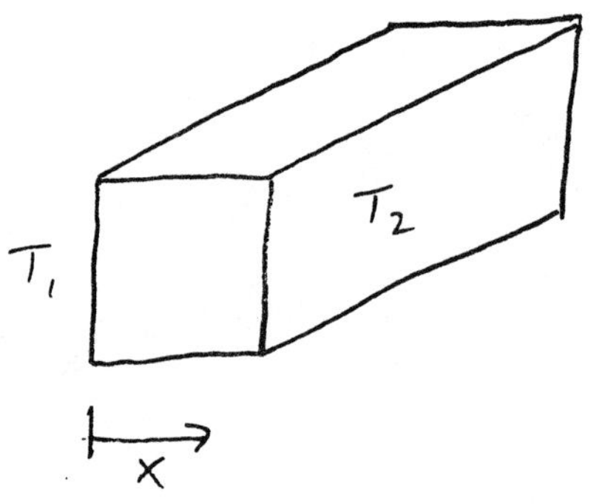

2-39 Use solution from Prob. 2-28

$T = T_0$ at $x = 0$

$$T_0 = \frac{\dot{q}}{2k} L^2 + \frac{T_1 + T_2}{2}$$

$$= \frac{(5 \times 10^5)(0.015)^2}{(2)(20)} + \frac{200 + 60}{2}$$

$$= 127.8\ °C$$

2-40 Use solution from Prob. 2-28

$$T_0 = \frac{\dot{q}}{2k} L^2 + \frac{T_1 + T_2}{2}$$

$$= \frac{(500 \times 10^6)(0.005)^2}{(2)(20)} + \frac{100 + 200}{2}$$

$$= 462.5\ °C$$

2-41 Behaves like half a plate having a thickness of 8 mm. Max Temp is at $x = 0$

$$T_0 = \frac{\dot{q} L^2}{2k} + T_w$$

$L = 0.004$ m $\quad T_w = 100\ °C$

$$T_0 = \frac{(200 \times 10^6)(0.004)}{(2)(25)} + 100$$

$$= 164\ °C$$

2-42

$$q \pi r^2 L = \frac{E^2}{R} = \frac{(10)^2 \pi (0.16)^2}{(7\times10^{-6})(30)}$$

$$\dot{q} = \frac{(10)^2 \pi (0.16)^2}{(7\times10^{-6})(30)(0.3)\pi(1.6\times10^{-3})^2} = 1587 \ MW/m^3$$

$$T_0 = \frac{\dot{q} r^2}{4k} + T_w = \frac{(1.587\times10^9)(1.6\times10^{-3})^2}{(4)(22.5)} + 93$$

$$T_0 = 138.1 \ °C$$

2-43

$$(200)^2 (0.099) = (5700)\pi(3\times10^{-3})(1)(T_w - 93)$$

$T_w = 166.7°C$ $\quad T_0 = 16.6 + 166.7 = 183.3°C$

2-44

$q = EI = \dot{q}\pi(r_o^2 - r_i^2)L = h\,2\pi L\,(T_i - T_f)$

$$\frac{d^2T}{dr^2} + \frac{1}{r}\frac{dT}{dr} + \frac{\dot{q}}{k} = 0 \qquad T = \frac{-\dot{q}r}{4k} + c_1 \ln r + c_2$$

$T = T_o$ at $r = r_o$ $\quad \frac{dT}{dr} = 0$ at $r = r_o$ $\quad T_o = \frac{-\dot{q}r}{4k} + c_1 \ln r_o + c_2$

$$\frac{dT}{dr} = 0 = \frac{-\dot{q}r}{2k} + \frac{c_1}{r_o} \qquad c_1 = \frac{\dot{q}r_o^2}{2k} \qquad T_i = \frac{-\dot{q}r}{4k} + \frac{\dot{q}r_o^2}{2k}\ln r_i + c_2$$

$$T_o = \frac{-\dot{q}r_o^2}{4k} + \frac{\dot{q}r_o^2}{2k}\ln r_o + c_2 \qquad T_i = T_o - \frac{\dot{q}}{4k}(r_i^2 - r_o^2) + \frac{\dot{q}r_o^2}{2k}\ln\left(\frac{r_i}{r_o}\right) \text{(a)}$$

$$\dot{q} = \frac{EI}{\pi(r_o^2 - r_i^2)L} \quad \text{(b)}$$

Insert (a) and (b) in
$EI = 2\pi r L h(T_i - T_f)$
and solve for h.

2-45

$\dot{q}$ uniform $T = T_w$ @ $r = R$ steady state, T varies only with r.

$$\frac{1}{r}\frac{\partial^2 (rT)}{\partial r^2} + \frac{1}{r^2 \sin\theta}\frac{\partial}{\partial\theta}\left(\sin\theta \frac{\partial T}{\partial\theta}\right) + \frac{1}{r^2\sin^2\theta}\frac{\partial^2 T}{\partial\theta^2} + \frac{\dot{q}}{k} = \frac{1}{\alpha}\frac{\partial T}{\partial t}$$

this reduces to:

$$\frac{1}{r}\frac{d^2(rT)}{dr^2} + \frac{\dot{q}}{k} = 0 \qquad \frac{d^2(rT)}{dr^2} = -\frac{\dot{q}r}{k} \qquad \text{Integrating } T = -\frac{\dot{q}r^2}{6k} + C_1 + \frac{C_2}{r}$$

Boundary conditions: $\dot{q}\frac{4}{3}\pi R^3 = -k\, 4\pi R^2 \left.\frac{dT}{dr}\right)_{r=R}$

① $\frac{dT}{dr} = -\frac{\dot{q}r}{3k}$ ② $T = T_w$ @ $r = R$

③ $\left.\frac{dT}{dr}\right)_{r=R} = 0$ then $C_1 = T_w + \frac{\dot{q}R^2}{6k}$ $C_2 = 0$

$$T - T_w = \frac{\dot{q}}{6k}(R^2 - r^2)$$

2-46

From Prob. 2-45

$$T - T_w = \frac{\dot{q}}{6k}(R^2 - r^2)$$

$$T_0 - T_w = \frac{(1\times10^6)(0.02)^2}{(6)(16)} = 4.17\ ^\circ C$$

$$q = \dot{q}V = \dot{q}\frac{4}{3}\pi R^3 = h\, 4\pi r^2 (T_w - T_\infty)$$

$$T_w - T_\infty = \frac{(1\times10^6)(0.02)}{(3)(15)} = 444.4\ ^\circ C$$

$$T_0 = 444.4 + 4.17 = 448.6\ ^\circ C$$

2-47 $R/L = \rho/\pi r_0^2 = \dfrac{2.9\times10^{-6}}{\pi(1.5)^2} = 4.1\times10^{-7}\ \Omega/km = 4.1\times10^{-5}\ \Omega/m$

$$\dot{q} = \frac{I^2R}{V} = \frac{(230)^2(4.1\times10^{-5})}{\pi(0.015)^2} = 3.07\times10^3\ W/m^3$$

$$T_0 = \frac{(3.07\times10^3)(0.015)^2}{(4)(190)} + 180 = 180.0009\ °C$$

2-48 $\dot{q} = a + br \qquad \dfrac{d}{dr}\left(r\dfrac{dT}{dr}\right) = -\dfrac{\dot{q}r}{k} = -\dfrac{(ar + br^2)}{k}$

$$r\frac{dT}{dr} = -\frac{1}{k}\left(\frac{ar^2}{2} + \frac{br^3}{3}\right) + C_1 \qquad T = -\frac{1}{k}\left(\frac{ar^2}{4} + \frac{br^3}{9}\right) + C_1 \ln r + C_2$$

$$\frac{dT}{dr} = -\frac{1}{k}\left(\frac{ar}{2} + \frac{br^2}{3}\right) + \frac{C_1}{r}$$

$T = T_i$ at $r = r_i$ $\qquad T = T_0$ at $r = r_0$

Solving for constants gives:

$$C_1 = \frac{T_i - T_0 - \frac{1}{k}\left(\frac{a(r_0^2 - r_i^2)}{4} + \frac{b(r_0^3 - r_i^3)}{9}\right)}{\ln\left(\frac{r_i}{r_0}\right)}$$

$$C_2 = T_0 + \frac{1}{k}\left(\frac{ar_0^2}{4} + \frac{br_0^3}{9}\right) + \frac{T_i - T_0 - \frac{1}{k}\left(\frac{a(r_0^2 - r_i^2)}{4} + \frac{b(r_0^3 - r_i^3)}{9}\right)}{\ln r_0}$$

Also $\dot{q} = \dot{q}_i$ at $r = r_i = a + br_i$

2-50 $r_i = 0.0125$ m $\quad r_o = 0.0129$ m

Assume inner surface is insulated

$$\frac{dT}{dr} = -\frac{\dot{q}r}{2k} + \frac{C_1}{r} = 0 \quad \text{at} \quad r = r_i$$

$$C_1 = \frac{\dot{q}r_i^2}{2k} \qquad (a)$$

$$T = \frac{-\dot{q}r^2}{4k} + C_1 \ln r + C_2 \qquad T_i = -\frac{\dot{q}r_i^2}{4k} C_1 \ln r_i + C_2$$

$$T_o = \frac{-\dot{q}r_o^2}{4k} + C_1 \ln r_o + C_2$$

$$T_i - T_o = \frac{\dot{q}}{4k}(r_o^2 - r_i^2) + C_1 \ln\left(\frac{r_i}{r_o}\right) \qquad (b)$$

Heat Transfer is:

$$q = \dot{q}V = \dot{q}\pi(r_o^2 - r_i^2) = h\pi(2r_o)(T_o - T_\infty) \qquad (c)$$

Inserting (a) in (b) gives

$$T_i - T_o = \frac{\dot{q}}{4k}(r_o^2 - r_i^2) + \frac{\dot{q}r_i^2}{2k}\ln\left(\frac{r_i}{r_o}\right) \qquad (d)$$

We take: $T_i = 250$ °C $\quad h = 100$ W/m² °C $\quad T_\infty = 40$ °C

$k = 24$ W/m·°C

Inserting the numerical values in Equations (c) and (d) and solving gives.

$\dot{q} = 53.26$ MW/m3

$T_o = 249.76$ °C

2-51 $k = 43 \frac{W}{m \cdot °C}$

$$U_i = \frac{1}{\frac{1}{500} + \frac{\ln(1.45/1.25)\pi(0.025)}{(2\pi)(43)} + \frac{0.025}{0.029}\left(\frac{1}{12}\right.}$$

$$= \frac{1}{2\times10^{-3} + 4.31\times10^{-5} + 71.84\times10^{-3}} = 13.54 \text{ W/m}^2\cdot°C$$

2-52 $r_o = \frac{k}{h} = \frac{0.18}{12} = 0.015 \text{ m} = 1.5 \text{ cm}$ a) $r_o = 1.25 + 0.05 = 1.3$ cm Increased

b) $r_o = 1.25 + 1.0 = 2.25$ cm Decreased.

2-53 $U = 1/R = \frac{1}{3.114\times10^{-2}} = 32.11 \text{ W/m}^2\cdot°C$

2-54 $A = 130.4 \text{ m}^2$

$$U = q/A\Delta T = \frac{44000}{(130.4)(260-38)} = 1.52 \text{ W/m}^2\cdot°C$$

2-55 For $L = 1$ m $\frac{1}{h_i A_i} = \frac{1}{(65)\pi(0.025)} = 0.1959$

$$\frac{\ln(r_o/r_i)}{2\pi k} = \frac{\ln(2.58/2.5)}{2\pi(18)} = 2.79\times10^{-4}$$

$$\frac{1}{h_o A_o} = \frac{1}{(6.5)\pi(0.0258)} = 1.898$$

$$UA = \frac{1}{\Sigma R} = 2.094$$

$$q/L = (2.094)(120-15) = 219.9 \text{ W/m}$$

2-56 $A = 1\ m^2$ $R_{glass} = \frac{\Delta x}{k} = \frac{0.005}{0.78} = 6.41 \times 10^{-3}$

$R_{air} = \frac{\Delta x}{k} = \frac{0.004}{0.026} = 0.1538$

$R_{conv_1} = \frac{1}{h} = \frac{1}{12} = 0.0833$

$R_{conv_2} = \frac{1}{50} = 0.02$

$$U = \frac{1}{(2)(6.41 \times 10^{-3}) + 0.1538 + 0.0833 + 0.02}$$

$$= \frac{1}{0.2699} = 3.705\ W/m^2 \cdot {}^\circ C \qquad R = 0.2699$$

Single glass plate: $R = 6.41 \times 10^{-3} + 0.0833 + 0.02 = 0.1097$

$$U = \frac{1}{R} = 9.11\ W/m^2\ {}^\circ C$$

2-57

$R_{cu} = \frac{0.001}{386} = 2.59 \times 10^{-6}$ $\Delta T_{cu} = (52)(2.59 \times 10^{-6}) = 1.35 \times 10^{-4}\ {}^\circ C$

$R_{st} = \frac{0.004}{43} = 9.3 \times 10^{-5}$ $\Delta T_{st} = (52)(9.3 \times 10^{-5}) = 4.84 \times 10^{-3}\ {}^\circ C$

$R_{As} = \frac{0.01}{0.166} = 0.0602$ $\Delta T_{As} = (52)(0.0602) = 3.13\ {}^\circ C$

$R_F = \frac{0.1}{0.038} = 2.632$ $\Delta T_F = (52)(2.632) = 136.9\ {}^\circ C$

$\Sigma R = 2.692$ $U = \frac{1}{R} = 0.371\ W/m^2 \cdot {}^\circ C$

$q = U \Delta T = (0.371)(150 - 10) = 52\ W/m^2$

Inside of copper: 150 °C

2-58

The general solution of eq. 2-19 (b) is:

$$\theta = T - T_\infty = C_1 e^{-mx} + C_2 e^{mx} \qquad (1)$$

boundary conditions:

① at $x = 0$ $\quad \theta = \theta_1 = T_1 - T_\infty$

② at $x = L$ $\quad \theta = \theta_2 = T_2 - T_\infty$

from:

① $\theta_1 = C_1 + C_2 \qquad C_1 = \theta_1 - C_2$

② $\theta_2 = C_1 e^{-mL} + C_2 e^{mL}$

$$\theta_2 = (\theta_1 - C_2) e^{-mL} + C_2 e^{mL}$$

$$= \theta_1 e^{-mL} - C_2 (e^{-mL} - e^{mL})$$

$$\theta_2 = \left[\frac{\theta_2 - \theta_1 e^{-mL}}{e^{mL} - e^{-mL}}\right] \qquad (2)$$

$$C_1 = \theta_1 - C_2 = \theta_1 - \frac{\theta_2 - \theta_1 e^{-mL}}{e^{mL} - e^{-mL}}$$

$$C_1 = \left[\frac{\theta_2 - \theta_1 e^{mL}}{e^{-mL} - e^{mL}}\right] \qquad (3)$$

Then eq. (1) becomes

$$\theta = \frac{\theta_2 - \theta_1 e^{mL}}{e^{-mL} - e^{mL}} e^{-mx} + \frac{\theta_2 - \theta_1 e^{-mL}}{e^{mL} - e^{-mL}} e^{mx}$$

$$\theta = \frac{e^{-mx}(\theta_2 - \theta_1 e^{mL}) + e^{mx}(\theta_1 e^{-mL} - \theta_2)}{e^{-mL} - e^{mL}}$$

Part heat lost by rod:

$$q = -kA \left.\frac{d\theta}{dx}\right]_{x=0} + kA \left.\frac{d\theta}{dx}\right]_{x=L}$$

2-58 (cont'd)

$$\frac{d\theta}{dx} = m\left[\frac{-e^{-mx}(\theta_2 - \theta_1 e^{mL}) + e^{mx}(\theta_1 e^{-mL} - \theta_2)}{e^{-mL} - e^{mL}}\right]$$

$$q = \frac{kAm\left[-e^{-mL}(\theta_2 - \theta_1 e^{mL}) + e^{mL}(\theta_1 e^{-mL} - \theta_2)\right]}{e^{-mL} - e^{mL}}$$

$$\frac{kAm\left[-(\theta_2 - \theta_1 e^{mL}) + (\theta_1 e^{-mL} - \theta_2)\right]}{e^{-mL} - e^{mL}}$$

$$q = \frac{kAm\left[(\theta_2 - \theta_1 e^{mL})(1 - e^{-mL}) + (\theta_2 - \theta_1 e^{-mL})(1 - e^{mL})\right]}{e^{-mL} - e^{mL}}$$

2-59

PART A:

$$\dot{q}A = -kA\,\frac{d^2T}{dx^2} + hP(T-T_\infty)$$

$$\frac{d^2T}{dx^2} - \frac{hP}{kA}(T-T_\infty) + \frac{\dot{q}}{k} = 0$$

let $\theta = T - T_\infty$

$$(D^2 - hP/kA)\,\theta = -\dot{q}/k$$

$$\theta = c_1 e^{\sqrt{hP/kA}\,x} + c_2 e^{-\sqrt{hP/kA}\,x} + \dot{q}A/hP$$

let $\sqrt{hP/kA} = m$

$\theta = \theta$ at $x=0$ $\quad \therefore c_1 = \theta_0 - \dfrac{\dot{q}A}{hP} - c_2$

$$-kA\,dT/dx\,\big]_{x=L} = hA(T-T_\infty)\big]_{x=L} = hA\theta_L$$

$$\theta = \frac{\left[e^{-mL}\left(\theta_0 - \frac{\dot{q}A}{hP}\right) - \frac{h\theta_L}{km}\right]e^{mx}}{e^{mL}+e^{-mL}}$$

$$+\ \frac{\left[e^{mL}\left(\theta_0 - \frac{\dot{q}A}{hP}\right) + \frac{h\theta_L}{km}\right]e^{-mx}}{e^{mL}+e^{-mL}} + \frac{\dot{q}A}{hP}$$

PART B:

$$q = \int_0^L [hP(T-T_\infty)]\,dx + hA\theta_L$$

$$q = \frac{hP}{m}\left\{-\frac{h\theta_L}{kL} + \frac{\left(\theta_0 - \dot{q}A/hP\right)\left(e^{mL}-e^{-mL}\right) + \frac{2h\theta_L}{km}}{e^{mL}+e^{-mL}}\right\}$$

$$+\dot{q}AL + hA\theta_L$$

2-59 (cont'd)

PART C:

$$-kA\frac{dT}{dx}\Big]_{x=0} = 0 = q_o$$

$$0 = \left[c_1 m e^{mx} - c_2 m e^{-mx}\right]_{x=0}$$

$$\therefore\ c_1 = c_2$$

$$2c_1 = \theta_o - \frac{\dot{q}A}{hP}$$

$$\dot{q} = \frac{hP}{A}\left[\theta_o + \frac{2h\theta_L}{km(e^{mL} - e^{-mL})}\right.$$

2-60 $\dfrac{d^2\theta}{dx^2} - \dfrac{hP}{kA}\theta = 0$ let $m = \sqrt{hP/kA}$

$T_\infty = 38$ $d = 12.5\,mm$ $L = 30\,cm$ $h = 17$

$\theta = c_1 e^{mx} + C_2 e^{-mx}$ at $x = 0$ $\theta = 200 - 38 = 162$

$k = 386$ $P = \pi d$ $A = \pi d/4$ $x = 0.3$ $\theta = 93 - 38 = 55$

$m = \left[\dfrac{(17)\pi(0.0125)(4)}{(386)\pi(0.0125)^2}\right]^{\frac{1}{2}} = 3.754$ $162 = C_1 + C_2$

$55 = 3.084c_1 + 0.324c_2$ $C_1 = 0.91$ $C_2 = 161.09$

$\theta = 0.91e^{mx} + 161.09e^{-mx}$

$q\int_0^L hP\theta\,dx = hP\dfrac{1}{m}\left[0.91e^{mx} - 161.09e^{-mx}\right]_0^L$

$= \sqrt{hPkA}\left[0.91e^{mx} - 161.09e^{-mx}\right]_0^{0.3}$

$= \left[(17)\pi(0.0125)(386)\pi(0.0125)^2\right]^{\frac{1}{2}} \times \left[0.91e^{mx} - 161.09e^{-mx}\right]_0^{0.3}$

$= 122.7\,W$

2-62 $k = 204 \frac{W}{m \cdot °C}$ $L_c = L + \frac{d}{2} = 15 + \frac{2.5}{2} = 16.25$ cm $T_0 = 260°C$ $T_\infty = 16°C$

$h = 15\ W/m^2 \cdot °C$ $A = \frac{\pi d^2}{4}$ $P = \pi d$

$$m = \sqrt{\frac{hP}{kA}} = \left[\frac{(15)\pi(0.025)(4)}{(204)\pi(0.025)^2}\right]^{1/2} = 3.43$$

$mL_c = (3.43)(0.1625) = 0.5573$ $q = \sqrt{hPkA}\,\theta_0 \tanh(mL_c)$

$$q = \left[(15)\pi(0.025)(204)\pi\frac{(0.025)^2}{4}\right]^{1/2}(260-16)\tanh(0.5573)$$

$= 42.41$ W

2-63

$$q = \int_0^\infty hP(T-T_\infty)\,dx = \int_0^\infty hP\theta\,dx = hP\int_0^\infty e^{-mx}\,dx$$

$$= \frac{hP\theta_0}{-m}\left[e^{-mx}\right]_0^\infty = \sqrt{hPkA}\,\theta_0 = \frac{hP\theta_0}{-\sqrt{hP/kA}}\left[e^{-\sqrt{hP/kA}\,L} - 1\right]$$

($e^{-\sqrt{hP/kA}\,L} \to 0$)

$$q = \frac{hP\theta_0}{\sqrt{\frac{hP}{kA}}} = \sqrt{hPkA}\,\theta_0$$

2-64

$\theta = C_1 e^{-mx} + C_2 e^{mx}$ In case II the end of the fin is insulated

$\left.\frac{dT}{dx}\right]_{x=L} = 0$ the boundary conditions are

$\theta = \theta_0$ @ $x = 0$

$\frac{d\theta}{dx} = 0$ @ $x = L$ then $\theta = \theta_0 \frac{\cosh[m(L-x)]}{\cosh(mL)}$

$dq_{conv} = hP\,dx\,(T - T_\infty)$ or $hP\,dx\,\theta$

$$q_{conv} = \int_0^L hP\,dx = \int_0^L \frac{hP\theta \cosh[m(L-x)]\,dx}{\cosh(mL)}$$

$$q_{conv} = \frac{hP\theta_0}{\cosh(mL)}\frac{1}{m}\left[\sinh(m(L-x))\right]_0^L = \sqrt{hPkA}\,\theta_0 \tanh(mL)$$

2-65

$$q = \sqrt{hPkA}\,\theta_0 = \left[\frac{(3.5)(\pi)(0.025)(372)(\pi)(0.025)^2}{4}\right]^{1/2}(93-38)$$

$$= 12.32 \text{ W}$$

2-66

$$q = \sqrt{hPkA}\,\theta_0 = \left[\frac{(3.5)\pi(0.025)(372)\pi(0.025)^2}{4}\right]^{1/2}(90-40)$$

$$= 11.2 \text{ W}$$

2-67

$T_0 = 150°C$ $T_\infty = 15°C$ $r_1 = 1.25$ cm $L = 6.4$ mm $t = 1.6$ mm $h = 23$ W/m^2·°C

$k = 210$ W/m·°C

$$L_c = L + \frac{t}{2} = 6.4 + 0.8 = 7.2 \text{ mm}$$

$$r_{2c} = r_1 + L_c = 1.25 + 0.72 = 1.97 \qquad \frac{r_{2c}}{r_1} = 1.576$$

$$A_m = t(r_{2c} - r_1) = (0.0016)(0.0072) = 1.152 \times 10^{-5} \text{ m}^2$$

$$L_c^{3/2}\left(\frac{h}{kA_m}\right)^{1/2} = (0.0072)^{3/2}\left[\frac{23}{(210)(1.152\times10^{-5})}\right]^{1/2} = 0.0596$$

From Fig 2-11 $\eta_f = 97\%$

$$q_{max} = 2h\pi(r_{2c}^2 - r_1^2)(T_0 - T_\infty) = 4.523 \text{ W}$$

$$q = (0.97)(4.523) = 4.387 \text{ W}$$

2-68

η = total efficiency A_f = Surface area of all fins η_f = fin efficiency

A = Total heat transfer Area including fins and exposed tube or other surface.

T_0 = Base Temp T_∞ = Environment Temp

$$q_{act.} = h(A - A_f)(T_0 - T_\infty) + \eta_f A_f h(T_0 - T_\infty)$$

$$q_{ideal} = hA(T_0 - T_\infty)$$

$$\eta_t = \frac{q_{act}}{q_{ideal}} = \frac{A - A_f + A_f\eta_f}{A} = 1 - \frac{A_f}{A}(1 - \eta_f)$$

2-69 $T_0 = 460°C$ $t = 6.4 mm$ $L = 2.5 cm$ $T_\infty = 93°C$ $h = 28$ $k = 16.3$

$A_m = \frac{t}{2} L = (0.0032)(0.025) = 8 \times 10^{-5} m^2$ $L_c^{3/2} \left(\frac{h}{kA_m}\right)^{1/2} = 0.579$

$\eta_f = 85\%$ for 1 m depth

$q = (0.85)(28)(1)(2)(0.025)(460-93) = 436.7 W$

2-70 $T_0 = 200°C$ $T_\infty = 93°C$ $L = 12.5 mm$ $t = 0.8 mm$ $r_1 = 1.25 cm$ $k = 204$ $L_c = 12.9 mm$

$h = 110$ $r_{2c} = 2.54 cm$ $r_{2c}/r_1 = 2.03$ $A_m = (0.0008)(0.0129) = 1.032 \times 10^{-5} m^2$

$L_c^{3/2} \left(\frac{h}{kA_m}\right)^{1/2} = 0.335$ $\eta_f = 0.87$ Each fin uses 9.5 mm of Length

# of fins $= \frac{1.0}{0.0095} = 105.3$ fins

Tube surface area $= (105.3)(\pi)(0.025)(9.5-0.8)(10^{-3}) = 0.0719 \ m^2/m$ Length

Tube heat transfer $= (110)(0.0719)(200-93) = 846.6 W$

Heat transfer per fin $= (0.87)(2)\pi(110)(0.0254^2 - 0.0125^2)(200-93) = 31.46$ W/fin

Total fin heat transfer $= (31.46)(105.3) = 3312 W$

Total heat transfer $= 846.6 + 3312 = 4159$ W/m

2-71

$r_1 = 1.25 cm$ $L = 6.4 mm$ $t = 3.2 mm$ $h = 28$ $T_0 = 260°C$ $T_\infty = 93°C$ $k = 43$

$L_c = 8 mm$ $r_{2c} = 2.05 cm$ $r_{2c}/r_1 = 1.64$ $A_m = (0.0032)(0.008) = 2.56 \times 10^{-5} m^2$

$L_c^{3/2} \left(\frac{h}{kA_m}\right)^{1/2} = 0.114$ $\eta_f = 97\%$

$q = (0.97)(28)(2)\pi(0.0205^2 - 0.0125^2)(260-93) = 7.52 W$

2-72 $k = 43$ $t = 2 cm$ $h = 20$ $L_c = 15 cm$ $L_c^{3/2} \left(\frac{h}{kA_m}\right)^{1/2} = 0.723$

$\eta_f = 0.75$

$q = (0.75)(20)(2)(0.15)(200-15) = 833$ W/m depth

2-73 $t = 1.6\ mm \quad r_1 = 1.25\ cm \quad L = 12.5\ mm \quad T_0 = 200°C \quad T_\infty = 20°C \quad h = 60\ \frac{W}{m^2 \cdot °C}$

$k = 204\ \frac{W}{m \cdot °C} \quad L_c = 13.3\ mm \quad r_{2c} = 2.58\ cm \quad r_{2c}/r_1 = 2.064$

$A_m = (0.0016)(0.0133) = 2.128 \times 10^{-5}\ m^2 \quad L_c^{3/2}\left(\frac{h}{kA_m}\right)^{1/2} = 0.18 \quad \eta_f = 95\%$

$$q = (0.95)(60)(2)\pi(0.0258^2 - 0.0125^2)(200 - 20) = 32.84\ W$$

2-75 $A = 2y = \frac{tx}{L} \qquad y = \frac{\frac{t}{2}}{L}x = \frac{tx}{2L}$

$$-kA\frac{dT}{dx} = hP\,dx\,(T - T_\infty) - \left[kA\frac{dT}{dx} + \frac{d}{dx}\left(kA\frac{dT}{dx}\right)dx\right]$$

$$\frac{d}{dx}\left(kA\frac{dT}{dx}\right) - hP(T - T_\infty) = 0 \qquad \theta = T - T_\infty$$

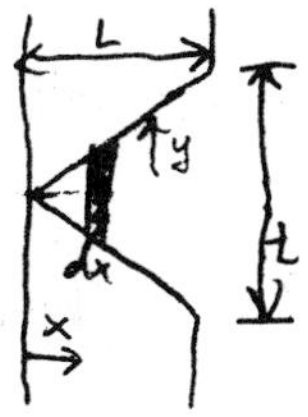

$$\frac{kt\,x}{L}\frac{d^2\theta}{dx^2} + \frac{kt}{L}\frac{d\theta}{dx} - hP\theta = 0$$

$$x\frac{d^2\theta}{dx^2} + \frac{d\theta}{dx} - \frac{hPL}{kt}\theta = 0$$

2-76 $k = 16 \quad h = 40 \quad T_0 = 250°C \quad T_\infty = 90°C \qquad P = (4)(0.0125) = 0.05\ m$

$A = (0.0125)^2 = 1.565 \times 10^{-4}\ m^2$

$$q = \sqrt{hPkA}\,\theta_0 = \left[(40)(0.05)(16)(1.565 \times 10^{-4})\right]^{1/2}(250 - 90)$$

$$= 11.31\ W$$

2-77 $t = 2.4\ mm \quad L = 19\ mm \quad h = 85 \quad k = 164 \quad T_0 = 90°C \quad T_\infty = 25°C \quad L_c = 20.2\ mm$

$A_m = (0.0012)(0.0202) = 2.424 \times 10^{-5}\ m^2 \quad L_c^{3/2}\left(\frac{h}{kA_m}\right)^{1/2} = 0.420 \quad \eta_f = 87\%$

$$q = (0.87)(85)(2)(0.0202)(90 - 25) = 194.2\ W/m$$

2-78 $L_c = 0.0614\ ft \quad r_{2c} = 2.629\ in. \quad A_m = (0.125)(0.688) = 0.081\ in^2 = 5.97 \times 10^{-4}\ ft^2$

$\frac{r_{2c}}{r_1} = 1.34 \quad L_c^{3/2}\left(\frac{h}{kA_m}\right)^{1/2} = 0.329 \quad \eta_f = 87\%$

$$q_{max} = 2h\pi(r_{2c}^2 - r_1^2)(450 - 100) = 591\ \frac{BTU}{hr.}$$

$$q = (0.87)(591) = 514\ \frac{BTU}{hr.}$$

2-79 Assume Insulated Tip Solution

$$\frac{\theta}{\theta_0} = \frac{e^{-mx}}{1+e^{-2mL}} + \frac{e^{mx}}{1+e^{2mL}} \qquad m^2 = \frac{hP}{kA} = \frac{(570)\pi(0.0016)(4)}{(22)\pi(0.0016)^2} = 6.48\times10^4$$

$m = 245.5\ m^{-1}$ $L = 12.5$ mm $= 0.0125$ m $mL = 3.18$ $2mL = 6.363$

$$\frac{\theta}{\theta_0} = 0.0829 \quad \theta = (0.0829)(49-25) = 1.99 \quad T = 1.99 + 25 = 26.99\,°C$$

For $h = 1200$ $m = 369.3$ $mL = 4.616$ $2mL = 9.232$ $\frac{\theta}{\theta_0} = 0.0198$

$T = (0.0198)(49-25) + 25 = 25.47\,°C$

For $h = 200$ $m = 150.8$ $mL = 1.884$ $2mL = 3.769$ $\frac{\theta}{\theta_0} = 0.297$

$T = (0.297)(49-25) + 25 = 32.13\,°C$

2-81 $L_c = 2.075$ cm $q_{max} = (500)(2)(0.02075)(200-20) = 3735$ W/m

$$L_c^{3/2}\left(\frac{h}{kA_m}\right)^{1/2} = 1.615 \qquad \eta_f = 42\%$$

$q_{act} = (0.42)(3735) = 1569$ W/m

2-82

$L = 3.5$ cm $t = 1.4$ mm $L_c = 3.57$ cm $k = 55$

$q_{max} = hA\theta_0 = (500)(2)(0.0357)(150-20) = 4641$ W/m

$$mL_c = \left(\frac{2h}{kA_m}\right)^{1/2} L_c^{3/2} = 4.068$$

$$\eta_f = \frac{\tanh(mL_c)}{mL_c} = 0.246 \qquad q_{act} = (0.246)(4641) = 1140\ \text{W/m}$$

2-83 $k = 43$ $h = 100$ $r_1 = 2.5$ cm $r_2 = 7.5$ cm $L = 5$ cm $T_\infty = 20\,°C$ $t = 2$ mm $r_{2c} = 7.51$ cm

$L_c = 5.1$ cm

$$L_c^{3/2}\left(\frac{h}{kA_m}\right)^{1/2} = 1.74 \qquad \frac{r_{2c}}{r_1} \cong 3 \qquad \eta_f = 0.27$$

$$q = \eta_f\, 2h\pi(r_{2c}^2 - r_1^2)\,\theta_0 = (0.27)(100)(2)\pi(0.0751^2 - 0.025^2)(150-20)$$

$= 110.6$ W

2-84 $r_1 = 1.5$ cm $L = 2$ cm $r_2 = 3.5$ cm $t = 1$ mm $h = 80$ $k = 200$ $L_c = 2.05$ cm

$r_{2c} = 3.55$ cm $L_c^{3/2}\left(\frac{h}{kA_m}\right)^{1/2} = 0.41$ $\frac{r_{2c}}{r_1} = 2.37$ $\eta_f = 0.81$

$$q = (80)\pi(0.0355^2 - 0.015^2)(2)(200-20)(0.81) = 75.9 \text{ W}$$

2-85 From Prob. 2-58 the solution is available:

$m = \left(\frac{hP}{kA}\right)^{\frac{1}{2}} = 41.23$ $L = 20$ cm $x = 10$ cm $mL = 8.25$

$\theta_1 = 50 - 20 = 30$ $\theta_2 = 100 - 20 = 80$ $e^{mL} = 3813$ $e^{-mL} = 2.262 \times 10^{-4}$

$e^{mx} = 61.74$ $e^{-mx} = 0.0162$

Inserting in soln. from Prob. 2-58

$$\theta(x = 10\text{cm}) = \frac{(0.0162)[80 - (30)(3813)] + (61.74)[(30)(2.262\times10^{-4}) - 80]}{2.262\times10^{-4} - 3813}$$

$$= 1.78$$

$$T = 1.78 + 20 = 21.78\ °\text{C}$$

2-86 $t = 2$ cm $L = 17$ cm $k = 43$ $h = 23$ $L_c = 18$ cm $L_c^{3/2}\left(\frac{h}{kA_m}\right)^{1/2} = 0.93$

$\eta_f = 0.64$

$$q = (0.64)(2)(23)(0.18)(230 - 25) = 1086 \text{ W/m}$$

2-87 $L = 5$ cm $L_c = 5$ cm $t = 4$ mm $k = 23$ $h = 20$ $L_c^{3/2}\left(\frac{h}{kA_m}\right)^{1/2} = 1.042$

$\eta_f = 0.68$ $q = \eta_f h A \theta_0$ $A = (2)(0.002^2 + 0.05^2)^{1/2} = 0.10008 \ \frac{\text{m}^2}{\text{m depth}}$

$$q = (0.68)(20)(0.10008)(200 - 40) = 217.8 \text{ W/m}$$

2-88 $t = 1.0$ mm $r_1 = 0.5$ in $= 1.27$ cm $L = 1.27$ cm $L_c = 1.32$ cm $r_2 = 2.54$ cm

$r_{2c} = 2.59$ cm $h = 56$ $k = 204$ $L_c^{3/2}\left(\frac{h}{kA_m}\right)^{1/2} = 0.219$ $\frac{r_{2c}}{r_1} = 2.04$

$\eta_f = 0.93$

$$q = (0.93)(2)\pi(0.0259^2 - 0.0127^2)(56)(125 - 30) = 15.84 \text{ W}$$

2-89 $t = 2$ mm $r_1 = 2.0$ cm $r_2 = 10.0$ cm $L = 8$ cm $L_c = 8.1$ cm $r_{2c} = 10.2$ cm

$h = 20$ $k = 17$ $L_c^{3/2}\left(\frac{h}{kA_m}\right)^{1/2} = 1.96$ $\frac{r_{2c}}{r_1} = 5.1$ $\eta_f = 0.19$

$$q = (0.19)(20)(\pi)(0.102^2 - 0.02^2)(2)(135 - 15) = 28.7 \text{ W}$$

2-90 $L = 2.5$ cm $t = 1.1$ mm $k = 55$ $h = 500$ $L_c = 2.555$ cm $L_c^{3/2}\left(\frac{h}{kA_m}\right)^{1/2} = 2.32$

$\eta_f = 0.33$ $q = (0.33)(2)(0.02555)(500)(125 - 20) = 885$ W/m

2-91 $t = 1.0$ mm $r_1 = 1.25$ cm $r_2 = 2.5$ cm $r_{2c} = 2.55$ cm $h = 25$ $k = 204$

$L = 1.25$ cm $L_c = 1.3$ cm $L_c^{3/2}\left(\frac{h}{kA_m}\right)^{1/2} = 0.249$ $\eta_f = 0.91$

$$q = (0.91)(2)(75)\pi(0.0255^2 - 0.0125^2)(100 - 30) = 4.94 \text{ W}$$

2-92 $d = 1$ cm $L = 5$ cm $h = 15$ $k = 0.78$ $L_c = 5.25$ cm $m = \left[\frac{hP}{kA}\right]^{1/2} = 87.7$

$mL_c = 4.605$ $\eta_f = \frac{\tanh(mL_c)}{mL_c} = 0.217$

$$q = (0.217)(15)\pi(0.01)(0.0525)(180 - 20) = 0.859 \text{ W}$$

2-93 $k = 1$ cm $L = 8$ cm $L_c = 8.5$ cm $m = \left[\frac{hP}{kA}\right]^{1/2} = 31.62$ $mL_c = 2.688$

$\eta_f = \frac{\tanh(mL_c)}{mL_c} = 0.369$ $q = (0.369)(45)(0.085)(4)(0.01)(300 - 50)$
$= 14.11$ W

2-94 $t = 1.0$ mm $\quad r_1 = 1.25$ cm $\quad L = 12$ mm $\quad h = 120 \quad L_c = 12.5$ mm $\quad r_{2c} = 2.5$ cm

$k = 386 \quad L_c^{3/2}\left(\frac{h}{kA_m}\right)^{1/2} = 0.659 \quad \frac{r_{2c}}{r_1} = 2 \quad \eta_f = 0.71$

$$q = (0.71)(120)\pi(0.025^2 - 0.0125^2)(2)(250-30) = 55.2 \text{ W}$$

2-95 $k = 17 \quad h = 47 \quad L = 5$ cm $\quad t = 2.5$ cm $\quad L_c = 6.25$ cm $\quad L_c^{3/2}\left(\frac{h}{kA_m}\right)^{1/2} = 0.657$

$\eta_f = 0.8$

$$q = (0.8)(47)(2)(0.0625)(100-20) = 376 \text{ W/m}$$

2-96 $r_1 = 1.5$ cm $\quad L = 2$ cm $\quad r_2 = 3.5$ cm $\quad t = 1$ mm $\quad r_{2c} = 3.55$ cm $\quad h = 80$

$k = 204 \quad L_c = 2.05$ cm $\quad L_c^{3/2}\left(\frac{h}{kA_m}\right)^{1/2} = 0.453 \quad \frac{r_{2c}}{r_1} = 2.37 \quad \eta_f = 0.79$

$$q = (0.79)(80)(2)\pi(0.0355^2 - 0.015^2)(200-20) = 74 \text{ W}$$

2-97 $r_1 = 1.5$ cm $\quad r_2 = 4.5$ cm $\quad t = 1.0$ mm $\quad h = 50 \quad r_{2c} = 4.55$ cm $\quad L_c = 3.05$ cm

$k = 204 \quad \eta_f = 0.6 \quad \frac{r_{2c}}{r_1} = 3 \quad L_c^{3/2}\left(\frac{h}{kA_m}\right)^{1/2} = 0.78$

$$k = \frac{1}{(0.001)(0.0305)}\left[0.0305\right]^3 \frac{50}{(0.78)^2} = 76.5 \text{ W/m}\cdot{}^\circ\text{C}$$

2-98 $t = 1.0$ mm $\quad L = 2.0$ cm $\quad r_1 = 1.0$ cm $\quad h = 200 \quad k = 204 \quad L_c = 2.05$ cm

$r_{2c} = 3.05$ cm $\quad L_c^{3/2}\left(\frac{h}{kA_m}\right)^{1/2} = 0.642 \quad \frac{r_{2c}}{r_1} = 3.05 \quad \eta_f = 0.68$

$$q = (0.68)(200)(2)\pi(0.0305^2 - 0.01^2)(150-20) = 92.2 \text{ W}$$

2-99 $k_A = k_B = 17 \quad A_c = 0.001A \quad \Delta T = 300^\circ\text{C} \quad A = \frac{\pi d^2}{4} = 5.067 \times 10^{-4}\ \text{m}^2 \quad \frac{L_g}{2} = 1.3 \times 10^{-6}$ m

$k_f = 0.035 \quad L_A = L_B = 7.5$ cm $\quad h_c = \frac{1}{L_g}\left[\frac{A_c}{A}\left(\frac{2k_A k_B}{k_A + k_B}\right) + \frac{A_v}{A}k_f\right] = 19986 \text{ W/m}^2\cdot{}^\circ\text{C}$

$$\frac{1}{h_c A} = \frac{1}{(19986)(5.067 \times 10^{-4})} = 0.0987\ {}^\circ\text{C/W}$$

$$q = \frac{300}{\frac{(0.075)(2)}{(17)(5.067 \times 10^{-4})} + 0.0987} = 17.31 \text{ W}$$

w/ no contact resistance

$$q = kA\frac{\Delta T}{\Delta x} = \frac{(17)(5.067 \times 10^{-4})(300)}{0.15} = 17.228 \text{ W}$$

2-101 $R_{th} = \frac{\Delta x}{kA} = \frac{5\times10^{-3}}{204} = 2.45\times10^{-5}$ $R_c = \frac{1}{h_c A} = 0.88\times10^{-4}$

$\Sigma R_{th} = (2)(2.45\times10^{-5}) + 0.88\times10^{-4} = 1.37\times10^{-4}$

$\Delta T_c = \frac{(80)(0.88\times10^{-4})}{1.37\times10^{-4}} = 51.4°C$

2-102 $t = 1\times10^{-3}$ $r_i = 0.0125$ $L = 0.0125$ $L_c = 0.0130$ $r_{2c} = 0.0255$ $\frac{r_o}{r_i} = 2.04$

$A_m = (0.001)(0.0130) = 1.3\times10^{-5}\ m^2$ $L_c^{3/2}\left(\frac{h}{kA_m}\right)^{1/2} = 0.322$ $\eta_f = 0.86$

$\frac{1}{h_c} = 0.88\times10^{-4}$ $T_o = 200°C$ $T_\infty = 20°C$

$R_{fin} = \frac{1}{2\eta_f \pi (r_{2c}^2 - r_1^2) h} = 2.997\ °C/W$

$R_c = \frac{1}{h_c A} = 1.1205\ °C/W$

$q = \frac{\theta_o}{R_{fin} + R_c} = 43.72\ W$ w/o contact resistance

$q = \frac{\theta_o}{R_{fin}} = 60.06\ W$ % Reduction $= \frac{60.06 - 43.72}{60.06} \times 100\% = 27.2\%$

2-103 $1/h_c = 0.9\times10^{-4}$ $A_c = 0.5\ cm^2$ $q = 300\ mW$

Assume fin at 27°C

$q = (300\times10^{-3}) = \frac{1}{0.9\times10^{-4}}(0.5)(10^{-4})(T_t - 27)$

$T_t = 27.54°C$

2-104 $t = 3.0\ mm$ $k = 204\ W/m \cdot °C$

$h = 50\ W/m^2 - °C$, $\rho = 2707\ kg/m^3$

Take $L = 2cm$

Rect. Fin $L_c = 2 + 0.15 = 2.15\ cm$

$$L_c^{3/2}\left[\frac{h}{kA_m}\right]^{1/2} = (0.0215)^{3/2}\left[\frac{50}{(204)(0.003)(0.0215)}\right]^{1/2}$$

$$= 0.1943$$

$\eta_f = 0.96$ $A_{surf} = (2)L_c = 0.043$

For same weight A_m = same

$$t(0.0215) = \frac{t}{2}L\ (\text{triang.})$$

$$L(\text{triang}) = 0.043$$

$$(\text{triang})\ L_c^{3/2}\left(\frac{h}{kA_m}\right)^{1/2} = 0.5496$$

$$\eta_f = 0.85$$

$$q \sim A_{surf}\,\eta_f$$

$$A_{surf}(\text{triang}) = 2(4.3^2 + 0.15^2)^{1/2} = 8.6 = 0.086\ m^2$$

$$A_{surf}\,\eta\ (\text{triangle}) = (0.086)(0.85) = 0.0731$$

$$A_{surf}\,\eta\ (\text{rect}) = (0.043)(0.96) = 0.0413$$

Triangle fin produces more heat transfer for given weight.

2-105 $r_1 = 1.0$, r_2 2.0 cm

$h = 160\ \text{W/m}^2\text{-}^\circ\text{C} \qquad k = 204\ \text{W/m-}^\circ\text{C}$

1.0 mm Fin $L_c = 1.05$ cm

$$L_c^{3/2}\left[\frac{h}{kA_m}\right]^{1/2} = (0.0105)^{3/2}\left[\frac{160}{(204)(0.001)(0.0105)}\right]^{1/2}$$

$$= 0.294$$

$\eta_f = 0.88$

$$q = (6)(160)\pi\left[0.0205^2 - 0.01^2\right](2)(\Delta T)(0.88) = 1.7\Delta T$$

2.0 mm Fin $L_c = 1.1$ cm

$$L_c^{3/2}\left[\frac{h}{kA_m}\right]^{1/2} = 0.218$$

$\eta_f = 0.92$

$$q = (3)(160)\pi\left[0.021^2 - 0.01^2\right](2)\Delta T\,(0.92) = 0.95\Delta T$$

3.0 mm Fin $L_c = 1.15$ cm

$$L_c^{3/2}\left[\frac{h}{kA_m}\right]^{1/2} = 0.186$$

$\eta_f = 0.95$

$$q = (2)(160)\pi\left[0.0215^2 - 0.01^2\right](2)\,\Delta T\,(0.95) = 0.69\,\Delta T$$

CONCLUSION: Several thin fins are better than a few thick fins. More heat transfer for the same weight of fins.

2-106 $2L = 20\text{cm} = 0.2\text{ m}$

$$q/A = \dot{q}(2L) = (2\times10^5)(-0.2) = 40{,}000\ \text{W/m}^2$$

$$= h(T_w - T_\infty) = (400)(T_w - 50)$$

$$T_w = 150°\text{C}$$

$$T_0 - T_w = \frac{\dot{q}L^2}{2k} = \frac{(2\times10^5)(0.1)^2}{(2)(20)} = 50°\text{C}$$

$$T_0 = 200°\text{C}$$

2-107 1/2 heat generated, max temperature at insulated surface which is the same as in Problem 2-106.

2-108 $L_c = 0.008$ $k = 204\ \text{W/m-°C}$

$h = 45\ \text{W/m}^2\text{-°C}$

$$L_c^{3/2}\left[\frac{h}{kA_m}\right]^{1/2} = (0.008)^{3/2}\left[\frac{(45)(2)}{(204)(0.008)(0.002)}\right]^{1/2}$$

$$= 0.119$$

$$\eta_f = 0.97$$

$$q = (2)(0.97)\left[0.008^2 + 0.001^2\right]^{1/2}(45)(200-25)$$

$$= 123.2\ \text{W/m}$$

2-109

$r_1 = 1.25\text{cm}$ $r_2 = 2.25\text{cm}$

$t = 2.0\text{mm}$ $r_{2c} = 2.35\text{cm}$ $r_{2c}/r_1 = 1.88$

$L_c = 1.1\text{ cm}$ $T_o = 180°C$ $T_\infty = 20°C$

$h = 50\text{ W/m}^2\text{-°C}$ $k = 204\text{ W/m-°C}$

L of tube $= 1.0\text{m}$

bare length $= 1.0 - (100)(0.002) = 0.8\text{ m}$

$$L_c^{3/2}\left[\frac{h}{kA_m}\right]^{1/2} = (0.011)^{3/2}\left[\frac{50}{(204)(0.002)(0.011)}\right]^{1/2}$$

$$= 0.122$$

$$\eta_f = 0.95$$

$$q\,(100\text{ fins}) = (100)(2)(0.95)\pi\left[0.0235^2 - 0.0125^2\right](50)(180-20)$$
$$= 1869\text{ W}$$

$$q\,(\text{bare tube}) = (50)\pi(0.025)(0.8)(180-20)$$
$$= 503\text{ W}$$

$$q\,(\text{total}) = 1869 + 503 = 2372\text{ W}$$

2-110

$L = 5\text{cm}$

$d = 2, 5, 10 \text{ mm}$

$T_\infty = 20°C$, $h = 40 \text{ W/m}^2\text{-°C}$

$k = 204 \text{ W/m}^2\text{-°C}$

$T_0 = 200°C$

$L_c = L + d/4$

2 mm pin $L_c = 5 + 0.05 = 5.05 \text{ cm} = 0.0505 \text{ m}$

$$m = \left[\frac{hP}{kA}\right]^{1/2} = \left[\frac{h\pi d}{k\pi d^2/4}\right]^{1/2}$$

$$= \left[\frac{4h}{kd}\right]^{1/2} = \left[\frac{(4)(40)}{(204)(0.002)}\right]^{1/2} = 19.8$$

$$mL_c = (19.8)(0.0505) = 1.0$$

$$\eta = \frac{\tanh(mL_c)}{mL_c} = 0.762$$

$$q = (0.762)(40)\pi(0.002)(0.0505)(200-20) = 1.74\text{W}$$

5 mm pin $L_c = 5 + 0.125 \text{ cm} = 0.05125 \text{ m}$

$m = 12.52$ $mL_c = 0.6419$

$$\eta = \frac{\tanh(mL_c)}{mL_c} = 0.882$$

$$q = (0.882)(40)\pi(0.005)(0.05125)(200-20) = 5.11 \text{ W}$$

2-110 (contd)

10 mm pin $L_c = 5 + 0.25 \text{ cm} = 0.0525 \text{ m}$

$m = 8.856$

$mL_c = 0.46495$

$$\eta = \frac{\tanh(mL_c)}{mL_c} = 0.934$$

$$q = (0.934)(40)\pi(0.01)(0.0525)(200-20) = 11.09 \text{ W}$$

d (mm)	q (W)	q/d (mm)	q/d² (per weight)
2	1.74	0.87	0.435
5	5.11	1.022	0.2044
10	11.09	1.109	0.1109

Conclusion: Smaller pins produce more heat transfer per unit weight.

2-111

See conclusion at end of Problem 2-110.

2-112

$k = 100\ W/m-°C$

$q = kA\,\frac{\Delta T}{\Delta x}$

$A = (1.7 - 1.5)\text{ cm} = \frac{0.2}{100} = 0.002\ m^2/m$

Δx = circumference at $r = 1.6$ cm for $\pi/4$

$\Delta x = (2)(1.6)(\pi/4) = 2.513\text{ cm} = 0.02513\text{ m}$

$R = \Delta x/kA = \frac{0.02513}{(100)(0.002)} = 0.1256\ \cdot °C^{-1}$

$q/L = \frac{50}{0.1256} = 398\ W/m$

2-115

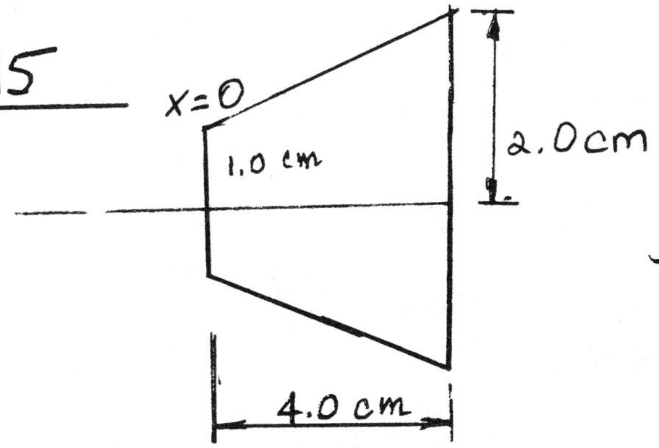

$k = 386\ W/m-°C$

$r = ax + b$

$r = 0.01$ at $x = 0$

$r = 0.02$ at $x = 0.04$

$0.01 = b$ $\quad a = 0.25$

$0.02 = (0.04)a + 0.01$

$r = 0.25x + 0.01$

$A = 2\pi r\,(0.0005)$

2-115 (contd)

$$q = -k\,2\pi(0.25x + 0.01)(0.0005)\frac{dT}{dx}$$

$$\int_0^{0.04} \frac{dx}{0.25x + 0.01} = \int -2\pi k(0.0005)\frac{dT}{q}$$

$$\frac{1}{0.25}\ln\left[\frac{(0.25)(0.04) + 0.01}{0.01}\right] =$$

$$2.773 = 2\pi(386)(0.0005)\frac{\Delta T}{q}$$

$$= 1.213\,\frac{\Delta T}{q}$$

$$q = \frac{\Delta T}{2.287} \qquad R = 2.287$$

For $\Delta T = 300°C$ $\quad q = \frac{300}{2.287} = 131.2\text{ W}$

3-1

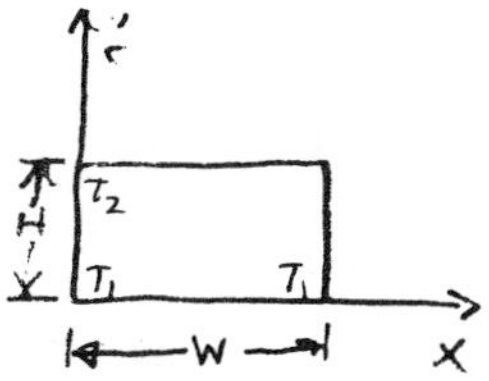

$T=T_1$ at $y=0$ $T=T_1$ at $x=0$ $T=T_1$ at $x=W$
$T=T_2$ at $y=H$ $T=XY$ For $\lambda^2=0$ $X=C_1+C_2x$
$Y=C_3+C_4y$ at $y=0$

$\therefore T-T_1=(C_1+C_2x)[C_3+0(C_4)]=\theta=0$ hence $C_3=0$

at $x=0$ $C_1=0$ also $C_2=0$ $T_2-T_1=[0+Ax][0+C_4H]=0$ which isn't true

For $\lambda^2<0$: $\theta=T-T_1$

$$T-T_1=[C_5e^{-\lambda x}+C_6e^{\lambda x}][C_7\cos(\lambda y)+C_8\sin(\lambda y)]$$

$\theta=0$ @ $y=0$ $0=(C_5e^{-\lambda x}+C_6e^{\lambda x})(C_7)$ $\therefore C_7=0$

$\theta=0$ @ $x=0$ $0=(C_5+C_6)(C_8\sin(\lambda y))$ $\therefore$ either $C_5+C_6=0$ or $C_8=0$ If $C_8=0$ then have trivial soln. then, $C_5+C_6=0$ and $C_5=-C_6$ $\theta=0$ @ $x=w$

$0=(C_5e^{-\lambda w}+C_6e^{\lambda w})(C_8\sin(\lambda y))$ $\therefore$ $\lambda=0$ but it was stated that $\lambda^2<0$

3-2 $$\frac{T-T_1}{T_2-T_1}=\frac{2}{\pi}\sum\frac{(-1)^{n+1}}{n}\sin\frac{n\pi x}{W}\frac{\sinh\left(\frac{n\pi y}{W}\right)}{\sinh\left(\frac{n\pi H}{W}\right)}$$ at $y=H$ $x=\frac{W}{2}$

$\frac{T-T_1}{T_2-T_1}=1$ (Ref Fig 3-2) Also, non zero terms for $n=1,3,5\ldots$ First four terms: $n=1,3,5,7$

$$\frac{T-T_1}{T_2-T_1}=\frac{2}{\pi}\left[2\sin\frac{\pi}{2}+\frac{2}{3}\sin\frac{3\pi}{2}+\frac{2}{5}\sin\frac{5\pi}{2}+\frac{2}{7}\sin\frac{7\pi}{2}\right]$$

$\frac{T-T_1}{T_2-T_1}=0.92$ Error is $(1.00-0.92)\times 100\% = 8\%$

3-4 $k_m = 11$ Btu/hr·ft·°F $= 19.04$ W/m·°C

$k_{glass} = 0.038$ W/m·°C

$$q = \frac{300 - 40}{\frac{\ln(2.5/1.5)}{2\pi(19.04)} + \frac{\ln(5/2.5)}{2\pi(0.038)}} = 89.38 \text{ W/m}$$

3-6 $k = 0.5$ Btu/hr ft°F $L = 6''$ wall thickness $T_i = 1000$°F $T_o = 200$°F

shape factors:

Walls	Edges	Corners
$S_1 = \frac{A}{L} = 6$	$S_1 = 0.54D = 1.62$	$S = 0.15L$
$S_2 = 4$	$S_2 = 1.08$	$= 0.075$
$S_3 = 12$	$S_3 = 0.54$	

$S_{tot} = 2\sum S_{walls} + 4\sum_{edges} + 8\sum_{corners} = 57.56$

$q = kS\Delta T = 23000 \ \frac{BTU}{hr}$

3-7 $D = 35 - 10 = 25$ cm $L = 5$ cm $A = (0.25)^2 = 0.0625 \text{ m}^2$ $k = 1.04$ W/m·°C

S_{wall} $\frac{A}{L} = 1.25$ $S_{edge} = 0.54D = 0.135$ $S_{corner} = 0.15L = 7.5\times10^{-3}$

$S = 2(1.25) + 4(0.135) + 8(7.5\times10^{-3}) = 9.18$

$q = kS\Delta T = (1.04)(9.18)(500-80) = 4009.8$ W

3-8 $r_1 = 4$ cm $r_2 = 150$ cm $D = 10$ cm $k = 1.4$ W/m °C

$$S = \frac{2\pi}{\cosh^{-1}\left[\frac{10^2 - 4^2 - 1.5^2}{(2)(4)(1.5)}\right]} = 2.411$$

$q = kS\Delta T = (1.4)(2.411)(200 - 35) = 556.8$ W/m

3-9 $k = 1.7$ W/m·°C $D = 2.4$ m $r = 0.5$ m $T_s = 4$°C $T_w = 30$°C

$$S = \frac{4\pi r}{1 - r/2D} = 7.014 \text{ m} \qquad q = kS\Delta T = (1.7)(7.014)(30-4) = 310 \text{ W}$$

3-10 $S = 4\pi r = 1.257$ m $k = 0.038$ W/m·°C $q = (1.257)(0.038)(170-20) = 7.165$ W

3-11 $r = 1$ m $k = 1.5$ W/m·°C $S = 4\pi r = 12.56$ m $q = (1.5)(12.56)(20-0) = 377$ W

3-13 $D > 3r$

$$q/L = \frac{2\pi(1.2)(110-25)}{\ln\left[\frac{(2)(1.3)}{0.04}\right]} = 153.5 \text{ W/m} = I^2R$$

$$= I^2(1.1\times10^{-4}) \qquad I = 1181 \text{ amp.}$$

3-14 $S = 4\pi r$ $q = (1.3)(4\pi)(0.02)(70-12) = 18.95$ W

3-15

$$S = \frac{2\pi}{\cosh^{-1}\left[\frac{25^2 - 7.5^2 - 2^2}{(2)(7.5)(2)}\right]} = 1.024$$

$$q/L = (1.024)(3.0)(100-20) = 245.8 \text{ W/m}$$

3-16

$$S = \frac{2\pi}{\cosh^{-1}\left[\frac{40^2 - 8^2 - 4^2}{(2)(8)(4)}\right]} = 1.628$$

$$q/L = (1.628)(0.7)(300-125) = 199.4 \text{ W/m}$$

3-17

$$S = \frac{4\pi(0.75)}{1 - \frac{0.75}{(2)(3.75)}} = 10.47$$

$$q = (10.47)(1.2)(300-30) = 3393 \text{ W}$$

3-18 $r = 1.25$ cm $L = 1$ m $T_r = 55°C$ $T_s = 10°C$ $k = 1.7$ W/m·°C

$$S = \frac{2\pi L}{\ln\left(\frac{2L}{r}\right)} = \frac{2\pi(1.0)}{\ln\left[\frac{(2)(1.0)}{0.0125}\right]} = 1.238 \qquad q = (1.238)(1.7)(55-10) = 94.71 \text{ W}$$

3-19

$k = 0.8$ W/m·°C $r_1 = 5$ cm $r_2 = 1.4$ cm $D = 12$ cm

$$S = \frac{2\pi}{\cosh^{-1}\left[\frac{12^2 - 1.4^2 - 5^2}{(2)(1.4)(5)}\right]} = 2.813$$

$$q = (2.813)(0.8)(300-15) = 641.4 \text{ W/m}$$

3-20 $D = 5$ m $r = 1.0$ m $k = 1.5$ W/m·°C

$$S = \frac{4\pi r}{1 - \frac{r}{2D}} = \frac{(4\pi)(1.0)}{1 - 1/10} = 13.96$$

$$q = (13.96)(1.5)(25-5) = 418.9 \text{ W}$$

3-21 $r = 2$ in $= 0.0508$ m $L = 100$ yd $= 91.44$ m $D = 9$ in $= 0.2286$ m

$$S = \frac{2\pi L}{\cosh^{-1}(D/r)} = \frac{2\pi(91.44)}{\cosh^{-1}(9/2)} = 262.98$$

$$q = (262.98)(1.2)(300-60)(5/9) = 42078 \text{ W}$$

3-22 $r_1 = 5$ cm $r_2 = 2.5$ cm $D = 20$ cm $k = 0.15$ W/m·°C

$$S = \frac{2\pi}{\cosh^{-1}\left[\frac{20^2 - 5^2 - 2.5^2}{(2)(2.5)(5)}\right]} = 1.857 \qquad q = (0.15)(1.857)(110-3) = 29.81 \text{ W/m}$$

3-23 $r = 1.5$ cm $D = 5$ cm $k = 15.5$ W/m·°C

$$S = \frac{2\pi}{\cosh^{-1}(D/r)} = \frac{2\pi}{\cosh^{-1}(5/1.5)} = 3.353 \qquad q = (15.5)(3.353)(135-46) = 4626 \text{ W/m}$$

3-24 $k = 0.2$ W/m·°C $D = 8.5$ m $r = 1.5$ m $S = \frac{4\pi r}{1 - r/2D} = 20.67$

$$q = (0.2)(20.67)(30-0) = 124 \text{ W}$$

3-25 $k = 0.74$ W/m·°C $D = 0$ $W = 50$ cm $L = 100$ cm

$$S = \frac{\pi W}{\ln(4W/L)} = \frac{(0.5)\pi}{\ln[(4)(0.5/1.0)]} = 2.266 \qquad q = (0.74)(2.266)(120-15) = 176.1 \text{ W}$$

3-26 $L = 5$ cm

$$S_{walls} = [(2)(0.6)(0.7) + 2(0.6)(0.8) + (0.7)(0.8)]/0.05 = 58.4$$

$$S_{edges} = (4)(0.54)(0.6 + 0.7 + 0.8) = 4.536$$

$$S_{corners} = (8)(0.15)(0.05) = 0.06$$

$$S_{tot} = 62.996$$

3-27 $r_1 = 7.5$ cm $T_1 = 150$°C $T_2 = 5$°C $k = 0.7$ W/m·°C $r_2 = 2.5$ cm $D = 15$ cm

$$S = \frac{2\pi}{\cosh^{-1}\left[\frac{15^2 - 7.5^2 - 2.5^2}{(2)(7.5)(2.5)}\right]} = 2.928$$

$$q = (0.7)(2.928)(150-5) = 297.2 \text{ W/m}$$

3-28 The temperature gradients are written:

$$\left.\frac{\partial T}{\partial x}\right|_{m+\frac{1}{2},n,p} \approx \frac{T_{m+1,n,p} - T_{m,n,p}}{\Delta x} \qquad \left.\frac{\partial T}{\partial x}\right|_{m-\frac{1}{2},n,p} \approx \frac{T_{m,n,p} - T_{m-1,n,p}}{\Delta x}$$

$$\left.\frac{\partial T}{\partial y}\right|_{m,n+\frac{1}{2},p} \approx \frac{T_{m,n+1,p} - T_{m,n,p}}{\Delta y} \qquad \left.\frac{\partial T}{\partial y}\right|_{m,n-\frac{1}{2},p} \approx \frac{T_{m,n,p} - T_{m,n-1,p}}{\Delta y}$$

$$\left.\frac{\partial T}{\partial z}\right|_{m,n,p+\frac{1}{2}} \approx \frac{T_{m,n,p+1} - T_{m,n,p}}{\Delta z} \qquad \left.\frac{\partial T}{\partial z}\right|_{m,n,p-\frac{1}{2}} \approx \frac{T_{m,n,p} - T_{m,n,p-1}}{\Delta z}$$

For 3 dimensions Laplaces eqm. is

$$\frac{\partial^2 T}{\partial x^2} + \frac{\partial^2 T}{\partial y^2} + \frac{\partial^2 T}{\partial z^2} = 0$$ also if $\Delta x = \Delta y = \Delta z$ the finite difference approximations become:

$$T_{m+1,n,p} + T_{m-1,n,p} + T_{m,n+1,p} + T_{m,n-1,p} + T_{m,n,p+1} + T_{m,n,p-1} - 6T_{m,n,p} = 0$$

3-29 the Temp. gradient is written:

$$\left.\frac{dT}{dx}\right|_{m+\frac{1}{2},n} \approx \frac{T_{m+1,n} - T_{m,n}}{\Delta x} \qquad \left.\frac{dT}{dx}\right|_{m-\frac{1}{2},n} = \frac{T_{m,n} - T_{m-1,n}}{\Delta x}$$

for 1 dimension $\frac{d^2T}{dx^2} = 0$ and the finite difference approximation is:

$$T_{m+1,n} + T_{m-1,n} - 2T_{m,n} = 0$$

3-30 $$kA\frac{(T_{m-1} - T_m)}{\Delta x} - hA(T_m - T_\infty) = 0 \qquad -kA\frac{dT}{dx} \quad hA\Delta T$$

then:

$$T_m\left[\frac{h}{k}(\Delta x) + 1\right] - T_\infty\left[\frac{h}{k}(\Delta x)\right] - T_{m-1} = 0$$

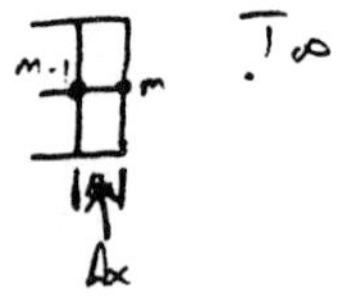

3-31

$$\frac{d^2T}{dx^2} - \frac{hP}{kA}(T - T_\infty) = 0 \qquad \left.\frac{d^2T}{dx^2}\right|_{m,n} \approx \frac{T_{m+1,n} - T_{m,n} - T_{m,n} + T_{m-1,n}}{(\Delta x)^2}$$

Substituting T_m for T above:

$$T_m\left[\frac{hP(\Delta x)^2}{kA} + 2\right] - \left[\frac{hP(\Delta x)^2}{kA}\right]T_\infty - \left[T_{m-1} + T_{m+1}\right] = \bar{q}_m$$

3-32

$$T_{m,n+1} + T_{m,n-1} + 2T_{m-1,n} - 4T_{m,n} = \bar{q}_{m,n}$$

$$\frac{hA(T_{m-1,n} - T_{m,n})}{\Delta x} + \frac{hA(T_{m,n+1} - T_{m,n})}{2\Delta y} + \frac{hA(T_{m,n-1} - T_{m,n})}{2\Delta y} = \bar{q}_{m,n}$$

for Δx Δy : $T_{m,n+1} + T_{m,n-1} + 2T_{m-1,n} - 4T_{m,n} = \bar{q}_{m,n}$

3-33

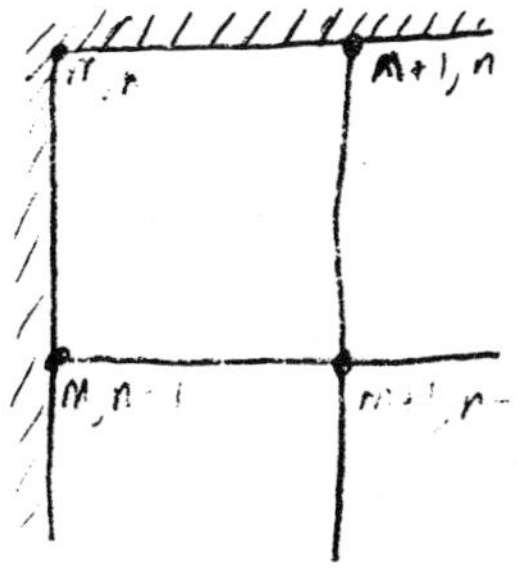

$$T_{m,n-1} + T_{m+1,n} - 2T_{m,n} = \bar{q}_{m,n}$$

3-35

$$k(T_{m-1,n} - T_{m,n}) + \frac{k}{2}(T_{m,n+1} - T_{m,n}) + \frac{k}{2}(T_{m,n-1} - T_{m,n}) + q''\Delta x = \bar{q}_{m,n}$$

3-38

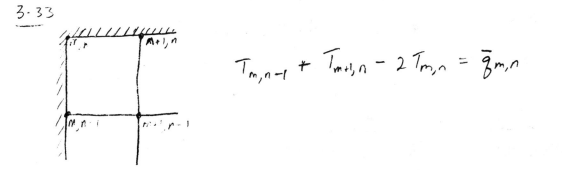

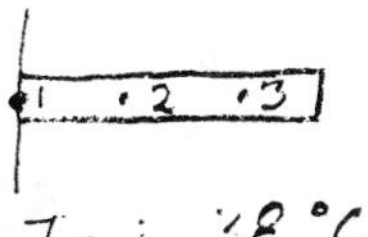

$k_{Al} = 204$ $\quad A = \dfrac{\pi(0.025)^2}{4} = 4.91 \times 10^{-4}\ m^2$ $\quad T_0 = 300°C$

$T_\infty = 38°C$ $\quad L = 15\ cm$ $\quad \Delta x = 0.05\ m$ $\quad h = 17$

$$\frac{hP(\Delta x)^2}{kA} = \frac{(17)\pi(0.025)(0.05)^2}{(204)(4.91\times10^{-4})} = 0.0333 \qquad \frac{h\Delta x}{k} = \frac{(17)(0.05)}{204} = 4.17\times10^{-3}$$

$2.033\,T_2 - 0.0333(38) - (300 + T_3) = q_1$; $\quad 2.033\,T_3 - (0.0333)(38) - (T_2 + T_4) = q_2$

$T_4(4.17\times10^{-3} + 0.01666 + 1) - (38)(4.17\times10^{-3} + 0.0166) - T_3 = q_4$

Solution: $T_2 = 279.7°C$ $\quad T_3 = 267.3°C$ $\quad T_4 = 262.6°C$

$$q = -\frac{kA}{\Delta x}(T_2 - 300) = 40.66\ W$$

3-40

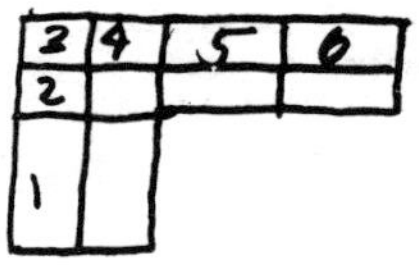

$\Delta x = \Delta y = 0.5\,m$ see internal nodes $k = 1.4$

$$4T_1 - 2T_2 = 800$$
$$-T_1 + 4T_2 - T_3 = 800$$
$$-T_2 + 4T_3 - T_4 = 300$$
$$-T_3 + 4T_4 - T_5 = 800$$
$$-T_4 + 4T_5 - T_6 = 800$$
$$-2T_5 + 4T_6 = 800$$

NODE	T
1	379.266
2	358.533
3	254.864
4	360.925
5	388.836
6	394.418

q(inside) = 6781.2 W/m depth

q(outside) = 6187.2 W/m

q(avg) = 6484.2 W/m

3-41

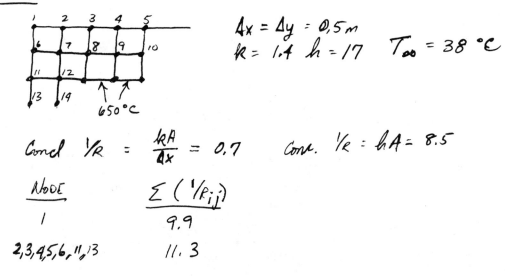

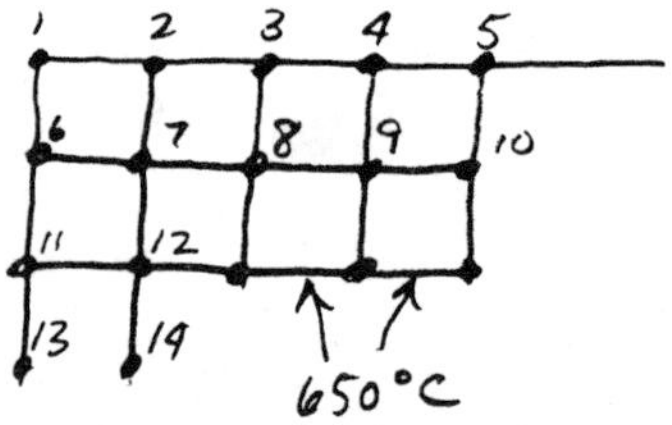

$\Delta x = \Delta y = 0.5\,m$

$k = 1.4 \quad h = 17 \quad T_\infty = 38\,°C$

Cond $1/R = \dfrac{kA}{\Delta x} = 0.7$ Conv. $1/R = hA = 8.5$

NODE	$\Sigma(1/R_{ij})$
1	9.9
2,3,4,5,6,11,13	11.3

USING THE GAUSS SEIDEL FORMAT:

3-41 cont.

The Equations:

```
T(1)  = ((.7)*(T(2)  + T(6))  + ((8.5)*(38)))/9.899999
T(2)  = ((.7)*(T(1)  + T(3))  + (8.5*38)  + 1.4*T(7))/11.3
T(3)  = ((.7)*(T(2)  + T(4))  + 1.4*T(8)  + (8.5*38))/11.3
T(4)  = ((.7)*(T(3)  + T(5))  + 1.4*T(9)  + (8.5*38))/11.3
T(5)  = (1.4*T(4)  + 1.4*T(10)  + (8.5*38))/11.3
T(6)  = ((.7)*(T(1)  + T(11))  + 1.4*T(7)  + (8.5*38))/11.3
T(7)  = (T(2)  + T(6)  + T(8)  + T(12))/4
T(8)  = (T(3)  + T(7)  + T(9)  + 650)/4
T(9)  = (T(4)  + T(8)  + T(10)  + 650)/4
T(10)  = (2*T(9)  + T(5)  + 650)/4
T(11)  = ((.7)*(T(6)  + T(13))  + 1.4*T(12)  + (8.5*38))/11.3
T(12)  = (T(7)  + T(11)  + T(14)  + 650)/4
T(13)  = (1.4*T(11)  + 1.4*T(14)  + (8.5*38))/11.3
T(14)  = (2*T(12)  + T(13)  + 650)/4
```

the solution:

```
T( 1 ) =  40.96808
T( 2 ) =  59.00557
T( 3 ) =  76.51266
T( 4 ) =  82.08206
T( 5 ) =  83.26541
T( 6 ) =  58.97151
T( 7 ) =  186.8046
T( 8 ) =  316.3084
T( 9 ) =  351.9162
T( 10 ) =  359.2745
T( 11 ) =  75.96281
T( 12 ) =  312.9329
T( 13 ) =  79.99095
T( 14 ) =  338.9642
```

3-43

$$1/R_{14} = 1/R_{12} = \frac{hA}{\Delta x} = 2.6$$

$$1/R_{1-\infty} = hA = 2.1$$

$$1/R_{2-\infty} = 4.2$$

$$1/R_{2-5} = 5.2$$

Node	$\Sigma(1/R_{ij})$
1	7.3
2	14.6
4	10.4

3-43 cont

the Equations:

```
T(1)  =  (2.6*T(2)  + 2.6*T(4))/7.3
T(2)  =  (2.6*T(1)  +  (2.6*10)  + 5.2*T(5))/14.6
T(4)  =  (2.6*T(1)  +  (2.6*38)  + 5.2*T(5))/10.4
T(5)  =  (10 + 38 + T(2)  + T(4))/4
```

the **Solution:**

```
T( 1 )  =   12.11863
T( 2 )  =   11.24257
T( 4 )  =   22.78283
T( 5 )  =   20.50635
```

3-45

the Equations:

```
T(1)  =  (1100 + T(3)  + T(4))/4
T(2)  =  (600 + T(3)  + T(4))/4
T(3)  =  (900 + T(1)  + T(2))/4
T(4)  =  (800 + T(1)  + T(2))/4
```

The Solution:

```
T( 1 )  =   487.5
T( 2 )  =   362.5
T( 3 )  =   437.5
T( 4 )  =   412.5
```

3-48

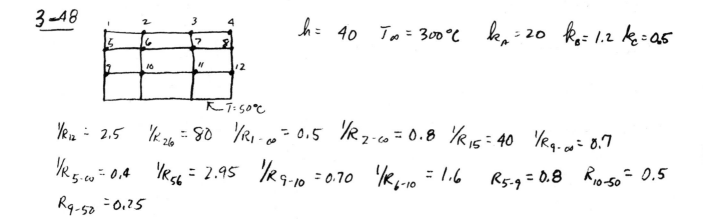

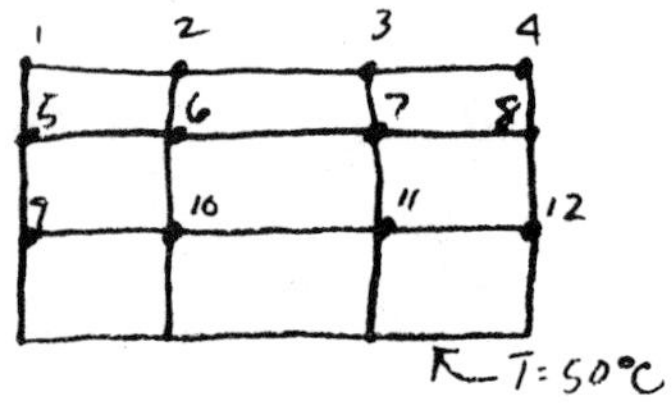

$h = 40$ $\quad T_\infty = 300°C$ $\quad k_A = 20$ $\quad k_B = 1.2$ $\quad k_C = 0.5$

$1/R_{12} = 2.5$ $\quad 1/R_{26} = 80$ $\quad 1/R_{1-\infty} = 0.5$ $\quad 1/R_{2-\infty} = 0.8$ $\quad 1/R_{15} = 40$ $\quad 1/R_{9-\infty} = 0.7$

$1/R_{5-\infty} = 0.4$ $\quad 1/R_{56} = 2.95$ $\quad 1/R_{9-10} = 0.70$ $\quad 1/R_{6-10} = 1.6$ $\quad R_{5-9} = 0.8$ $\quad R_{10-50} = 0.5$

$R_{9-50} = 0.25$

3-48 cont

The Equations

```
T(1)  = ((.5*300) + 2.5*T(2) + 40*T(5))/43
T(2)  = ((.8*300) + 2.5*T(1) + 2.5*T(3) + 80*T(6))/85.8
T(3)  = ((.8*300) + 2.5*T(2) + 2.5*T(4) + 80*T(7))/85.8
T(4)  = ((.5*300) + 2.5*T(3) + 40*T(8))/43
T(5)  = (40*T(1) + 2.95*T(6) + .8*T(9) + (.4*300))/44.15
T(6)  = (80*T(2) + 2.95*T(5) + 2.95*T(7) + 1.6*T(10))/87.5
T(7)  = (80*T(3) + 2.95*T(6) + 2.95*T(8) + 1.6*T(11))/87.5
T(8)  = (40*T(4) + 2.95*T(7) + .8*T(12) + (.4*300))/44.15
T(9)  = ((.25*50) + .8*T(5) + .7*T(10) + (.7*300))/2.45
T(10) = ((.5*50) + 1.6*T(6) + .7*T(9) + .7*T(11))/3.5
T(11) = ((.5*50) + 1.6*T(7) + .7*T(10) + .7*T(12))/3.5
T(12) = ((.25*50) + .7*T(11) + .8*T(8) + (.7*300))/2.45
```

The **Solution:**

```
T( 1 ) =  254.9767
T( 2 ) =  250.7219
T( 3 ) =  250.7219
T( 4 ) =  254.9767
T( 5 ) =  254.6798
T( 6 ) =  250.0962
T( 7 ) =  250.0962
T( 8 ) =  254.6798
T( 9 ) =  234.0802
T( 10 ) =  210.3607
T( 11 ) =  210.3607
T( 12 ) =  234.0802
```

3-49 $1/R_{1-6} = 0.25$ $1/R_{1-5} = 2.0$ $1/R_{8-\infty} = 0.4$ $1/R_{1-\infty} = 0.8$ $1/R_{4-\infty} = 0.2$ $\Sigma = 3.3$

Equations

```
T(1) = (75 + 2*T(5) + .25*T(2) + 16)/3.3
T(2) = (.25*T(1) + .25*T(3) + 2*T(6) + 16)/3.3
T(3) = (.25*T(2) + .25*T(4) + 2*T(7) + 16)/3.3
T(4) = (.25*T(3) + T(8) + 8 + 4)/1.85
T(5) = (4*T(1) + .5*T(6) + 150)/5
T(6) = (4*T(2) + .5*T(7) + .5*T(5))/5
T(7) = (4*T(3) + .5*T(6) + .5*T(8))/5
T(8) = (2*T(4) + .5*T(7) + 5)/2.9
```

Solution

```
T( 1 ) =  101.5548
T( 2 ) =  45.66036
T( 3 ) =  28.03639
T( 4 ) =  22.29107
T( 5 ) =  116.3579
T( 6 ) =  51.1407
T( 7 ) =  29.76612
T( 8 ) =  22.22938
```

3-50

THE EQUATIONS

```
T(1)=(1.5+2*(T(2)/.1)+(T(3)/.2))/25.15
T(2)=(2.25+(T(1)/.1)+(T(4)/.4))/12.725
T(3)=((T(1)/.2)+2*(T(4)/.05)+(T(5)/.2))/50!
T(4)=((T(2)/.4)+(T(3)/.05)+(T(6)/.4)+3!)/25.3
T(5)=((T(3)/.2)+2*(T(6)/.05)+(T(7)/.2))/50!
T(6)=((T(4)/.4)+(T(5)/.05)+(T(8)/.4)+3!)/25.3
T(7)=((T(5)/.2)+2*(T(8)/.05)+(T(9)/.2))/50!
T(8)=((T(6)/.4)+(T(7)/.05)+(T(10)/.4)+3!)/25.3
T(9)=((T(7)/.2)+2*(T(10)/.05)+(T(13)/.2))/50!
T(10)=((T(8)/.4)+(T(9)/.05)+(T(14)/.1)+(T(11)/.3)+3.75)/36.21
T(11)=((T(10)/.3)+(T(15)/.08)+(T(12)/.2)+3.75)/21.21
T(12)=((T(11)/.2)+(T(16)/.2)+1.5)/10.15
T(13)=((T(9)/.2)+2*(T(14)/.05)+1000)/50!
T(14)=((T(10)/.1)+(T(13)/.05)+(T(15)/.15)+2000)/46.67
T(15)=((T(11)/.08)+(T(14)/.15)+(T(16)/.1)+2500)/41.67
T(16)=((T(12)/.2)+(T(15)/.1)+1000)/20
```

THE SOLUTIONS

```
T( 1 ) =  109.7125
T( 2 ) =  108.9678
T( 3 ) =  115.6857
T( 4 ) =  114.8968
T( 5 ) =  127.9756
T( 6 ) =  127.1026
T( 7 ) =  147.253
T( 8 ) =  146.3778
T( 9 ) =  173.5348
T( 10 ) =  175.0166
T( 11 ) =  184.3022
T( 12 ) =  185.5953
T( 13 ) =  187.9632
T( 14 ) =  188.2622
T( 15 ) =  191.5148
T( 16 ) =  192.1562
TOTAL NUMBER OF ITERATIONS:  250
```

3-51

$$1/R_{12} = \frac{(2.3)(0.125)}{0.25} = 1.15 = 1/R_{14} \quad 1/R_{1-\infty} = (25)(0.125) = 3.125$$

$$\Sigma = 5.425$$

The Equations:

```
T(1) = ((5*3.125) + 1.15*T(2) + 1.15*T(4))/5.425
T(2) = (1.15*T(1) + 1.15*T(3) + (6.25*5) + 2.3*T(5))/10.85
T(3) = (1.15*T(2) + 115 + (6.25*5) + 2.3*T(6))/10.85
T(4) = (1.15*T(1) + 2.3*T(5) + 1.15*T(7))/4.6
T(5) = (T(2) + T(4) + T(6) + T(8))/4
T(6) = (T(3) + T(5) + T(9) + 100)/4
T(7) = (1.15*T(4) + 115 + 2.3*T(8))/4.6
T(8) = (T(5) + T(7) + T(9) + 100)/4
T(9) = (T(6) + T(8) + 200)/4
```

.21198

The Solution:

```
T( 1 ) =  17.86873
T( 2 ) =  19.51926
T( 3 ) =  29.93323
T( 4 ) =  51.18759
T( 5 ) =  54.59205
T( 6 ) =  67.86016
T( 7 ) =  77.69751
T( 8 ) =  79.80123
T( 9 ) =  86.91534
```

3-52

$$1/R_{1-\infty} = (12)(0.25) = 3 \quad 1/R_{12} = (1.5)(0.125)/0.25 = 0.75$$

The Equations:

```
T(1) = (.75*T(2) + 37.5 + 45 + 1.5*T(3))/6
T(2) = (.75*T(1) + 1.5*T(4) + 45 + 37.5)/6
T(3) = (T(1) + T(4) + T(5) + 50)/4
T(4) = (T(2) + T(3) + T(6) + 50)/4
T(5) = (T(3) + T(6) + 100)/4
T(6) = (T(4) + T(5) + 100)/4
```

The Solution:

```
T( 1 ) =  27.6
T( 2 ) =  27.6
T( 3 ) =  41.6
T( 4 ) =  41.6
T( 5 ) =  47.2
T( 6 ) =  47.2
```

3-53 $1/R_{12} = (10)(0.005)/(0.01) = 5$ $1/R_{1-\infty} = (125)(0.01) = 1.25$ $1/R_{1-5} = 10$

$\Sigma = 21.25$ $1/R_{59} = 40$ $1/R_{5-6} = (40)(0.005)/0.01 + \frac{10}{2} = 25$

The Equations:

```
T(1)  = (10*T(2) + 12.5 + 10*T(5))/21.25
T(2)  = (5*T(1) + 5*T(3) + 12.5 + 10*T(6))/21.25
T(3)  = (5*T(2) + 5*T(4) + 10*T(7) + 12.5)/21.25
T(4)  = (5*T(3) + 12.5 + 5*T(8))/11.25
T(5)  = (10*T(1) + 12000 + 50*T(6))/100
T(6)  = (25*T(5) + 12000 + 10*T(2) + 25*T(7))/100
T(7)  = (25*T(6) + 10*T(3) + 25*T(11) + 10*T(8))/70
T(8)  = (10*T(7) + 5*T(4) + 5*T(12) + 12.5)/21.25
T(11) = (12000 + 25*T(7) + 10*T(12) + 25*T(14))/100
T(12) = (10*T(11) + 12.5 + 5*T(8) + 5*T(15))/21.25
T(14) = (50*T(11) + 10*T(15) + 12000)/100
T(15) = (10*T(14) + 10*T(12) + 12.5)/21.25
```

The Solution:

```
T( 1 ) =  253.5477
T( 2 ) =  249.8672
T( 3 ) =  236.621
T( 4 ) =  211.4409
T( 5 ) =  287.6715
T( 6 ) =  284.6335
T( 7 ) =  270.9156
T( 8 ) =  236.621
T( 11 ) =  284.6335
T( 12 ) =  249.8672
T( 14 ) =  287.6715
T( 15 ) =  253.5476
```

3-54 $hA/\Delta x = 0.5498$ $1/R_{12} = 1/R_{1-\infty} = 0.5498$ $1/R_{1-\infty} = 0.05027$

$q_1 = (50\times10^6)\pi(0.01)^2(0.02) = 3.142$ $\Sigma 1/R_{ij} = 2(0.5498) + 0.05027 = 1.1499$

Equations:

```
T(1) = (109.96 + .5498*T(2) + 1.26 + 3.142)/1.1499
T(2) = (.5498*(T(1) + T(3)) + 1.26 + 3.142)/1.1499
T(3) = (.5498*(T(2) + T(4)) + 1.26 + 3.142)/1.1499
T(4) = (.5498*(T(3) + T(5)) + 1.26 + 3.142)/1.1499
T(5) = (.5498*T(4) + .63 + 1.571 + .314)/.5875
```

Solution

```
T( 1 ) =  172.9641
T( 2 ) =  153.7457
T( 3 ) =  140.5867
T( 4 ) =  132.2831
T( 5 ) =  128.0753
```

3-55 $1/R_{12} = 0.2$ $1/R_{16} = 0.05$ $1/R_{1-\infty} = 0.25$ $1/R_{1-\infty_6} = 0.5$

$\Sigma = 1.0$

Equations

```
T(1) = (10 + 5 + .2*T(2) + .05*T(6))/1!
T(2) = (.2*T(1) + .2*T(3) + .1*T(7) + 10)/1!
T(3) = (.2*T(2) + .1*T(4) + .15*T(8) + 15)/1.5
T(4) = (.1*T(3) + .1*T(5) + .2*T(9) + 20)/1.4
T(5) = (.2*T(4) + .2*T(10) + 20)/1.4
T(6) = (.05*T(1) + .05*T(11) + .4*T(7) + 20)/1.5
T(7) = (.4*T(6) + .4*T(8) + .1*T(2) + .1*T(12))/1!
T(8) = (.4*T(7) + .15*T(3) + 75 + .2*T(9))/.9
T(9) = (.2*T(8) + .2*T(10) + 100 + .2*T(4))/.8
T(10) = (.4*T(9) + .2*T(5) + 100)/.8
T(11) = (.4*T(12) + .05*T(6) + .05*T(13) + 20)/1.5
T(12) = (.4*T(11) + .1*T(7) + .1*T(14) + 200)/1!
T(13) = (.8*T(11) + .4*T(14) + 20)/1.5
T(14) = (.4*T(13) + .2*T(12) + 200)/1!
```

Solution

```
T( 1 ) =  25.01783
T( 2 ) =  36.62113
T( 3 ) =  39.62346
T( 4 ) =  58.51667
T( 5 ) =  61.20274
T( 6 ) =  53.87198
T( 7 ) =  136.9287
T( 8 ) =  208.3953
T( 9 ) =  259.2036
T( 10 ) =  269.9025
T( 11 ) =  95.7117
T( 12 ) =  283.5975
T( 13 ) =  148.6994
T( 14 ) =  316.1992
```

3-57 $1/R_{12} = (4.0)(0.00125)/0.01 = 0.5 \quad 1/R_{1-\infty} = (75)(0.01) = 0.75$

$1/R_{13} = (4.0)(0.01)/0.0025 = 16 \quad \Sigma = 17.75$

Equations

```
T(1)  =  (.5*T(2)  + 50.D + 16.D*T(3)  + 0)/17.75
T(2)  =  (.5*T(1)  + 16.D*T(4)  + 50.D)/17.75
T(3)  =  (1!*T(4)  + 100.D + 16.D*T(3)  + 16.D*T(1))/34
T(4)  =  (1!*T(3)  + 100.D + 16.D*T(6)  + 16.D*T(2))/34
T(5)  =  (.5*T(6)  + 50.D + 16.D*T(3))/17
T(6)  =  (.5*T(5)  + 50.D + 16.D*T(4))/17
```

Solution

```
T( 1 ) =   67.35356
T( 2 ) =   69.55632
T( 3 ) =   69.42171
T( 4 ) =   71.93425
T( 5 ) =   70.41794
T( 6 ) =   72.71512
```

3-58 $1/R_{12} = (20)(0.005)/0.01 = 10 = 1/R_{14}$

$q_1 = (90)(10^6)(0.005)(0.005) = 2250$

$1/R_{1-\infty} = (100)(0.005) = 0.5 \quad \Sigma = 20.5$

Equations

```
T(1)  =  (10*T(2)  + 10*T(4)  + 10  + 2250)/20.5
T(2)  =  (10*T(1)  + 20*T(5)  + 10*T(3)  + 4500  + 20)/41
T(3)  =  (10*T(2)  + 1000  + 20*T(6)  + 20  + 4500)/41
T(4)  =  (20*T(5)  + 10*T(1)  + 10*T(7)  + 4500)/40
T(5)  =  (T(2)  + T(4)  + T(6)  + T(8)  + 450)/4
T(6)  =  (T(3)  + T(5)  + T(9)  + 100  + 450)/4
T(7)  =  (20*T(8)  + 10*T(4)  + 10*T(10)  + 4500)/40
T(8)  =  (T(5)  + T(7)  + T(9)  + T(11)  + 450)/4
T(9)  =  (T(6)  + T(8)  + T(12)  + 100  + 450)/4
T(10)  =  (10*T(11)  + 10*T(7)  + 2250)/20
T(11)  =  (10*T(10)  + 10*T(12)  + 4500)/20
T(12)  =  (10*T(11)  + 1000  + 4500)/20
```

Solution

```
T( 1 ) =   1938.338
T( 2 ) =   1736.704
T( 3 ) =   1127.731
T( 4 ) =   2010.89
T( 5 ) =   1801.208
T( 6 ) =   1167.497
```

Solution

```
T( 7 ) =   2052.807
T( 8 ) =   1839.742
T( 9 ) =   1191.047
T( 10 ) =   2070.855
T( 11 ) =   1863.903
T( 12 ) =   1206.952
```

3-62

Equations

```
T(1) = (20*T(2) + 5*T(3) + 1.5)/25
T(2) = (10*T(1) + 2.5*T(4) + 2.25)/12.5
T(3) = (40*T(4) + 5*T(1) + 5*T(5))/50
T(4) = (20*T(3) + 2.5*T(2) + 2.5*T(6) + 3)/25
T(5) = (5*T(3) + 5*T(7) + 40*T(6))/50
T(6) = (20*T(5) + 2.5*T(4) + 2.5*T(8) + 3)/25
T(7) = (5*T(5) + 5*T(9) + 40*T(8))/50
T(8) = (20*T(7) + 2.5*T(6) + 2.5*T(10) + 3)/25
T(9) = (40*T(10) + 5*T(7) + 5*T(13))/50
T(10) = (20*T(9) + (10*T(11)/3) + 10*T(14) + 3.75)*3/100
T(11) = ((10*T(10)/3) + 5*T(12) + 12.5*T(15) + 3.75)/20.8333
T(12) = (5*T(11) + 2.5*T(16) + 1.5)/7.5
T(13) = (5*T(9) + 1000 + 40*T(14))/50
T(14) = (20*T(13) + 10*T(10) + 2000 + (10*T(15)/1.5))/46.6667
T(15) = ((10*T(14)/1.5) + 12.5*T(11) + 2500 + 10*T(16))/41.6667
T(16) = (10*T(15) + 2.5*T(12) + 500)/15
```

Solution

```
T( 1 ) =  207.2835
T( 2 ) =  207.3585
T( 3 ) =  206.6836
T( 4 ) =  206.7585
T( 5 ) =  205.4843
T( 6 ) =  205.5585
T( 7 ) =  203.6915
T( 8 ) =  203.752
T( 9 ) =  201.4148
T( 10 ) =  201.2298
T( 11 ) =  200.8928
T( 12 ) =  200.9544
T( 13 ) =  200.619
T( 14 ) =  200.5969
T( 15 ) =  200.4778
T( 16 ) =  200.4776
```

3-63 $1/R_{9\,10} = (40)(0.005)/0.01 = 20 = 1/R_{16-13}$ $1/R_{9-5} = (40)(0.01)/0.01 = 40$

Equations

```
T(1)  = (10*T(2) + 12.5 + 10*T(5))/21.25
T(2)  = (5*T(1) + 5*T(3) + 12.5 + 10*T(6))/21.25
T(3)  = (5*T(2) + 5*T(4) + 10*T(7) + 12.5)/21.25
T(4)  = (5*T(3) + 12.5 + 5*T(8))/11.25
T(5)  = (10*T(1) + 12000 + 50*T(6))/100
T(6)  = (25*T(5) + 12000 + 10*T(2) + 25*T(7))/100
T(7)  = (25*T(6) + 10*T(3) + 25*T(11) + 10*T(8))/70
T(8)  = (10*T(7) + 5*T(4) + 5*T(12) + 12.5)/21.25
T(9)  = (40*T(5) + 40*T(10) + 3)/80
T(10) = (40*T(6) + 20*T(9) + 20*T(13) + 40*T(11) + 3)/120
T(11) = (12000 + 25*T(7) + 10*T(12) + 25*T(14))/100
T(12) = (10*T(11) + 12.5 + 5*T(8) + 5*T(15))/21.25
T(13) = (40*T(10) + 40*T(14) + 3)/80
T(14) = (50*T(11) + 10*T(15) + 12000)/100
T(15) = (10*T(14) + 10*T(12) + 12.5)/21.25
```

Solution

```
T( 1 ) =  253.5477
T( 2 ) =  249.8672
T( 3 ) =  236.621
T( 4 ) =  211.4409
T( 5 ) =  287.6715
T( 6 ) =  284.6335
T( 7 ) =  270.9156
T( 8 ) =  236.621
T( 9 ) =  286.5163
T( 10 ) =  285.2861
T( 11 ) =  284.6335
T( 12 ) =  249.8672
T( 13 ) =  286.5163
T( 14 ) =  287.6715
T( 15 ) =  253.5476
```

3-65

Take resistance in terms of mean areas between nodes.

$k = 210\ W/m\text{-}^\circ C \quad h = 200\ W/m^2\text{-}^\circ C$

$r_0 = 0.5\ cm,\ r_4 = 0.25\ cm \quad T_\infty = 10^\circ C$

$r_2 = 0.375\ cm,\ r_1 = 0.4375\ cm,\ r_3 = 0.3125\ cm$

$A_0 = \pi(0.005)^2 = 7.85 \times 10^{-5}$

$A_1 = \pi(0.004375)^2 = 6.013 \times 10^{-5}\ m^2$

$A_2 = \pi(0.00375)^2 = 4.418 \times 10^{-5}$

$A_3 = \pi(0.003125)^2 = 3.068 \times 10^{-5}\ m^2$

$A_4 = \pi(0.0025)^2 = 1.963 \times 10^{-5}$

Assume convection areas are $\pi d \Delta x$ at node.

$1/R_{10} = kA/\Delta x = (210)(6.013 + 7.85)(10^{-5})/(2)(0.015) = 0.9704$

$1/R_{12} = (210)(6.013 + 4.418)(10^{-5})/(2)(0.015) = 0.7302$

Also,

$1/R_{23} = (210)(4.418 + 3.068)(10^{-5})/(2)(0.015) = 0.524$

$1/R_{34} = (210)(3.068 + 1.963)(10^{-5})/(2)(0.015) = 0.3522$

$1/R_{4+} = 0$

3-65 (contd)

$$1/R_{100} = \pi(2)(0.4375)(0.015)(200)(0.01) = 0.0825$$

$$1/R_{200} = \pi(2)(0.375)(0.015)(200)(0.01) = 0.0707$$

$$1/R_{300} = \pi(2)(0.3125)(0.015)(200)(0.01) = 0.0589$$

$$1/R_{4-\infty} = \pi(2)(0.25)(0.0075)(200)(0.01) = 0.0236$$

$$0.9704(200-T_1) + 0.7302(T_2-T_1) + 0.0825(10-T_1) = 0$$
$$0.7302(T_1-T_2) + 0.524(T_3-T_2) + 0.0707(10-T_2) = 0$$
$$0.524(T_2-T_3) + 0.3522(T_4-T_3) + 0.0589(10-T_3) = 0$$
$$0.3522(T_3-T_4) + 0.0236(10-T_4) = 0$$

$$-1.7831\,T_1 + 0.7302\,T_2 = -194.905$$
$$0.7302\,T_1 - 1.3249\,T_2 + 0.524\,T_3 = -0.707$$
$$0.524\,T_2 - 0.9351\,T_3 + 0.3522\,T_4 = -0.589$$
$$0.3522\,T_3 - 0.3758\,T_4 = -0.236$$

SOLUTIONS: $T_1 = 167.45°C$ $T_2 = 141.99°C$

$T_3 = 124.32°C$ $T_4 = 117.14°C$

3-66

$$Bi = \frac{h \Delta x}{k}$$

Node

① $Bi\, T_\infty + T_3 + T_2 - (Bi + 2) T_1 = 0$

② $2 Bi\, T_\infty + T_1 + T_4 - 2(Bi + 1) T_2 = 0$

③ $T_1 + 2T_4 + 100 - 4T_3 = 0$

④ $2 Bi\, T_\infty + (2)(100) + 2T_3 + T_2 - 2(Bi + 3) T_4 = 0$

⑤ $Bi\, T_\infty + \frac{1}{2}(200 + T_4 + T_6) - (Bi + 2) T_5 = 0$

⑥ $Bi\, T_\infty + \frac{1}{2}(200 + T_5 + T_7) - (Bi + 2) T_6 = 0$

⑦ $T_6 + 100 - (Bi + 2) T_7 = 0$

3-67 INITIAL EQUATIONS:

$$140 + T_2 + T_3 - 4T_1 = 0$$
$$T_1 + T_4 + T_5 + 100 - 4T_3 = 0$$
$$40 + T_1 + T_4 - 4T_2 = 0$$
$$T_2 + T_3 + T_6 - 4T_4 = 0$$
$$200 + T_3 + T_6 - 4T_5 = 0$$
$$100 + T_4 + T_5 - 4T_6 = 0$$

MATRIX FORM:

T_1	T_2	T_3	T_4	T_5	T_6	C
-4	1	1	0	0	0	-140
1	-4	0	1	0	0	-40
1	0	-4	1	1	0	-100
0	1	1	-4	0	1	0
0	0	1	0	-4	1	-200
0	0	0	1	1	-4	-100

SOLUTIONS:

$T_1 = 61.615$, $T_2 = 35.528$, $T_3 = 70.932$

$T_4 = 40.969$, $T_5 = 81.615$, $T_6 = 55.528\ ^\circ C$

3-68

NODE:

① $100 + 2T_2 + T_4 - 4T_1 = 0$

② $T_1 + 100 + T_3 + T_5 - 4T_2 = 0$

③ $T_2 + 150 + T_6 - 4T_3 = 0$

④ $T_1 + T_7 + 2T_5 - 4T_4 = 0$

⑤ $T_4 + T_2 + T_6 + T_8 - 4T_5 = 0$

⑥ $T_5 + T_3 + 50 + T_9 - 4T_6 = 0$

⑦ $\frac{k}{2}\left[T_4 + T_8 - 2T_7\right] + q'' \frac{\Delta x}{2} = 0$

⑧ $\frac{k}{2}\left[T_7 + 2T_5 + T_9 - 4T_8\right] + q'' \frac{\Delta x}{2} = 0$

⑨ $\frac{k}{2}\left[T_8 + 2T_6 + 50 - 4T_9\right] + q'' \frac{\Delta x}{2} = 0$

3-69

NODE:

① $k\frac{\Delta x/2}{\Delta y}(T_5 - T_1) + h\,\Delta x/2\,(T_\infty - T_1)$

$$+ \frac{k\,\Delta y/2}{\Delta x}(T_2 - T_1) = 0$$

$$T_5 - T_1 + \frac{h\Delta y}{k}(T_\infty - T_1) + T_2 - T_1 = 0$$

$$T_1 = \left(T_5 + T_2 + \frac{h\Delta y}{k}\right) \Big/ \left(2 + \frac{h\Delta y}{k}\right)$$

Let $Bi = h\Delta y/k$

② $T_2 = \dfrac{T_6 + (T_1 + T_3)/2 + Bi\,T_\infty}{2 + Bi}$

③ $T_3 = \dfrac{T_7 + (T_2 + T_4)/2 + Bi\,T_\infty}{2 + Bi}$

④ $T_4 = \dfrac{T_8 + (T_3 + 50)/2 + Bi\,T_\infty}{2 + Bi}$

⑤ $T_5 = (T_1 + T_9 + 2T_6)/4$

⑥ $T_6 = (T_5 + T_2 + T_7 + T_{10})/4$

⑦ $T_7 = (T_6 + T_3 + T_8 + T_{11})/4$

⑧ $T_8 = (T_7 + T_4 + 100 + T_{12})/4$

⑨ $T_9 = (T_5 + T_{10})/2$

3-69 (contd)

(10) $T_{10} = (T_9 + T_{11} + 2T_6)/4$

(11) $T_{11} = (T_{10} + T_{12} + 2T_7)/4$

(12) $T_{12} = (T_{11} + 150 + 2T_8)/4$

3-70

$$x^2 + y^2 = 100 \qquad \text{circle, } r = 10$$

@ $x = 6$, $y = 8$ cm

$6 + b\Delta y = 8 \qquad b = 2/3$

@ $y = 6$, $x = 8$ and $a = 2/3$

@ $y = 9$, $x = \sqrt{19} = 4.359$ cm

$9 - 8 = 1 = c\Delta y \qquad c = 1/3$

Using Table 3-2(f) for node 4

$$0 = \frac{2}{(2/3)(5/3)} T_3 + \frac{2}{4/3} T_6 + \frac{2}{5/3} T_5 + \frac{2}{(2/3)(5/3)} - 2(3/2 + 3/2) T_4$$

Using Table 3-2(g) for Node 3

$$\frac{h\Delta x}{k} = \frac{(30)(0.03)}{10} = 0.09$$

Continued next page

3-70 (contd)

Continuation of using Table 3-2(g)

$$\frac{2/3}{[(2/3)^2+(2/3)^2]^{1/2}} T_2 + \frac{2/3}{[(1/3)^2+1]} T_7$$

$$+ \frac{4/3}{2/3} T_4 + (0.09)[(1/3^2+1)^{1/2} + (2/3^2 + 2/3^2)^{1/2}](20)$$

$$-[\qquad\qquad\qquad] T_3$$

↑ Insert values of a, b, c, etc.

3-71

$b = 1/2 \qquad a = 1/2 \qquad c = 3/2$

Nomenclature fits Table 3-2 (f, g) exactly. Insert above values for final equations.

3-72

Same as Prob. 3-71 except that insulated surface is equivalent to $h = 0$, inserted in the equations.

3-73

The same as Prob. 3-71 except that T_∞ would be inserted for T_1 and T_3.

3-74 Assume $\Delta x = \Delta y$

Nodes 3 and 6 correspond to Node M, N of Table 3-2(f) but for different media. The correspondance with the equations in Table 3-2(f) is as follows:

Node 3, Material A

$$a = 0.5, \quad b = 0.5$$

$$T_{m,n} = T_3, \; T_{m,n-1} = T_2, \; T_{m+1,n} = T_1$$
$$T_2 = T_4, \quad T_1 = T_5$$

Node 6, Material A $\quad a = 0.5, \quad b = 0.5$

$$T_{m,n} = T_6, \quad T_{m,n-1} = T_8, \quad T_{m+1,n} = T_7$$
$$T_2 = T_4, \quad T_1 = T_9$$

For nodes 4 and 5, the connecting resistances between the two materials must be taken into account $a = 1/2$, $b = 1/2$, $c = 1/2$ for Material A.

Node 4, Material A

$$R_{43_A} = \frac{2b}{k_A(a+1)}, \quad R_{49})_A = \frac{2(c^2+1)^{1/2}}{b\,k_A}$$

$$R_{45_A} = \frac{2(a^2+b^2)^{1/2}}{b\,k_A}$$

3-74 (contd)

Node 4, Material B

$a = 1/2, \; b = 1/2, \; c = 1/2$

$$R_{46,B} = \frac{2a}{(1+b)\,k_B} \qquad R_{45_B} = \frac{2(c^2+1)^{1/2}}{b\,k_B}$$

$$R_{49_B} = \frac{2(a^2+b^2)^{1/2}}{b\,k_B}$$

$$R_{49}\Big|_{Tot} = \frac{1}{1/R_{49_A} + 1/R_{49_B}} = R_{49}$$

$$R_{45}\Big|_{Tot} = \frac{1}{1/R_{45_A} + 1/R_{45_B}} = R_{45}$$

Nodal Equation for Node 4:

$$\frac{T_6 - T_4}{R_{46_B}} + \frac{T_3 - T_4}{R_{43_A}} + \frac{T_9 - T_4}{R_{49}} + \frac{T_5 - T_4}{R_{45}} = 0$$

Similar equation for Node 5 in terms of nodes 3, 4, 10, and 11.

3-75

$L = 0.2\ m$

$S = 8.24\ L = 1.648$

$q = kS\Delta T = (2.3)(1.648)(80-10) = 265.3\ W$

Sphere $\quad S = 4\pi r = 4\pi(0.1) = 1.2567$

$q = kS\Delta T = (2.3)(1.2567)(80-10) = 202.32\ W$

$V(\text{cube}) = (0.2)^3 = 8 \times 10^{-3}\ m^3$

$V(\text{sphere}) = \frac{4}{3}\pi(0.1)^3 = 4.189 \times 10^{-3}\ m^3$

$(q/V)(\text{cube}) = 265.3/8\times10^{-3} = 33162\ W/m^3$

$(q/V)(\text{sphere}) = 202.32/4.189\times10^{-3} = 48300\ W/m^3$

3-76

$r = 5\ cm \qquad D = 15\ cm$

$$S/L = \frac{2\pi}{\ln(4D/r)} = \frac{2\pi}{\ln(60/5)} = 2.529$$

$q/L = kS\Delta T = (10)(2.529)(100-20) = 2023\ W$

3-78

$W = 100\ cm \qquad L = 10\ cm \qquad D = 2.0\ m$

$$S = \frac{2\pi W}{\ln(4W/L)} = \frac{2\pi(1.0)}{\ln\left[\frac{(4)(100)}{10}\right]} = 1.703$$

$q = kS\Delta T = (1.5)(1.703)(50-10) = 102.2\ W$

3-79

$$S = 4r = (4)(0.025) = 0.1$$

$$q = kS\Delta T = (3)(0.1)(75-15) = 18 \text{ W}$$

3-80

$$S = \frac{\pi W}{\ln(4W/L)} = \frac{\pi(0.05)}{\ln(4)} = 0.133$$

$$q = kS\Delta T = (3)(0.133)(75-15) = 20.4 \text{ W}$$

$$q/A)_{DISC} = 18/\pi(0.025)^2 = 9167 \text{ W/m}^2$$

$$q/A)_{SQUARE} = 20.4/(0.05)^2 = 8158 \text{ W/m}^2$$

3-81

$$S = \frac{2\pi L}{\ln(0.54W/r)} \qquad r = 5\text{ cm} \quad L = 20\text{ m} \quad W = 20\text{ cm}$$

$$S = \frac{2\pi(20)}{\ln[(0.54)(20)/5]} = 163.2$$

$$q = kS\Delta T = (50 \times 10^{-3})(163.2)(200-35)$$

$$= 1346 \text{ W}$$

4-1 $T_{0_{cu}}$ = Initial Sphere Temperature. $T_\infty = T_m + A\sin\omega\tau$

Heat Transferred = $q = hA(T - T_\infty) = -\rho cV \frac{dT}{d\tau}$

$$q = hA\left[T - (T_m + A\sin\omega\tau)\right] \quad \frac{dT}{d\tau} + kT = kA\sin\omega\tau + kT_m$$

where $k = \frac{hA}{\rho cV}$ the solution for T is:

$$T = C_1 e^{-k\tau} - \frac{Ak\omega}{\omega^2 + k^2}\cos\omega\tau + \frac{k^2 A}{\omega^2 + k^2}\sin\omega\tau + T_m$$ Solving for C:

$T = T_0$ at $\tau = 0$ $\quad C_1 = T_0 - T_m + \frac{Ak\omega}{\omega^2+k^2}$ then

$$T - T_m = \left(T_0 - T_m + \frac{Ak\omega}{\omega^2+k^2}\right)e^{-k\tau} - \frac{Ak\omega}{\omega^2+k^2}\cos\omega\tau + \frac{k^2A}{\omega^2+k^2}\sin\omega\tau$$

4-2 $\alpha = 1.8\times10^{-6}\ m^2/sec$ $\quad 2L = 2.5\ cm$ $\quad T_i = 150°C$ $\quad T_1 = 30°C$

$\tau = 1\ min = 60\ sec$ $\quad \frac{\pi x}{2L} = \pi/2$; $\left(\frac{\pi}{2L}\right)^2\alpha\tau = 1.705$ $\quad$ 1st four nonzero terms $n = 1,3,5,7$

$$\frac{T - T_1}{T_i - T_1} = \frac{4}{\pi}\left[0.1818 - 7.22\times10^{-8} + 6.15\times10^{-20}\right] = 0.231$$

$$T = 30 + (0.231)(150 - 30) = 57.8°C \qquad \frac{\alpha\tau}{L^2} = 0.69 \qquad \frac{\theta_0}{\theta_i} = 0.25$$

4-3 at $\tau = 0$ $\quad \frac{x}{2L} = \frac{1}{2}$ $\quad \frac{\pi x}{2L} = \pi/2$

$$\frac{T - T_1}{T_i - T_1} = \frac{4}{\pi}\left[\sin\frac{\pi}{2} + \frac{1}{3}\sin\frac{3\pi}{2} + \frac{1}{5}\sin\frac{5\pi}{2} + \frac{1}{7}\sin\frac{7\pi}{2}\right] = 0.9216$$

correct value is 1.0 $\quad$ Error = 7.84 %

4-4

$$q = \sigma A(T^4 - T_\infty^4) + hA(T - T_\infty) = -c\rho V\frac{dT}{d\tau}$$

4-5 $T_0 = 260°C$ $\quad T = 90°C$ $\quad T_\infty = 35°C$ $\quad R_{th} = \frac{1}{2}\left(\frac{\Delta x}{kA}\right) = 3.754\times10^{-4}$

$$C_{th} = \rho cV = (8900)(0.38\times10^3)(0.05)(0.3)^2 = 1.52\times10^4$$

$$1/R_{th}C_{th} = 0.175 \qquad \frac{T - T_\infty}{T_0 - T_\infty} = e^{-\frac{1}{R_{th}C_{th}}\tau} = e^{-0.175\tau} = \frac{90 - 35}{260 - 35}$$

$\tau = 8.05$ sec

4-6 $m = \rho V = 5.5$ kg $\rho = 2707$ kg/m³ $c = 0.896$ kJ/kg °C $h = 58$ W/m² °C

$\frac{4}{3}\pi r^3 (2707) = 5.5$ $r = 0.0786$ m $A = 4\pi r^2 = 0.0776$ m²

$$\frac{hA}{\rho c V} = \frac{(58)(0.0776)}{(5.5)(896)} = 9.137 \times 10^{-4} \qquad \frac{90-15}{290-15} = e^{-9.137\times10^{-4}\tau}$$

$\tau = 1422$ sec $= 0.395$ hr.

4-10 $T_o = 30°C$ $T_\infty = 150°C$ $h = 140$ W/m²·°C $d = 6.4$ mm $T = 120°C$ $\rho = 7817$ kg/m³

$c = 460$ J/kg·°C $A = \pi d L$ $V = \frac{\pi d^2}{4} L$ $A/V = 4/d$ $\frac{hA}{\rho c V} = 0.0243$

$$\frac{T_1 - T_\infty}{T_o - T_\infty} = e^{-0.0243\tau} = 0.25 \qquad \tau = 57 \text{ sec}$$

4-11 $d = 5$ cm $r_o = 2.5$ cm $T_o = 250°C$ $T_\infty = 30°C$ $h = 28$ W/m²·°C $T = 90°C$

$$\frac{A}{V} = \frac{3}{r_o} \qquad \frac{hA}{\rho c V} = 9.8\times10^{-4} \qquad \frac{90-30}{250-30} = 0.273 = e^{-9.8\times10^{-4}\tau} \qquad \tau = 1325 \text{ s} = 0.368 \text{ hr.}$$

4-13 Lumped Capacity $\rho = 8954$ $c = 383$ $\frac{hA}{\rho c V} = \frac{(5)(4\pi)(0.015)^2}{(8954)(383)(\frac{4}{3}\pi(0.015)^3)} = 8.75 \times 10^{-4}$

$$\frac{25-10}{50-10} = e^{-8.75\times10^{-4}\tau} \qquad \tau = 1121 \text{ sec}$$

4-14 $\rho = 2707$ $c = 896$ $\sigma A T^4 = -\rho c V \frac{dT}{d\tau}$ T in °K

$$\frac{dT}{T^4} = \frac{-\sigma A}{\rho c V} d\tau \qquad \frac{1}{T^3} - \frac{1}{T_o^3} = \frac{\sigma A \tau}{\rho c V}$$

$T = -240 + 273 = 33$ K

$T_o = 40 + 273 = 313$ K $\qquad \tau = 9.9 \times 10^6$ sec

4-15 $\rho = 999.8$ $c = 4225$ $L = 2d$ $A = 2.5\pi d^2$ $V = \frac{1}{2}\pi d^3$ $d = 6.06$ cm $A = 288.5$ cm²

$$\frac{hA}{\rho c V} = \frac{(15)(288.5)(10^{-4})}{(999.8)(4225)(350)(10^{-6})} = 2.927\times10^{-4}$$

$$\frac{15-22}{1-22} = e^{-2.927\times10^{-4}\tau} \qquad \tau = 375 \text{ sec}$$

4-16

$$\frac{h(V/A)}{k} = \frac{(10)(0.006)}{(3)(204)} = 9.8 \times 10^{-5}$$

lumped capacity $\rho = 2707$
$c = 896$

$$\frac{200-20}{400-20} = e^{-\left[\frac{(10)(3)\tau}{(0.006)(896)(2707)}\right]} = 0.4737$$

$\tau = 362$ sec.

4-17

$$\frac{h(V/A)}{k} = \frac{hV}{3k} = \frac{(20)(0.02)}{(3)(380)} = 3.5 \times 10^{-4}$$

Lumped capacity $c = 383$ $\rho = 8954$

$$\frac{80-30}{200-30} = e^{-\left[\frac{(20)(3)(\tau)}{(0.02)(383)(8954)}\right]} = 0.294$$

$\tau = 1399$ sec.

4-18 $A = \frac{90-35}{2} = 27.5°C$ $x = 5$ cm $= 0.05$ m

$k = 1.37$ W/m-°C $a = 7 \times 10^{-7}$ m²/sec

$n = 1$ cyc./15 min $= 1.1111 \times 10^{-3}$ cyc/sec.

$$x\sqrt{\frac{\pi n}{\alpha}} = 0.05\left[\frac{\pi(1.1111 \times 10^{-3})}{7 \times 10^{-7}}\right]^{1/2} = 3.531$$

$2n\pi\tau = 2\pi(1.1111 \times 10^{-3})(2)(3600) = 50.26$

$2\pi n\tau - x\sqrt{\frac{\pi n}{\alpha}} = 46.734$ radians $= 2677.66°$

$\cos(2677.66) = -0.925$ $\sin(2677.66) = 0.3801$

$$q/A = kAe^{-x\sqrt{\frac{\pi n}{\alpha}}}\left(\sqrt{\frac{\pi n}{\alpha}}\right)\left\{\cos(2677.66) + \sin(2677.66)\right\}$$

$$q/A = (1.37)(27.5)(e^{-3.531})\left[-0.925 + 0.3801\right] = -0.601 \text{ W/m}^2$$

4-19

Maximum points when sine function is max. ie.:

$$2\pi n\tau - x\sqrt{\frac{\pi n}{\alpha}} = \pi/2 \quad \text{at } x=0 \quad \tau = \frac{\pi}{4\pi n} = \frac{1}{4n}$$

$$\text{at } x = x_1 \quad \tau = \frac{\pi}{4\pi n} + \frac{x\sqrt{\frac{\pi n}{\alpha}}}{2\pi n} = \frac{1}{4n} + \frac{x}{2}\sqrt{1/\pi\alpha n}$$

$$\Delta\tau = \frac{x}{2}\sqrt{1/\pi\alpha n}$$

4-20

$T_i = 54°C \quad T_\infty = 10°C \quad h = 2.6 \text{ W/m}^2\cdot°C$

$x = 7 \text{ cm} \quad \tau = 30 \text{ min} \quad \alpha = 7\times10^{-7} \text{ m}^2/\text{sec}$

$k = 1.37 \text{ W/m}\cdot°C$

$$\frac{x}{2\sqrt{\alpha\tau}} = \frac{0.07}{2[(7\times10^{-7})(30)(60)]^{1/2}} = 0.986 \qquad \frac{T-T_i}{T_\infty - T_i} = 0.0062$$

$$\frac{h\sqrt{\alpha\tau}}{k} = \frac{(2.6)[(7.7\times10^{-7})(30)(60)]^{1/2}}{1.37} = 0.067 \qquad T = 53.73°C$$

4-21

$T_i = 300°C \quad T_o = 35°C \quad x = 7.5 \text{ cm} \quad \tau = 4 \text{ min} = 240 \text{ sec}$

$\alpha = 11.23\times10^{-5} \text{ m}^2/\text{s}$

$$X = \frac{x}{2\sqrt{\alpha\tau}} = \frac{0.075}{2[(11.23\times10^{-5})(240)]^{1/2}} = 0.2284$$

$$\text{erf } X = 0.2533 = \frac{T - T_o}{T_i - T_o} \qquad T = 102.1°C$$

4-22

$\alpha = 7\times10^{-7}\ m^2/s \qquad T_i = 55°C \qquad T_o = 15°C$

$T = 25°C \qquad x = 5\ cm$

$$\frac{T-T_o}{T_i-T_o} = \frac{25-15}{55-15} = 0.25 = \text{erf}\ x \qquad x = 0.2253 = \frac{x}{2\sqrt{\alpha\tau}}$$

$\tau = 17589$ sec $= 4.89$ hrs.

4-23

$q_o/A = 0.32\times10^6\ W/m^2 \qquad \tau = 300s \qquad T_i = 30°C$

$k = 386 \qquad \alpha = 11.23\times10^{-5}\ m^2/s$

$x = 0$

$$T = 30 + \frac{(2)(0.32\times10^6)\left[\frac{(11.23\times10^{-5})(300)}{\pi}\right]^{1/2}}{386}e^0 = 201.7°C$$

$x = 15\ cm$

$$x = \frac{x}{2\sqrt{\alpha\tau}} = \frac{0.15}{2[(11.23\times10^{-5})(300)]^{1/2}} = 0.4086$$

$$\frac{x^2}{4\alpha\tau} = 0.167 \qquad \text{erf}\ x = 0.4173$$

$$T-T_i = \frac{2(0.32\times10^6)\left[\frac{(11.23\times10^{-5})(300)}{\pi}\right]^{1/2}}{386}e^{-0.167} - \frac{(0.32\times10^6)(0.15)}{386}(1-0.4173)$$

$= 72.83 \qquad T = 102.83\ °C$

4-24

$T_i = 90°C \qquad T_o = 30°C \qquad x = 7.5\ cm \qquad \tau = 5\ sec$

$k = 386\ W/m\cdot°C \qquad \alpha = 11.23\times10^{-5}\ m^2/s$

$$q/A = \frac{-k(T_i-T_o)}{\sqrt{\pi\alpha\tau}}\ e^{-\frac{x^2}{4\alpha\tau}}$$

$$q/A = \frac{(-386)(90-30)}{[\pi(11.23\times10^{-5})(5)]^{1/2}}\exp\left[\frac{-(0.075)^2}{4(11.23\times10^{-5})(5)}\right] = -45.06\ kW/m^2$$

4-25 $T_i = 30°C$ $q_0/A = 15000\ W/m^2$ $x = 2.5\,cm$ $\tau = 120\,sec$

$k = 204$ $\alpha = 8.42 \times 10^{-5}$

$$T - T_i = \frac{15000}{204}\left\{2\left[\frac{(8.42\times10^{-5})(120)}{\pi}\right]^{1/2} \exp\left[\frac{-(0.025)^2}{4(8.42\times10^{-5})(120)}\right] - (0.025)\left[1 - \operatorname{erf}\left(\frac{0.025}{2\sqrt{8.42\times10^{-5}(120)}}\right)\right]\right\} = 6.59$$

$T = 36.59°C$

4-28

$$\frac{T - T_0}{T_i - T_0} = \frac{-1-(-1)}{-20-(-1)} = 0.5263 = \operatorname{erf}\frac{x}{2\sqrt{\alpha\tau}}$$

$$\frac{x}{2\sqrt{\alpha\tau}} = 0.5267$$

$$\tau = \frac{1}{0.048}\left[\frac{(0.015)(3.2808)}{2(0.5267)}\right]^2 = 0.04547\ hr = 163.7\ sec$$

4-29

$q/A = 900\ W/m^2$ $T_i = 20°C$ $x = 10\,cm$ $\tau = 9\,hr = 32400\,sec$

$k = 1.37\ W/m\cdot°C$ $\alpha = 7.5\times10^{-7}\ m^2/s$

$$T = 20 + \frac{(2)(900)[(7.5\times10^{-7})(32400)/\pi]^{1/2}}{1.37}\exp\left[\frac{-(0.1)^2}{4(7.57\times10^{-7})(32400)}\right] - \frac{(900)(0.1)}{1.37}\left[1 - \operatorname{erf}\left(\frac{0.1}{(2)(7.5\times10^{-7})(32400)^{1/2}}\right)\right]$$

$= 81.5\ °C$

4-30 $T_i = 300°C$ $T_0 = 100°C$ $X = 0.03\ m$ $T = 200°C$

$$\frac{200-100}{300-100} = 0.5 = \operatorname{erf}\left[\frac{x}{2\sqrt{\alpha\tau}}\right] \qquad \alpha = 0.444\times10^{-5}$$

$$\frac{x}{2\sqrt{\alpha\tau}} = 0.48 \qquad \tau = \left[\frac{0.03}{(2)(0.48)}\right]^2 \Big/ 0.444\times10^{-5} = 2200\,sec.$$

4-31 $T_i = 40°C$ $h = 25\ W/m^2\cdot°C$ $T_\infty = 2°C$ $x = 0.08\ m$ $T(x) = 20°C$

$\alpha = 5.2 \times 10^{-7}$ $k = 0.69\ W/m\cdot°C$

$$\frac{T - T_i}{T_\infty - T_i} = \frac{20 - 40}{2 - 40} = 0.5263 \qquad \frac{x}{2\sqrt{\alpha\tau}} \simeq 0.4$$

$$\tau = \left[\frac{0.08}{(2)(0.4)}\right]^2 \Big/ 5.2 \times 10^{-7} = 19231 \text{ sec}$$

4-32 $T_i = 30°C$ $q/A = 3 \times 10^4$ $\tau = 10 \text{ min} = 600 \text{ sec}$ $x = 3\ cm$

$k = 2.32\ W/m\cdot°C$ $\alpha = 9.2 \times 10^{-7}\ m^2/s$

$$T = 30 + \frac{(2)(3\times10^4)\left[(9.2\times10^{-7})(600)/\pi\right]^{1/2}}{2.32} \exp\left[\frac{-(0.03)^2}{(4)(9.2\times10^{-7})(600)}\right]$$

$$- \frac{(3\times10^4)(0.03)}{2.32}\left[1 - \text{erf}\left(\frac{0.03}{(2)(9.2\times10^{-7})(600)}\right)\right] = 30 + 228$$

$$= 258°C$$

4-33 From symmetry same as inf. plate 6 cm thick

$\tau = 360$ sec. $L = 3 cm$ $\alpha = 11.23 \times 10^{-5}$ $k = 370$ $\frac{\alpha\tau}{L^2} = 44.92$

$$\theta_{x=L}/\theta_i = \frac{150 - 100}{250 - 100} = 0.33$$

Iterative Solution:

k/hL	θ_0/θ_i	θ/θ_0	θ/θ_i	$\theta/\theta_i - 0.33$
100	0.65	1.0	0.65	0.32
50	0.42	0.98	0.41	0.08
45	0.38	0.98	0.37	0.04
40	0.34	0.98	0.33	0

$$h = \frac{370}{(40)(0.03)} = 308.3\ W/m^2\cdot°C$$

4-34 $L = 5\text{ cm}$ $h = 1400$ $k = 230$ $T_i = 400$ $\alpha = 8.42 \times 10^{-5}$ $T_\infty = 90$

$T_o = 180$ $k/hL = 3.29$ $\theta_o/\theta_i = 0.29$ $\dfrac{\alpha\tau}{L^2} = 5.0$

$$\tau = \frac{(0.05)^2(5)}{8.42 \times 10^{-5}} = 148 \text{ sec}$$

4-35 $T_i = 350°C$ $T_\infty = 100°C$ $T(x=L) = 150°C$ $\tau = 6\text{ min} = 360\text{ sec}$

$k = 374\ W/m \cdot °C$ $\alpha = 11.23 \times 10^{-5}$ $L = 0.1\text{ m}$ $\dfrac{\alpha\tau}{L^2} = 4.04$

$$\theta/\theta_i = \frac{150 - 100}{250 - 100} = \frac{50}{150} = 0.33 \qquad x/L = 1.0$$

$$k/hL \approx 3.5 \qquad h = \frac{374}{(3.5)(0.1)} = 1069\ W/m^2 \cdot °C$$

4-36 $L = 5\text{ cm}$ $T_i = 400°C$ $T_\infty = 90°C$ $h = 1400\ W/m^2 \cdot °C$ $k = 204\ W/m \cdot °C$

$\alpha = 8.4 \times 10^{-5}\ m^2/s$ $\theta/\theta_i = \dfrac{180 - 90}{400 - 90} = 0.29$

$$k/hL = \frac{204}{(1400)(0.05)} = 2.91 \qquad \frac{\alpha\tau}{L^2} = 4.2$$

$$\tau = (4.2)(0.05)^2 / 8.4 \times 10^{-5} = 125 \text{ sec}$$

4-37 $L = 0.015\text{ m}$ $T_i = 500°C$ $T_\infty = 40°C$; $h = 150\ W/m^2 \cdot °C$; $k = 16.3\ W/m \cdot °C$

$\alpha = 0.44 \times 10^{-5}\ m^2/s$ $k/hL = \dfrac{16.3}{(150)(0.015)} = 7.24$

at $\dfrac{x}{L} = 1.0$ $\theta/\theta_o = 0.93$

For $\theta_o/\theta_i = \dfrac{100 - 40}{500 - 40} = 0.13$ $\alpha\tau/L^2 = 16$

$$\tau = \frac{(16)(0.015)^2}{0.44 \times 10^{-5}} = 818 \text{ sec}$$

For $\theta/\theta_i = 0.13$ $\theta_o/\theta_i = \dfrac{0.13}{0.93} = 0.14$ $\alpha\tau/L^2 = 15$

$$\tau = \frac{(15)(0.015)^2}{0.44 \times 10^{-5}} = 767 \text{ sec.}$$

4-38 $r_0 = 5\,cm$ $L = 5\,cm$ $T_i = 250°C$ $T_\infty = 30°C$ $h = 280\,W/m^2\cdot°C$

$k = 43\,W/m\cdot°C$ $\alpha = 1.172 \times 10^{-5}\,m^2/sec$ $\tau = 2\,min = 120\,sec$

$$\frac{\alpha\tau}{L^2} = \frac{\alpha\tau}{r_0^2} = 0.563 \qquad \frac{k}{hL} = \frac{k}{hr_0} = 3.071$$

Cylinder: $\frac{\theta_0}{\theta_i} = 0.86$ Plate: $\frac{\theta_0}{\theta_i} = 0.93$ $\frac{\theta_z}{\theta_0} = 0.86$

Center: $\theta/\theta_i = (0.86)(0.93) = 0.8$ $T = 175°C$

End center: $\theta/\theta_i = (0.86)(0.93)(0.86) = 0.688$ $T = 151°C$

4-39 $r_0 = 5.5\,cm$ $T_i = 300°C$ $T_\infty = 50°C$ $h = 1200\,W/m^2\cdot°C$

$T_0 = 80°C$ $\rho = 2707$ $c = 896\,J/kg\cdot°C$ $k = 204\,W/m\cdot°C$

$\alpha = 8.4 \times 10^{-5}\,m^2/s$

$$k/hr_0 = \frac{204}{(1200)(0.055)} = 3.09 \qquad \theta_0/\theta_i = \frac{80-50}{300-50} = 0.12$$

$$\frac{\alpha\tau}{r_0^2} = 3.7 \qquad \tau = (3.7)(0.055)^2/8.4 \times 10^{-5} = 133\,sec.$$

$$\frac{h^2\alpha\tau}{k^2} = \frac{(1200)^2(8.4\times10^{-5})(133)}{(204)^2} = 0.386$$

$$\frac{hr_0}{k} = 0.324 \qquad \frac{Q}{Q_0} = 0.85$$

$Q_0 = \rho c V\theta_i = (2707)(896)(300-50)\,\pi(0.055)^2 = 5.76\,MJ$

$Q = (0.85)(5.76) = 4.9\,MJ$

4-40 $T_i = 70°F$ $T_\infty = 350°F$ $h = 2.5$ $k_w = 0.395$ $\rho = 59.6$

$c_p = 1.0$ $\alpha = 0.00663$ $V = 0.084\,ft^3$ (assume sperical roast)

$r = 0.271\,ft = r_0$ $k/hr_0 = 0.583$ $\theta_0/\theta_i = 0.536$ $\frac{\alpha\tau}{r_0^2} = 0.3$

$\tau = 3.32\,hrs.$

4-41 $\alpha = 9.5 \times 10^{-7}$ m²/s $r_0 = 1.25$ cm $k = 1.52$ W/m·°C $T_i = 25°C$
$T_\infty = 200°C$ $h = 110$ W/m²·°C $\tau = 4$ min $= 240$ sec.

$$\frac{k}{hr_0} = \frac{1.52}{(110)(0.0125)} = 1.105 \qquad r/r_0 = \frac{0.64}{1.25} = 0.51$$

$$\frac{\alpha\tau}{r_0^2} = \frac{(9.5\times10^{-7})(240)}{(0.0125)^2} = 1.46 \qquad \frac{\theta_0}{\theta_i} = 0.048 \qquad \frac{\theta_r}{\theta_0} = 0.89$$

center $T = (25-200)(0.048) + 200 = 191.6°C$ $r = 6.4$ mm

$T = (25-200)(0.048)(0.89) + 200 = 192.5°C$

4-42 $T_i = 300°C$ $T_0 = 120°C$ $d = 1.5$ mm $r_0 = 0.75$ mm $T_\infty = 100°C$
$h = 5000$ W/m²·°C $k = 35$ W/m·°C $\alpha = 2.34 \times 10^{-5}$ m²/s

$$\frac{k}{hr_0} = \frac{35}{(5000)(0.00075)} = 9.33 \qquad \frac{\theta_0}{\theta_i} = \frac{120-100}{300-100} = 0.1$$

$$\frac{\alpha\tau}{r_0^2} = 7.3 \qquad \tau = \frac{(7.3)(0.00075)^2}{2.34\times10^{-5}} = 0.175 \text{ sec}$$

4-43 $r_0 = 5$ cm $T_\infty = 10°C$ $T_i = 250°C$ $h = 280$ $\alpha = 1.172 \times 10^{-5}$ $T_0 = 150°C$
$k = 43$ $\frac{\theta_0}{\theta_i} = \frac{150-10}{250-10} = 0.583$

$$\frac{k}{hr_0} = \frac{43}{(280)(0.05)} = 3.07 \qquad \frac{\alpha\tau}{r_0^2} = 0.75 \qquad \tau = \frac{(0.75)(0.05)^2}{1.172\times10^{-5}} = 160 \text{ sec}$$
$= 2.67$ min

4-44 $T_i = 200°C$ $T_\infty = 20°C$ $h = 14$ W/m²°C $r_0 = 0.0075$ m
$T_0 = 35°C$ $k = 0.78$ W/m·°C $\alpha = 3.4 \times 10^{-7}$

$$k/hr_0 = \frac{0.78}{(14)(0.0075)} = 7.43 \qquad \frac{\theta_0}{\theta_i} = \frac{35-20}{200-20} = 0.083$$

$$\frac{\alpha\tau}{r_0^2} = 6.5 \qquad \tau = (6.5)(0.0075)^2/3.4\times10^{-7} = 1075 \text{ sec}$$

4-45 $r_o = 0.00075\ m$ $T_i = 200°C$ $h = 5000\ W/m^2\cdot°C$ $T_\infty = 100°C$ $T_o = 120°C$

$k = 35\ W/m\cdot°C$ $\alpha = 2.34 \times 10^{-5}$

$$k/hr_o = \frac{35}{(5000)(0.00075)} = 9.33 \qquad \frac{\theta_o}{\theta_i} = \frac{120-100}{200-100} = 0.2$$

$$\frac{\alpha\tau}{r_o^2} = 5.2 \qquad \tau = (5.2)(0.00075)^2/2.34\times10^{-5} = 0.125\ \text{sec.}$$

4-47 $T_i = 250°C$ $T_\infty = 30°C$ $h = 570$ $\tau = 120\ \text{sec}$ $L_1 = L_2 = 1.25\ cm$

$L_3 = 3.75\ cm$ $k = 43\ W/m\cdot°C$ $\alpha = 1.172\times10^{-5}\ m^2/\text{sec.}$

$$\left.\frac{k}{hL}\right|_{1,2} = 6.035 \quad \left.\frac{k}{hL}\right|_3 = 2.01 \quad \left.\frac{\alpha\tau}{L^2}\right|_{1,2} = 9.00 \quad \left.\frac{\alpha\tau}{L^2}\right|_3 = 1.00$$

$$\left.\frac{\theta_o}{\theta_i}\right|_{1,2} = 0.25 \quad \left.\frac{\theta_o}{\theta_i}\right|_3 = 0.7 \quad \text{Center } \frac{\theta_o}{\theta_i} = (0.25)^2(0.07) = 0.0438$$

$T = 39.6\ °C$

4-48 $L = 5\ cm$ $T_i = 350°C$ $T_\infty = 90°C$ $h = 1200\ W/m^2\cdot°C$ $\tau = 60\ \text{sec}$

$k = 204\ W/m\cdot°C$ $\alpha = 8.42\times10^{-5}\ m^2/s$ $\frac{k}{hL} = 3.4$ $\frac{\alpha\tau}{L^2} = 2.02$

$$\frac{\theta_o}{\theta_i} = 0.59 \quad \frac{\theta_L}{\theta_o} = 0.86 \quad \text{Center of face: } \theta/\theta_i = (0.59)^3(0.86) = 0.177$$

$T = 136\ °C$

4-49

$r_o = 7.5\ cm$ $L = 15\ cm$ $T_i = 25°C$ $T_\infty = 0°C$ $T_o = 6°C$ $h = 17\ W/m^2\cdot°C$

$\alpha = 7\times10^{-7}\ m^2/s$ $k = 1.37\ W/m\cdot°C$ $k/hL = 0.54$ $k/hr_o = 1.075$

$$\frac{\theta_o}{\theta_i} = \frac{6-0}{25-0} = 0.24 = \left(\frac{\theta_o}{\theta_i}\right)_{cyl.}\left(\frac{\theta_o}{\theta_i}\right)_{plate}$$

Iterative Solution:

τ	$\frac{\alpha\tau}{r_o^2}$	C	$\frac{\alpha\tau}{L^2}$	P	CP
3600	0.448	0.65	0.112	1.0	0.65
7200	0.896	0.28	0.224	0.9	0.252

$\tau \simeq 7200\ \text{sec} = 2\ \text{hrs.}$

4-50 $k = 240$ $\frac{h(V/A)}{k} = \frac{(20)(0.04)^3}{(6)(0.04)^2(240)} = 3.3 \times 10^{-3}$

Lumped Capacity: $\rho = 2707$ $c = 896$

$$\frac{hA}{\rho c V} = \frac{(20)(6)(0.04)^2}{(2707)(896)(0.04)^3} = 7.42 \times 10^{-3}$$

$$\frac{200-100}{450-100} = e^{-7.42 \times 10^{-3}\tau} \qquad \tau = 168.8 \text{ sec.}$$

4-51 $L = 5.5$ cm $T_i = 400\,°C$ $T_\infty = 85\,°C$ $h = 1100\ W/m^2 \cdot °C$ $\tau = 60$ sec

$k = 204\ W/m \cdot °C$ $\alpha = 8.4 \times 10^{-5}\ m^2/s$ $k/hL = \frac{204}{(1100)(0.055)} = 3.37$

$$\frac{\alpha \tau}{L^2} = \frac{(8.4 \times 10^{-5})(60)}{(0.055)^2} = 1.67 \qquad \frac{\theta_0}{\theta_i} = 0.7$$

at $\frac{x}{L} = 1.0$ $\theta/\theta_0 = 0.86$ $\frac{\theta}{\theta_i}$ at center of face $= (0.7)^3(0.86) = 0.295$

$T = (0.295)(400 - 85) + 85 = 178\ °C$

4-52

$L = 0.025$ $T_i = 100°C$ $T_\infty = 25°C$ $h = 20\ w/m^2 \cdot °C$

$T_0 = 50°C$ $k = 204\ w/m \cdot °C$ $\alpha = 8.4 \times 10^{-5}\ m^2/sec$

$$\frac{h(V/A)}{k} = \frac{(20)(0.025)^3}{(6)(0.025)^2(204)} = 4.08 \times 10^{-4}$$

Lumped Capacity:

$$\frac{50-25}{100-25} = \exp\left[\frac{-(20)(6)(0.025)^2\,\tau}{(2707)(896)(0.025)^3}\right]$$

$\tau = 555$ sec.

4-53 $T_i = 200\ °C$ $T_\infty = 30\ °C$ $\tau = 600$ sec. $h = 200\ W/m \cdot °C$ $d = 10$ cm

$L = 15$ cm $T_0 = 100\ °C$ $k = 16.3$ $\alpha = 0.44 \times 10^{-5}$ $\rho = 7817$ $c = 460$

$$k/hr_0 = \frac{16.3}{(200)(0.05)} = 1.63 \qquad k/hL = \frac{16.3}{(200)(0.075)} = 1.09$$

$$\frac{\alpha\tau}{r_0^2} = 1.056 \qquad \frac{\alpha\tau}{L^2} = 0.47$$

$$\left.\frac{\theta_0}{\theta_i}\right)_c = 0.45 \qquad \left.\frac{\theta_0}{\theta_i}\right)_p = 0.9 \qquad \frac{\theta_0}{\theta_i} = (0.45)(0.9) = 0.405$$

$$T = (0.405)(200 - 30) + 30 = 98.8\ °C$$

Plate $\frac{hL}{k} = 0.92$ $\quad h^2\alpha\tau/k^2 = 0.397$

Cyl $\frac{hr_0}{k} = 0.61$

$$Q/Q_0\big|_{cyl} = 0.55 \qquad Q/Q_0\big|_{plate} = 0.2$$

$$\left.\frac{Q}{Q_0}\right|_{total} = 0.55 + 0.2[1 - 0.55] = 0.64$$

$$Q_0 = \rho c V \theta_i = (7817)(460)\pi(0.05)^2(0.15)(200-30) = \mathbf{0.72\ MJ}$$

$$Q = (0.64)(0.72) = \mathbf{0.46\ MJ}$$

4-63 $\rho = 2000\ kg/m^3$ $c = 960\ J/kg\,°C$ $k = 1.04\ W/m \cdot °C$

$$1/R_{12} = \frac{kA}{\Delta x} = (1.04)(0.005)/0.02 = 0.26$$

$$1/R_{13} = (1.04)(0.02)/0.01 = 2.08$$

$$C_i = \rho_i c_i V_i$$

NODE	V_i	C_i	$\Sigma\, 1/R_{ij}$	$\Delta \tau_{max}$
1	0.0001	192	3.6	53.3
2	0.00005	96	2.05	46.8
3	0.0002	384	5.2	73.8
4	0.0001	192	3.1	61.9

$$1/R_{1-\infty} = (50)(0.02) = 1.0$$

$$1/R_{2-\infty} = (50)(0.015) = 0.75$$

$$1/R_{4-\infty} = (50)(0.01) = 0.5$$

$$1/R_{34} = (1.04)(0.01)/0.02 = 0.52$$

$$1/R_{24} = (1.04)(0.01)/0.01 = 1.04$$

4-64 $1/R_{31} = \frac{kA}{\Delta x} = (2.32)(0.01)/0.01 = 2.32$

$1/R_{34} = (2.32)(0.005)/0.02 + (0.48)(0.005)/0.02 = 0.7$

$1/R_{35} = (0.48)(0.01)/0.01 = 0.48$

$1/R_{3-\infty} = hA = (50)(0.01) = 0.5$

$\Sigma\, 1/R_{ij} = 4.0$ $C_3 = \Sigma \rho c \Delta V = (3000)(840)(0.005)(0.01) + (1440)(1000)(0.005)(0.01) = 198$

$$\Delta \tau_{max} = 198/4.0 = 49.5 \text{ sec}$$

$$T_3^{p+1} = \frac{\Delta \tau}{C_3}\left[2.32\, T_1^p + 0.7\, T_4^p + 0.48\, T_5^p + (0.5)(40)\right] + T_3^p$$

4-65 $1/R_{m+} = \frac{kA}{\Delta x} = \frac{1}{2}k = \frac{1}{R_{m-}}$

$1/R_{n-} = k \qquad 1/R_{n+} = hA = h\Delta x$

$$T_{m,n}^{p+1} = \frac{\Delta\tau}{C_{m,n}}\left[\frac{k}{2}\left(T_{m-1,n} + T_{m+1,n} + 2T_{m,n-1}\right) + h\Delta x T_\infty\right] + T_{m,n}^{p}$$

4-66 $1/R_{12} = \frac{kA}{\Delta x} = (20)(0.02)/0.02 = 20 \qquad 1/R_{13} = (20)(0.01)/0.04 = 5$

$1/R_{14} = (1.2)(0.01)/0.04 = 0.3 \qquad 1/R_{1-\infty} = hA = (40)(0.02) = 0.8$

$\Sigma\, 1/R_{ij} = 26.1 \qquad C_1 = \Sigma \rho c V = (7800)(460)(0.01)(0.04) + (1600)(850)(0.01)(0.04)$

$= 1979.2$

$$\Delta\tau_{max} = \frac{1979.2}{26.1} = 75.83 \text{ sec}$$

$$T_1^{p+1} = \frac{\Delta\tau}{C_1}\left[5T_3^p + 20T_2^p + 0.3T_4^p + (0.8)(40)\right] + T_1^p$$

4-68

	ρ	c	k
A	1440	840	0.48
B	2787	883	164

$$1/R_{54} = \frac{(0.48)(0.005)}{0.02} + \frac{(164)(0.005)}{0.02} = 41.12$$

$1/R_{52} = \frac{(0.48)(0.02)}{0.01} = 0.96 \qquad 1/R_{56} = 0.12 \qquad 1/R_{58} = \frac{(164)(0.005)}{0.02} = 41$

$1/R_{5-\infty} = (35)(0.005 + 0.01) = 0.525 \qquad \Sigma\, 1/R = 83.725$

$C_5 = (1440)(840)(0.02)(0.005) + (2787)(883)(0.01)(0.005) = 244.006$

$$\Delta\tau_{max} = \frac{244.006}{83.725} = 2.914 \text{ sec.}$$

$$T_5^{p+1} = \left[41.12\,T_4^p + 0.96\,T_2^p + 0.12\,T_6^p + 41\,T_8^p + (55)(0.525)\right]\frac{\Delta\tau}{C_5} + \left(1 - \Delta\tau/2.914\right)T_5^p$$

4-69

Node 1

$$1/R_{12} = \frac{(1.04)(0.02)}{0.01} = 2.08$$

$$1/R_{13} = (1.04)(0.01)/0.02) = 0.52$$

$$\sum 1/R = (2)(2.08) + (2)(.52) = 5.2$$

$$C_1 = (2000)(960)(0.01)(0.02) = 384$$

Node 2

$$C_2 = 384/2 = 192$$

$$1/R_{2-\infty} = (60)(0.02) = 1.2$$

$$1/R_{24} = (1.04)(0.005)/0.02 = 0.26$$

$$\sum 1/R = 2.08 + (2)(0.26) + 1.2 = 3.8$$

Node 3

$$C_3 = (2000)(960)(0.01)(0.015) = 288$$

$$1/R_{34} = (1.04)(0.015)/0.01 = 1.56$$

$$1/R_{35} = (1.04)(0.01)/0.01 = 1.04$$

$$\sum 1/R = (2)(1.56) + 1.04 + 0.52 = 4.68$$

4-69 (con't)

Node 4

$C_4 = (2000)(960)[(0.005)(0.015) + (0.005)(0.01)]$

$= 240$

$1/R_{4+} = (1.04)(0.005)/0.02 = 0.26$

$1/R_{4\infty} = (60)(0.02) = 1.2$

$1/R_{46} = (1.04)(0.015)/0.01 = 1.56$

$\sum 1/R = 0.26 + 0.26 + 1.56 + 1.56 + 1.2 = 4.84$

Node 5

Take all $1/R = (1.04)(0.01)/0.01) = 1.04$

$\sum 1/R = 4.16$

$C_5 = (2000)(960)(0.01)(0.01) = 192$

Node	C	$\sum 1/R$	$\Delta \tau$ max, sec.
1	384	5.2	73.85
2	192	3.8	50.53
3	288	4.68	61.54
4	240	4.84	49.59
5	192	4.16	46.15

4-69 (contd)

Node	C	$\Sigma 1/R$	$\Delta\tau_{MAX}$, s
1	384	5.2	73.85
2	192	3.8	50.53
3	288	4.68	61.54
4	240	4.58	52.4
5	192	4.16	46.15

4-70

$$1/R_{35} = (15)(0.0025)/0.02 = 1/R_{31} = 1.875$$

$$1/R_{34} = (15)(0.02)/0.005 = 60 \quad 1/R_{4-\infty} = (25)(0.02) = 0.5$$

$$\Sigma 1/R = 64.25$$

$$T_3 = \left[(1.875)(T_1 + T_5) + 60T_4 + 0.5\,T_\infty\right]/64.25$$

4-71 $1/R_{21} = (10)(0.005)/0.02 = 2.5 \qquad 1/R_{23} = (2)/4 = 0.5$

$1/R_{25} = (10) + 2 = 12 \qquad 1/R_{2-\infty} = (40)(0.02) = 0.8$

$\Sigma 1/R = 15.8 \qquad C_2 = (6500)(300)(0.01)(0.005) + (2000)(700)(0.01)(0.005)$

$= 167.5$

$\Delta\tau_{max} = 10.601$ sec.

$$T_2^{p+1} = \left[2.5\,T_1^p + 0.5\,T_3^p + 12\,T_5^p + (0.8)(20)\right]\frac{\Delta\tau}{167.5} + \left(1 - \Delta\tau/10.601\right)\Big]\,T_2^p$$

4-72

	ρ	c	k
A	1440	840	0.48
B	2787	883	164
C	7817	460	16.3

$$1/R_{52} = \frac{(0.48)(0.01)}{0.01} = 0.48$$

$$1/R_{54} = \frac{(0.48)(0.005)}{0.01} + (164)(0.005)/0.01 = 82.24$$

$$1/R_{56} = \frac{(0.48)(0.005)}{0.01} + (16.3)(0.005)/0.01 = 8.39$$

$$1/R_{58} = [164 + 16.3](0.005)/0.01 = 90.15$$

$\Sigma 1/R = 181.26$

$C_5 = (1440)(840)(0.01)(0.005) + [(2787)(883) + (7817)(460)](0.005)(0.005)$

$= 211.90$

$$\Delta\tau_{max} = \frac{211.90}{181.26} = 1.169 \text{ sec.}$$

4-73

$1/R_{42} = (20)(0.005)/0.01 = 10 \qquad 1/R_{43} = (20)(0.01)/0.01 = 20$

$1/R_{45} = (2)(0.005)/0.02 = 0.5 \qquad 1/R_{47} = 10 + (2)(0.01)/0.01 = 12$

$1/R_{4-\infty} = (50)(0.01 + 0.005) = 0.75 \qquad \Sigma\, 1/R = 43.25$

$C_4 = (7800)(500)(0.01)(0.005) + (1600)(800)(0.005)(0.01) = 259$

$\Delta\tau_{max} = \dfrac{259}{43.25} = 5.988$ sec.

$$T_4^{p+1} = \left[10T_2^p + 20T_3^p + 0.5\,T_5^p + 12T_7^p + (0.75)(50)\right]\Delta\tau/C_4 + \left[1 - \Delta\tau/5.988\right]T_4^p$$

4-74 $k = 16.3 \quad \rho = 7817 \quad c = 460$ see Table 4-2(d)

$Bi = \dfrac{h\,\Delta x}{k} = (60)(0.01)/16.3 = 0.0368$

$Fo = \alpha\,\Delta\tau/(\Delta x)^2 \qquad Fo\,(3 + Bi) \le 3/4$

$\alpha = 16.3/(7817)(460) = 4.53 \times 10^{-6}$

$Fo < (3/4)/3.0368 = 0.24697$

$\Delta\tau \le (0.24697)(0.01)^2/4.53 \times 10^{-6} = 5.452$ sec.

4-75 $c_1 = (3000)(840)(0.075)^2 = 14175$

$c_2 = (2)(14175) = 28350$

$c_4 = (2)(14175) = 28350$

$c_5 = 56700$

NODE	$\Delta\tau_{MAX}$
1	1942
2	1942
4	2726
5	2726

Equations and Solutions on next page

4-75 (cont)

The equations:

```
TP(1) = ((2.6*T(2) + 2.6*T(4))*DT)/14175 + (1 - DT/1942)*T(1)
TP(2) = ((2.6*T(1) + 26 + 5.2*T(5))*DT)/28350  + (1 - DT/1942)*T(2)
TP(4) = ((2.6*T(1) + 2.6*38 + 5.2*T(5))*DT)/28350 + (1 - DT/2726)*T(4)
TP(5) = 5.2*((T(2) + T(4) + 38 + 10)*DT)/56700! + (1 - DT/2726)*T(5)
```

the Solution:

```
T( 1 ) =  12.12136
T( 2 ) =  11.24476
  .
  .
T( 4 ) =  22.78445
T( 5 ) =  20.50759
```

4-76

The equations:

```
TP(1) = (1100 + T(3) + T(4))/4
TP(2) = (600 + T(3) + T(4))/4
TP(3) = (900 + T(1) + T(2))/4
TP(4) = (800 + T(1) + T(2))/4
```

The Solution:

```
T( 1 ) =  487.5
T( 2 ) =  362.5
T( 3 ) =  437.5
T( 4 ) =  412.5
```

4-78

	ΔT_{max}
$C_1 = 89.7$	3.5666
$C_2 = 44.85$	3.8114
$C_3 = 179.4$	3.588
$C_4 = 89.7$	3.5455
$C_5 = 179.4$	3.588
$C_6 = 89.7$	3.5455
$C_7 = 179.4$	3.588
$C_8 = 89.7$	3.5455
$C_9 = 179.4$	3.588
$C_{10} = 224.25$	6.6527
$C_{11} = 224.25$	10.5737
$C_{12} = 89.7$	8.8374
$C_{13} = 179.4$	3.588
$C_{14} = 358.8$	7.6885
$C_{15} = 448.5$	10.7639
$C_{16} = 179.4$	8.97

equations and solution on next page

THE EQUATIONS

```
170 TP(1)=T(1)+(DT/89.7)*(((T(3)-T(1))/.2)+2*((T(2)-T(1))/.1)+((10-T(1))/6.67))
175 TP(2)=T(2)+(DT/44.85)*(((T(1)-T(2))/.1)+((T(4)-T(2))/.4)+((10-T(2))/4.44))
180 TP(3)=T(3)+(DT/179.4)*(((T(1)-T(3))/.2)+2*((T(4)-T(3))/.05)+((T(5)-T(3))/.2)
)
185 TP(4)=T(4)+(DT/89.7)*(((T(2)-T(4))/.4)+((T(3)-T(4))/.05)+((T(6)-T(4))/.4)+((
10-T(4))/3.33))
190 TP(5)=T(5)+(DT/179.4)*(((T(3)-T(5))/.2)+2*((T(6)-T(5))/.05)+((T(7)-T(5))/.2)
)
195 TP(6)=T(6)+(DT/89.7)*(((T(4)-T(6))/.4)+((T(5)-T(6))/.05)+((T(8)-T(6))/.4)+((
10-T(6))/3.33))
200 TP(7)=T(7)+(DT/179.4)*(((T(5)-T(7))/.2)+2*((T(8)-T(7))/.05)+((T(9)-T(7))/.2)
)
205 TP(8)=T(8)+(DT/89.7)*(((T(6)-T(8))/.4)+((T(7)-T(8))/.05)+((T(10)-T(8))/.4)+(
(10-T(8))/3.33))
210 TP(9)=T(9)+(DT/179.4)*(((T(7)-T(9))/.2)+2*((T(10)-T(9))/.05)+((T(13)-T(9))/.
2))
215 TP(10)=T(10)+(DT/224.25)*(((T(8)-T(10))/.4)+((T(9)-T(10))/.05)+((T(14)-T(10)
)/.1)+((T(11)-T(10))/.3)+((10-T(10))/2.67))
220 TP(11)=T(11)+(DT/224.25)*(((T(10)-T(11))/.3)+((T(15)-T(11))/.08)+((T(12)-T(1
1))/.2)+((10-T(11))/2.67))
225 TP(12)=T(12)+(DT/89.7)*(((T(11)-T(12))/.2)+((T(16)-T(12))/.2)+((10-T(12))/6.
67))
230 TP(13)=T(13)+(DT/179.4)*(((T(9)-T(13))/.2)+2*((T(14)-T(13))/.05)+((200-T(13)
)/.2))
235 TP(14)=T(14)+(DT/358.8)*(((T(10)-T(14))/.1)+((T(13)-T(14))/.05)+((T(15)-T(14
))/.15)+((200-T(14))/.1))
240 TP(15)=T(15)+(DT/448.5)*(((T(11)-T(15))/.08)+((T(14)-T(15))/.15)+((T(16)-T(1
5))/.1)+((200-T(15))/.08))
245 TP(16)=T(16)+(DT/179.4)*(((T(12)-T(16))/.2)+((T(15)-T(16))/.1)+((200-T(16))/
.2))
```

THE SOLUTIONS FOR
TIME INCREMENT = 3.525 sec = DT

```
T( 1 ) =  184.2047
T( 2 ) =  182.9081
T( 3 ) =  187.9496
T( 4 ) =  186.6328
T( 5 ) =  189.6357
T( 6 ) =  188.2666
T( 7 ) =  191.0745
T( 8 ) =  189.8256
T( 9 ) =  194.7192
T( 10 ) =  194.1308
T( 11 ) =  194.1903
T( 12 ) =  194.1661
T( 13 ) =  198.3055
T( 14 ) =  198.2499
T( 15 ) =  198.1529
T( 16 ) =  198.1328
TOTAL NUMBER OF TIME INCREMENTS =  10
```

```
T( 1 ) =  124.4725
T( 2 ) =  123.6152
T( 3 ) =  129.9779
T( 4 ) =  129.0814
T( 5 ) =  140.0393
T( 6 ) =  139.0735
T( 7 ) =  155.6121
T( 8 ) =  154.6524
T( 9 ) =  177.3351
T( 10 ) =  178.4133
T( 11 ) =  185.7976
T( 12 ) =  186.7947
T( 13 ) =  189.744
T( 14 ) =  189.9733
T( 15 ) =  192.5038
T( 16 ) =  192.9872
TOTAL NUMBER OF TIME INCREMENTS =  100
```

4-79

	$\Delta\tau_{max}$
$c_1 = 37500$	6912
$c_2 = 75000$	6912
$c_3 = 75000$	6912
$c_4 = 75000$	16304
$c_5 = 150000$	16304
$c_6 = 150000$	16304
$c_7 = 75000$	16304
$c_8 = 150000$	16304
$c_9 = 150000$	16304

Equations and Solutions on next page

4-79 cont

The Equations:

```
130 TP(1) = ((5*(5 - T(1)) + 1.15*(T(2) - T(1)) + 1.15*(T(4) - T(1)))*DT)/37500!
+ T(1)
140 TP(2) = ((1.15*(T(1) - T(2)) + 1.15*(T(3) - T(2)) + 6.25*(5 - T(2)) + 2.3*(T
(5) - T(2)))*DT)/75000! + T(2)
150 TP(3) = ((1.15*(T(2) - T(3)) + 1.15*(100 - T(3)) + 6.25*(5 - T(3)) + 2.3*(T(
6) - T(3)))*DT)/75000! + T(3)
160 TP(4) = ((1.15*(T(1) - T(4)) + 2.3*(T(5) - T(4)) + 1.15*(T(7) - T(4)))*DT)/7
5000! + T(4)
170 TP(5) = ((2.3*(T(2) + T(4) + T(6) + T(8) - 4*T(5)))*DT)/150000! + T(5)
180 TP(6) = ((2.3*(T(3) + T(5) + T(9) + 100 - 4*T(6)))*DT)/150000! + T(6)
190 TP(7) = ((1.15*(T(4) - T(7)) + 1.15*(100 - T(7)) + 2.3*(T(8) - T(7)))*DT)/75
000! + T(7)
200 TP(8) = ((2.3*(T(5) + T(7) + T(9) + 100 - 4*T(8)))*DT)/150000! + T(8)
210 TP(9) = ((2.3*(T(6) + T(8) + 200 - 4*T(9)))*DT)/150000! + T(9)
```

The Solution.

```
Number of time increments: 10
T( 1 ) =  17.9077
T( 2 ) =  23.40188
T( 3 ) =  32.57268
T( 4 ) =  66.37385
T( 5 ) =  68.88091
T( 6 ) =  76.62771
T( 7 ) =  90.58165
T( 8 ) =  91.35402
T( 9 ) =  93.88974
```

```
Number of time increments: 20
T( 1 ) =  15.45997
T( 2 ) =  20.4147
T( 3 ) =  30.69091
T( 4 ) =  55.31455
T( 5 ) =  58.8419
T( 6 ) =  70.49835
T( 7 ) =  81.98696
T( 8 ) =  83.67387
T( 9 ) =  89.22552
```

4-80

$C_1 = 70312$

$C_2 = 70312$

$C_3 = 140625$

$C_4 = 140625$

$C_5 = 140625$

$C_6 = 140625$

NODE	ΔT_{max}
1	11719
2	11719
3	23438
4	23438
5	23438
6	23438

4-80 cont

The Equations

```
130 TP(1) = ((.75*(T(2) - T(1)) + 3*(15 - T(1)) + .75*(50 - T(1)) + 1.5*(T(3) -
T(1)))*DT)/70312! + T(1)
140 TP(2) = ((.75*(T(1) - T(2)) + 1.5*(T(4) - T(2)) + 3*(15 - T(2)) + .75*(50 -
T(2)))*DT)/70312! + T(2)
150 TP(3) = (1.5*(T(1) + T(4) + T(5) + 50 - 4*T(3))*DT)/140625! + T(3)
160 TP(4) = (1.5*(T(2) + T(3) + T(6) + 50 - 4*T(4))*DT)/140625! + T(4)
170 TP(5) = (1.5*(T(3) + T(6) + 100 - 4*T(5))*DT)/140625! + T(5)
180 TP(6) = (1.5*(T(4) + T(5) + 100 - 4*T(6))*DT)/140625! + T(6)
```

The Solution

```
Number of time increments: 10
T( 1 ) =  28.00879
T( 2 ) =  28.00879
T( 3 ) =  42.65092
T( 4 ) =  42.65092
T( 5 ) =  48.01731
T( 6 ) =  48.01731
```

4-82

$C_1 = (7600)(450)\pi(0.01)^2(0.02) = 21.488$

$C_2 = 21.488$

$C_3 = 21.488$

$C_4 = 21.488$

$C_5 = 10.744$

NODE	$\Delta \tau_{max}$
1	18.687
2	18.687
3	18.687
4	18.687
5	18.288

4-82 cont.

The Equations

```
130 TP(1) = ((.5498*(T(2) - T(1)) + .5498*(200 - T(1)) + .05027*(25 - T(1)) + 3.
142)*DT)/21.488 + T(1)
140 TP(2) = ((.5498*(T(1) + T(3) - 2*T(2)) + .05027*(25 - T(2)) + 3.142)*DT)/21.
488 + T(2)
150 TP(3) = ((.5498*(T(2) + T(4) - 2*T(3)) + .05027*(25 - T(3)) + 3.142)*DT)/21.
488 + T(3)
160 TP(4) = ((.5498*(T(3) + T(5) - 2*T(4)) + .05027*(25 - T(4)) + 3.142)*DT)/21.
488 + T(4)
170 TP(5) = ((.5498*(T(4) - T(1)) + .0377*(25 - T(5)) + 1.571)*DT)/10.744 + T(5)
```

The Solution

```
Number of time increments: 10
T( 1 ) =  182.0792
T( 2 ) =  168.3011
T( 3 ) =  150.4356
T( 4 ) =  112.7128
T( 5 ) =  19.68562
```

4-83

$C_1 = 600$

$C_2 = 1200$

$C_3 = 1800$

$C_4 = 2400$

$C_5 = 2400$

$C_6 = 1200$

$C_7 = 2400$

$C_8 = 3600$

$C_9 = 4800$

$C_{10} = 4800$

$C_{11} = 1200$

$C_{12} = 2400$

$C_{13} = 1200$

$C_{14} = 2400$

NODE	$\Delta\tau_{max}$
1	600
2	1200
3	1500
4	1714
5	1714
6	800
7	2400
8	4000
9	6000
10	6000
11	800
12	2400
13	800
14	2400

4-83 cont

The Equations

```
130 TP(1) = ((.2*(T(2) - T(1)) + .05*(T(6) - T(1)) + .75*(20 - T(1)))*DT)/600 +
T(1)
140 TP(2) = ((.2*(T(1) - T(2)) + .2*(T(4) - T(2)) + .1*(T(7) - T(2)) + .5*(20 -
T(2)))*DT)/1200 + T(2)
150 TP(3) = ((.2*(T(2) - T(3)) + .1*(T(4) - T(3)) + .15*(T(8) - T(3)) + .75*(20
- T(3)))*DT)/1800 + T(3)
160 TP(4) = ((.1*(T(3) - T(4)) + .1*(T(5) - T(4)) + .2*(T(9) - T(4)) + (20 - T(4
)))*DT)/2400 + T(4)
170 TP(5) = ((.2*(T(4) - T(5)) + .2*(T(10) - T(5)) + (20 - T(5)))*DT)/2400 + T(5
)
180 TP(6) = ((.05*(T(1) - T(6)) + .05*(T(11) - T(6)) + .4*(T(7) - T(6)) + (20 -
T(6)))*DT)/1200 + T(6)
190 TP(7) = ((.4*(T(6) - T(7)) + .4*(T(8) - T(7)) + .1*(T(2) - T(7)) + .1*(T(12)
 - T(7)))*DT)/2400 + T(7)
200 TP(8) = ((.4*(T(7) - T(8)) + .15*(T(3) - T(8)) + .2*(T(9) - T(8)) + .15*(500
 - T(8)))*DT)/3600 + T(8)
210 TP(9) = ((.2*(T(8) - T(9)) + .2*(T(10) - T(9)) + .2*(T(4) - T(9)) + .2*(500
- T(9)))*DT)/4800 + T(9)
220 TP(10) = ((.4*(T(9) - T(10)) + .2*(T(5) - T(10)) + .2*(500 - T(10)))*DT)/480
0 + T(10)
230 TP(11) = ((.4*(T(12) - T(11)) + .05*(T(6) - T(11)) + .05*(T(13) - T(11)) + 2
0 - T(11))*DT)/1200 + T(11)
240 TP(12) = ((.4*(T(11) - T(12)) + .1*(T(7) - T(12)) + .1*(T(14) - T(12)) + .4*
(500 - T(12)))*DT)/2400 + T(12)
250 TP(13) = ((.8*(T(11) - T(13)) + .4*(T(14) - T(13)) + 20 - T(13))*DT)/1200 +
T(13)
260 TP(14) = ((.4*(T(13) - T(14)) + .2*(T(12) - T(14)) + .4*(500 - T(14)))*DT)/2
400 + T(14)
```

4-83 cont

The Solution

```
Number of time increments: 10
T( 1 ) =  40.42402
T( 2 ) =  88.08902
T( 3 ) =  111.5521
T( 4 ) =  118.0645
T( 5 ) =  118.4374
T( 6 ) =  111.3538
T( 7 ) =  321.7555
T( 8 ) =  428.1198
T( 9 ) =  453.0409
T( 10 ) =  454.2848
T( 11 ) =  121.5945
T( 12 ) =  357.7431
T( 13 ) =  122.0525
T( 14 ) =  359.5641
```

```
Number of time increments: 20
T( 1 ) =  31.48851
T( 2 ) =  59.641
T( 3 ) =  76.75708
T( 4 ) =  86.62274
T( 5 ) =  87.81902
T( 6 ) =  82.65151
T( 7 ) =  232.7442
T( 8 ) =  340.468
T( 9 ) =  393.3104
T( 10 ) =  399.4332
T( 11 ) =  102.848
T( 12 ) =  307.0833
T( 13 ) =  104.2127
T( 14 ) =  312.7511
```

4-85

$C_1 = 412.5$

$C_2 = 412.5$

$C_3 = 825$

$C_4 = 825$

$C_5 = 412.5$

$C_6 = 412.5$

NODE	$\Delta\tau_{max}$
1	23.23
2	23.23
3	24.26
4	24.26
5	24.26
6	24.26

4-85 cont.

The Equations

```
130 TP(1) = ((.5*(T(2) - T(1)) + 16*(T(3) - T(1)) + .5*(100 - T(1)) + .75*(0 - T
(1)))*DT)/412.5 + T(1)
140 TP(2) = ((.5*(T(1) - T(2)) + 16*(T(4) - T(2)) + .5*(100 - T(2)) + .75*(0 - T
(2)))*DT)/412.5 + T(2)
150 TP(3) = ((T(4) - T(3) + 100 - T(3) + 16*(T(5) - T(3)) + 16*(T(1) - T(3)))*DT
)/825 + T(3)
160 TP(4) = ((T(3) - T(4) + 100 - T(4) + 16*(T(6) - T(4)) + 16*(T(2) - T(4)))*DT
)/825 + T(4)
170 TP(5) = ((.5*(T(6) - T(5)) + .5*(100 - T(5)) + 16*(T(3) - T(5)))*DT)/412.5 +
 T(5)
180 TP(6) = ((.5*(T(5) - T(6)) + .5*(100 - T(6)) + 16*(T(4) - T(6)))*DT)/412.5 +
 T(6)
```

The Solutions

```
Number of time increments: 10
T( 1 ) =  89.77267
T( 2 ) =  89.77267
T( 3 ) =  92.86581
T( 4 ) =  92.86581
T( 5 ) =  93.88928
T( 6 ) =  93.88928
```

4-86

$C_1 = 87.4$

$C_2 = 174.8$

$C_3 = 174.8$

$C_4 = 174.8$

$C_5 = 349.6$

$C_6 = 349.6$

$C_7 = 174.8$

$C_8 = 349.6$

$C_9 = 349.6$

$C_{10} = 87.4$

$C_{11} = 174.8$

$C_{12} = 174.8$

NODE	ΔT_{max}
1	4.263
2	4.263
3	4.263
4	4.37
5	4.37
6	4.37
7	4.37
8	4.37
9	4.37
10	4.37
11	4.37
12	4.37

4-86 cont

the Equations

```
130 TP(1) = ((10*(T(2) - T(1)) + 10*(T(4) - T(1)) + .5*(20 - T(1)) + 2250)*DT)/8
7.4 + T(1)
140 TP(2) = ((10*(T(1) - T(2)) + 10*(T(3) - T(2)) + 20*(T(5) - T(2)) + (20 - T(2
)) + 4500)*DT)/174.8 + T(2)
150 TP(3) = ((10*(T(2) - T(3)) + 10*(100 - T(3)) + (20 - T(3)) + 4500 + 20*(T(6)
 - T(3)))*DT)/174.8 + T(3)
160 TP(4) = ((20*(T(5) - T(4)) + 10*(T(1) - T(4)) + 10*(T(7) - T(4)) + 4500)*DT)
/174.8 + T(4)
170 TP(5) = (20*(T(2) + T(4) + T(6) + T(8) - 4*T(5) + 450)*DT)/349.6 + T(5)
180 TP(6) = (20*(T(3) + T(5) + T(9) + 100 - 4*T(6) + 450)*DT)/349.6 + T(6)
190 TP(7) = ((20*(T(8) - T(7)) + 10*(T(4) - T(7)) + 10*(T(10) - T(7)) + 4500)*DT
)/174.8 + T(7)
200 TP(8) = (20*(T(5) + T(7) + T(9) + T(11) - 4*T(8) + 450)*DT)/349.6 + T(8)
210 TP(9) = (20*(T(6) + T(8) + T(12) - 4*T(9) + 550)*DT)/349.6 + T(9)
220 TP(10) = ((10*(T(11) - T(10)) + 10*(T(7) - T(10)) + 2250)*DT)/87.4 + T(10)
230 TP(11) = ((10*(T(10) - T(11)) + 10*(T(12) - T(11)) + 4500)*DT)/174.8 + T(11)
240 TP(12) = ((10*(T(11) - T(12)) + 10*(100 - T(12)) + 4500)*DT)/174.8 + T(12)
```

The Solution

```
Number of time increments: 10
T( 1 ) =  966.5926
T( 2 ) =  894.7732
T( 3 ) =  641.3108
T( 4 ) =  996.7071
T( 5 ) =  922.2206
T( 6 ) =  659.5205
T( 7 ) =  1009.09
T( 8 ) =  933.5721
T( 9 ) =  666.7418
T( 10 ) =  1012.862
T( 11 ) =  938.3229
T( 12 ) =  669.8716
```

```
Number of time increments: 20
T( 1 ) =  1436.131
T( 2 ) =  1301.619
T( 3 ) =  876.4111
T( 4 ) =  1486.383
T( 5 ) =  1346.631
T( 6 ) =  904.8259
T( 7 ) =  1512.15
T( 8 ) =  1370.23
T( 9 ) =  919.3698
T( 10 ) =  1522.108
T( 11 ) =  1383.305
T( 12 ) =  927.9388
```

4-89

$$\frac{\alpha \Delta \tau}{(\Delta x)^2} = \frac{1}{4} \qquad \Delta \tau_{max} = \frac{(0.15)^2}{4} \Big/ 1.29 \times 10^{-5} = 436 \text{ sec.}$$

$$c_1 = c_2 = c_3 = c_4 = 78488$$

The Equations

```
TP(1) = (45*(T(2) + T(3) + 100 + 30 - 4*T(1))*DT)/78488! + T(1)
TP(2) = (45*(T(1) + T(4) + 30 + 100 - 4*T(2))*DT)/78488! + T(2)
TP(3) = (45*(T(1) + T(4) + 200 - 4*T(3))*DT)/78488! + T(3)
TP(4) = (45*(T(2) + T(3) + 200 - 4*T(4))*DT)/78488! + T(4)
```

The Solution

```
Number of time increments: 10
T( 1 ) =  85.67889
T( 2 ) =  85.67889
T( 3 ) =  102.6063
T( 4 ) =  102.6063
```

4-90

$c_1 = 350$ $\qquad c_5 = 700$

$c_2 = 700$ $\qquad c_6 = 1400$

$c_3 = 700$ $\qquad c_7 = 1400$

$c_4 = 350$ $\qquad c_8 = 700$

NODE	$\Delta \tau_{max}$
1	16.83
2	17.16
3	17.16
4	16.83
5	17.16
6	17.5
7	17.5
8	17.16

The Equations

```
130 TP(1) = ((10*(T(2) - T(1)) + 10*(T(5) - T(1)) + .8*(300 - T(1)))*DT)/350 + T
(1)
140 TP(2) = ((10*(T(1) - T(2)) + 20*(T(5) - T(2)) + .8*(300 - T(2)))*DT)/700 + T
(2)
150 TP(5) = ((10*(T(1) - T(5)) + 10*(50 - T(5)) + 20*(T(6) - T(5)))*DT)/700 + T(
5)
160 TP(6) = (20*(T(2) + T(5) + 50 - 3*T(6))*DT)/1400 + T(6)
```

The Solutions

DT = 0.25

```
Number of time increments: 4
DT =  .25
T( 1 ) =  50.56186
T( 2 ) =  50.28407

T( 5 ) =  50.00302
T( 6 ) =  50.00152
Number of time increments: 40
DT =  .25
T( 1 ) =  54.67344
T( 2 ) =  52.65144

T( 5 ) =  50.30585
T( 6 ) =  50.17934
```

DT = 5

```
Number of time increments: 12
DT =  5
T( 1 ) =  65.24096
T( 2 ) =  61.52972

T( 5 ) =  54.24393
T( 6 ) =  53.51868
```

DT = 15

```
Number of time increments: 100
DT =  15
T( 1 ) =  77.22425
T( 2 ) =  73.71596

T( 5 ) =  62.91046
T( 6 ) =  62.20881
```

4-90 cont

4-97 Wall thickness = 0.25 m $T_\infty = 600\,°C$ $h = 100\ W/m^2\,°C$

$k = 0.16\ W/m\,°C$ $\alpha = 3.5 \times 10^{-7}\ m^2/s$

Approximate as inf. plate with $2L = 0.5$ Center plane is insulated

$$k/hL = \frac{0.16}{(100)(0.25)} = 6.4 \times 10^{-3} \qquad A = (6)(1.0) = 6.0\ m^2$$

$$V = 1.0 - (0.5)^3 = 0.875\ m^3$$

$$\frac{h(V/A)}{k} = 546.9 \quad \text{\underline{Not} lumped capacity}$$

$$\frac{\theta_0}{\theta_i} = \frac{150-600}{30-600} = 0.789$$

$$\frac{\alpha\tau}{L^2} = 0.2 \qquad \tau = \frac{(0.25)^2(0.2)}{3.5\times10^{-7}} = 3.57\times10^4\ \text{sec}$$

$$= 9.92\ \text{hr.}$$

4-99 $k = 1.07$ $\alpha = 5.4\times10^{-7}$ $T_i = 20°C$ $x = 2.0\ cm$ $q/A = 4500\ W/m^2$

$$T_0 = T_i + \frac{(2q_0/A)\sqrt{\frac{\alpha\tau}{\pi}}}{k}$$

τ	T_0
0	20
100	54.87
300	80.4
900	124.61
1500	155.05
1400	150.47
1390	150
2000	175.94

$$q/A = -k\frac{\partial T}{\partial x} \quad \text{at } x = 2cm,\ \tau = 1390$$

$$= \frac{2q_0}{A}\sqrt{\frac{\alpha\tau}{\pi}}\left(\frac{-x^2}{4\alpha\tau}\right)\exp\left(\frac{-x^2}{4\alpha\tau}\right)\left(\frac{-2x}{4\alpha\tau}\right)$$

$$-\frac{q_0}{A}\,x\left[-\exp\left(\frac{-x^2}{4\alpha\tau}\right)\left(\frac{1}{2\sqrt{\alpha\tau}}\right)\right]$$

$$-\frac{q_0}{A}\left[1-\text{erf}\left(\frac{x}{2\sqrt{\alpha\tau}}\right)\right]$$

$$= 1180\ W/m^2$$

4-107

$$\frac{k}{hr_0} = \frac{3.2}{(350)(0.075)} = 0.122$$

$$\frac{\alpha\tau}{r_0^2} = \frac{(13\times10^{-7})(21)(60)}{(0.075)^2} = 0.291$$

$$r/r_0 = 4.5/7.5 = 0.6$$

$$\frac{\theta_0}{\theta_i} = 0.65 \qquad \frac{\theta}{\theta_0} = 0.59$$

$$T = (0.65)(0.59)(120-30) + 30$$
$$= 64.5\ °C$$

$$hr_0/k = 8.2$$

$$\frac{h^2\alpha\tau}{k^2} = \frac{(350)^2(13\times10^{-7})(21)(60)}{(3.2)^2} = 19.6$$

$$Q/Q_0 = 0.92$$

$$\rho c = \frac{k}{\alpha}$$

$$Q = \frac{(0.92)(3.2)}{13\times10^{-7}}(120-30)\frac{4}{3}\pi(0.075)^3$$
$$= 3.6\times10^5\ J$$

4-108

$r_o = 1.5$ in $= 3.81$ cm

$k = 0.585$

$\alpha = 1.4 \times 10^{-7}$

$\rho = 999$

$c = 4195$

Take T = 3°C at outside of orange to prevent frost bite

$$\frac{k}{hr_o} = \frac{0.585}{(45)(0.0381)} = 0.341 \qquad \frac{\theta}{\theta_o} = 0.33$$

$$\frac{\theta}{\theta_i} = \frac{3-0}{25-0} = 0.12 = \frac{\theta_o}{\theta_i}(0.33)$$

$$\frac{\theta_o}{\theta_i} = 0.364 \rightarrow \frac{\alpha\tau}{r_o^2} = 0.35$$

$$\tau = \frac{(0.35)(0.0381)^2}{1.3 \times 10^{-7}} = 3888 \text{ sec}$$

$$\frac{hr_o}{k} = 2.93 \qquad \frac{h^2\alpha\tau}{k^2} = \frac{(45)^2(1.3\times10^{-7})(3888)}{(0.585)^2}$$

$$= 2.99$$

$$\frac{Q}{Q_o} = 0.84$$

For 100 oranges:

$$Q = (100)(999)(4195)\tfrac{4}{3}\pi(0.0381)^3(0.84)(25-0)$$

$$= 1.19 \times 10^6 \text{ J} = 1125 \text{ Btu}$$

5-2 $\frac{u}{u_\infty} = \frac{y}{\delta} \qquad \rho \frac{d}{dx}\int_0^\delta (u_\infty - u)\,u\,dy = \mu \frac{du}{dy}\Big)_{y=0}$

$$\rho \frac{d}{dx}\int_0^\delta \left[u_\infty - u_\infty \frac{u}{\delta}\right] u_\infty \frac{y}{\delta}\,dy = \mu \frac{u_\infty}{\delta}\Big]_{y=0} \qquad \frac{d}{dx}\,\frac{1}{6}\rho\, u_\infty^2 \delta = \frac{\mu u_\infty}{\delta}$$

$$\delta\, d\delta = \frac{6\nu}{u_\infty}dx \qquad \delta^2 = \frac{12\nu x}{u_\infty} \qquad \delta = \frac{3.47\,x}{\sqrt{Re_x}}$$

5-3 $\delta = \frac{4.64x}{\sqrt{Re_x}} = 4.64\,x^{1/2}\left(\frac{\mu}{u_\infty \rho}\right)^{1/2} \qquad u = \frac{3u_\infty}{2} - \frac{u_\infty}{2}\left(\frac{y}{\delta}\right)^3$

$$u = \frac{3u_\infty y\,x^{1/2}}{2\left(4.64\left(\frac{\mu}{u_\infty\rho}\right)^{1/2}\right)} - \frac{u_\infty y^3 x^{1/2}}{2\left[4.64\left(\frac{\mu}{u_\infty\rho}\right)^{1/2}\right]^3} \qquad \frac{\partial u}{\partial x} + \frac{\partial v}{\partial y} = 0 \qquad \frac{\partial u}{\partial x} = -\frac{\partial v}{\partial y}$$

$$\frac{\partial u}{\partial x} = -\frac{3u_\infty x^{-1} y}{4\delta} + \frac{3u_\infty x^{-1} y^3}{4\delta^3} \qquad v = \int\left[\frac{-3u_\infty\,4.64}{4\sqrt{Re_x}}\left(\frac{y}{\delta^2}\right) + \frac{3u_\infty\,4.64}{4\delta\sqrt{Re_x}}\left(\frac{y^3}{\delta^3}\right)\right]dy$$

$$v = \frac{2.32\,u_\infty}{\sqrt{Re_x}}\left[\frac{3}{4}\left(\frac{y}{\delta}\right)^2 - \frac{3}{8}\left(\frac{y}{\delta}\right)^4\right] \text{ at } y=\delta \quad v = \frac{2.32\,u_\infty(3/8)}{\sqrt{Re_x}} \quad \text{at } x = 6\text{ in} = 15.24\text{ cm}$$

$$Re = \frac{(0.99)(30)(0.1524)}{2.13\times 10^{-5}} = 213\,000 \quad @\ x = 12\text{ in} \quad Re = 426\,000$$

$$v = \frac{(2.32)(30)(3/8)}{(213\,000)^{1/2}} = 0.0566 \text{ m/sec} \qquad v = \frac{(2.32)(30)(3/8)}{(426\,000)^{1/2}} = 0.0400 \text{ m/sec}$$

5-4 $\frac{u}{u_\infty} = \frac{y}{\delta} \qquad \frac{\partial u}{\partial x} + \frac{\partial v}{\partial y} = 0 \qquad \delta = \frac{3.47x}{\sqrt{Re_x}}$

$$u = u_\infty \frac{y}{3.47}\sqrt{\frac{\rho u_\infty}{\mu}}\,x^{-1/2} \qquad \frac{\partial u}{\partial x} = -u_\infty \frac{y}{2(3.47)}\sqrt{Re_x}\,x^{-2}$$

$$\frac{\partial v}{\partial y} = u_\infty \frac{y}{2(3.47)}\frac{1}{x^2}\sqrt{Re_x} \qquad v = u_\infty \frac{y^2}{4(3.47)}\frac{1}{x^2}\sqrt{Re_x}$$

at $x = 6$ in $\quad v = \frac{(3.47)(30)}{(4)(213000)^{1/2}} = 0.0563$ m/sec

at $x = 12$ in. $\quad v = \frac{(3.47)(30)}{4(426000)^{1/2}} = 0.040$ m/sec

5-5

$$\frac{u}{u_\infty} = \frac{y}{\delta} \qquad \frac{\theta}{\theta_\infty} = \frac{T-T_w}{T_\infty - T_w} = \frac{3}{2}\frac{y}{\delta_t} - \frac{1}{2}\left(\frac{y}{\delta_t}\right)^3$$

$$\frac{d}{dx}\int_0^H (T_\infty - T)u\,dy = \alpha\left(\frac{dT}{dy}\right)_w$$

$$\frac{d}{dx}\int_0^H (\theta_\infty - \theta)u\,dy = \alpha\left(\frac{dT}{dy}\right)_w$$

$$\frac{d}{dx}\int_0^H \left[\theta_\infty - \theta_\infty \frac{3}{2}\frac{y}{\delta_t} + \frac{\theta_\infty}{2}\left(y/\delta_t\right)^3\right]u\,dy = \alpha\left(\frac{dT}{dy}\right)_w$$

$$\frac{\theta_\infty u_\infty}{10}\frac{d}{dx}\left(\frac{\delta_t^2}{\delta}\right) = \alpha\left(\frac{dT}{dy}\right)_w \qquad \zeta = \frac{\delta_t}{\delta}$$

$$\frac{\theta_\infty u_\infty}{10}\frac{d}{dx}\left(\delta\,\zeta^2\right) = \alpha\left(\frac{dT}{dy}\right)_w$$

$$\alpha\left(\frac{dT}{dy}\right)_w = \alpha\theta_0\left[\frac{3}{2\delta_t} - \frac{3y^2}{2\delta_t^3}\right]_0 = \frac{3\alpha\theta_\infty}{2\delta_t}$$

$$\frac{\theta_\infty u_\infty}{10}\frac{d}{dx}\left(\delta\,\zeta^2\right) = \frac{3\alpha\,\theta_\infty}{2\delta_t}$$

$$\frac{1}{15}u_\infty\frac{d}{dx}\left(\delta\,\zeta^2\right) = \frac{\alpha}{\zeta\delta}$$

$$\frac{1}{15}u_\infty\left(2\delta^2\zeta^2\frac{d\zeta}{dx} + \zeta^3\delta\frac{d\delta}{dx}\right) = \alpha$$

$$\rho\frac{d}{dx}\int_0^\delta (u_\infty - u)u\,dy = \mu\left.\frac{du}{dy}\right)_{y=0}$$

$$\rho\frac{d}{dx}\left(u_\infty^2\frac{\delta}{6}\right) = \mu\frac{d}{dy}\left(\frac{u_\infty y}{\delta}\right)_{y=0}$$

$$\delta\,d\delta = \frac{6\mu}{\rho u_\infty}dx \qquad \delta^2 = \frac{12\mu}{\rho u_\infty}x$$

5-5 contd

$$\frac{1}{15} u_\infty \left(\frac{24\mu}{\rho u_\infty} x \zeta^2 \frac{d\zeta}{dx} + \frac{6\mu}{\rho u_\infty} \zeta^3 \right) = \alpha$$

$$4x \zeta^2 \frac{d\zeta}{dx} + \zeta^3 = \frac{15\alpha}{6\nu} \qquad \zeta^2 \frac{d\zeta}{dx} = \frac{1}{3}\frac{d}{dx}(\zeta^3)$$

$$\frac{d}{dx}(\zeta^3) + \frac{3}{4x}(\zeta^3) = \frac{15\alpha}{6\nu}\frac{3}{4x}$$

$$x^{3/4}\zeta^3 = \frac{45}{24}\frac{\alpha}{\nu}\int x^{-1/4}dx + c \qquad \zeta^3 = \frac{15}{6}\frac{\alpha}{\nu} + cx^{-3/4}$$

$$\zeta = 0 \text{ at } x = x_0 \qquad c = -\frac{15}{6}\frac{\alpha}{\nu} x_0^{3/4}$$

$$\zeta = \frac{\delta_t}{\delta} = \left\{ \frac{15}{6}\frac{\alpha}{\nu}\left[1 - \left(\frac{x_0}{x}\right)^{3/4}\right]\right\}^{1/3}$$

$$\delta = \left(\frac{12\mu x}{\rho u_\infty}\right)^{1/2} = \frac{3.47x}{\sqrt{Re_x}}$$

$$h = \frac{3}{2}\frac{k}{\delta\zeta} = \frac{3}{2}\frac{k\sqrt{Re_x}}{3.47x}\frac{Pr^{1/3}}{\left\{\frac{15}{6}\left[1-\left(\frac{x_0}{x}\right)^{3/4}\right]\right\}^{1/3}}$$

$$h_x = 0.319\frac{k}{x}\sqrt{Re_x}\, Pr^{1/3}\left[1 - \left(\frac{x_0}{x}\right)^{3/4}\right]^{-1/3}$$

5-6

$p = 2 \times 10^4$ N/m^2 $T = 5°C = 278°K$ air

$\delta = 1.25\text{cm} = 0.0125\text{m}$ $u = 1.5\text{m/sec}$

$$\rho = \frac{2 \times 10^4}{(287)(278)} = 0.251 \text{ kg/m}^3 \qquad \mu = 1.864 \times 10^{-5} \text{ kg/m-sec}$$

$$x = \frac{\rho u_\infty}{\mu}\left(\frac{\delta}{4.64}\right)^2 = \frac{(0.25)(1.5)}{1.864 \times 10^{-5}}\left(\frac{0.0125}{4.64}\right)^2 = 0.147\text{m}$$

5-7

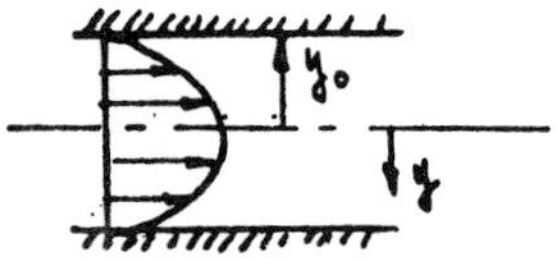

$$2yp - 2yp - 2yp = -2\tau dx$$
$$= -2\left(\mu \frac{du}{dy}\right) dx$$
$$y\,dp = \mu\,dx \frac{du}{dy}$$

$$u = \frac{1}{2\mu}\frac{dp}{dx} y^2 + c \quad \text{at } y = y_0 \quad u = 0 \quad c = \frac{-1}{2\mu}\frac{dp}{dx} y_0^2$$

$$u = \frac{1}{2\mu}\frac{dp}{dx}\left(y^2 - y_0^2\right)$$

$$\text{at } y = 0 \quad u = u_0 \quad u_0 = \frac{1}{2\mu}\frac{dp}{dx}\left(-y_0^2\right)$$

$$\frac{u}{u_0} = \left(1 - \frac{y^2}{y_0^2}\right)$$

5-10

$$\frac{\partial^2 u}{\partial y^2} = \frac{1}{\nu}\left(u\frac{\partial u}{\partial x} + v\frac{\partial u}{\partial y}\right) \qquad \frac{\partial^3 u}{\partial y^3} = \frac{1}{\nu}\left(u\frac{\partial^2 u}{\partial y \partial x} + \frac{\partial u}{\partial y}\cdot\frac{\partial y}{\partial x}\right.$$
$$\left. + v\frac{\partial^2 u}{\partial y^2} + \frac{\partial v}{\partial y}\frac{\partial u}{\partial y}\right) - \frac{1}{\nu}\left[u\frac{\partial^2 u}{\partial y \partial x} + \frac{\partial u}{\partial y}\left(\frac{\partial y}{\partial x} + \frac{\partial v}{\partial y}\right) + v\frac{\partial^2 u}{\partial y^2}\right]$$

$$u = v = 0 \ @ \ y = 0 \qquad \frac{\partial u}{\partial x} + \frac{\partial v}{\partial y} = 0 \quad \therefore \frac{\partial^3 u}{\partial y^3} = 0 \ @ \ y = 0$$

5-12

$$\frac{\delta_t}{\delta} = \frac{1}{1.026\, Pr^{1/3}}$$

	Pr	δ_t/δ
Air	0.709	1.093
H_2O	7	0.5095
He	0.705	1.095
NH_3	2.02	0.771
Glycerine	12.5	0.42

5-13 $T_\infty = 15°C$ $u_\infty = 3$ m/s $x = 5$ cm $\mu = 1.13\times10^{-3}$ kg/m·s $\rho = 999.8$ kg/m³

$$Re = \frac{(999.8)(3)(0.05)}{1.13\times10^{-3}} = 1.33\times10^5 \qquad \delta = \frac{(4.64)(0.05)}{(1.33\times10^5)^{1/2}} = 6.37\times10^{-4}\text{ m}$$

$$\text{mass flow} = \int_0^\delta \rho u\,dy = \frac{5}{8}\rho u_\infty \delta = 1.19\text{ kg/sec for 1m depth.}$$

5-14 $$\rho = \frac{p}{RT} = \frac{1.032\times10^5}{(287)(363)} = 0.99\text{ kg/m}^3 \quad T_\infty = 363°K \quad \mu = 2.13\times10^{-5}\text{ kg/m-sec}$$

$$Re = \frac{(0.99)(30)(0.025)}{2.13\times10^{-5}} = 34860 \qquad \delta = \frac{(4.64)(0.025)}{(34860)^{1/2}} = 6.21\times10^{-4}\text{ m}$$

5-15 $$T_f = \frac{75+20}{2} = 47.5°C = 320.5\text{ K} \qquad \mu = 1.94\times10^{-5}$$

$$\rho = \frac{2\times10^4}{(287)(320.5)} = 0.217 \qquad k = 0.028 \quad Pr = 0.7$$

$$Re_x = \frac{(0.217)(20)\,x}{1.94\times10^{-5}} = 2.24\times10^5\,x$$

$$h_x = (0.332)(0.028)(0.7)^{1/3}(2.24\times10^5)^{1/2}\,x^{-1/2}\left[1-\left(\frac{x_0}{x}\right)^{3/4}\right]^{-1/3}$$

$$h_x = \frac{3.904}{x^{1/2}\left[1-\left(\frac{x_0}{x}\right)^{3/4}\right]^{1/3}}$$

x	h_x W/m²·°C
0.08	38.18
0.19	11.27
0.3	8.24

$$q = \int_{x_0}^{x} h_x\,dx\,(T_w - T_\infty)$$

$$\int h_x\,dx = 3.79 \qquad q = (3.79)(75-20) = 208.6\text{ W}$$

5-16

$\delta = 7.5$ mm $= 0.0075$ m $\rho = 1000$

$\mu = 1.5\times10^{-3}$ kg/m-sec

$$x = \frac{\rho u_\infty}{\mu}\left(\frac{\delta}{4.64}\right)^2 = \frac{(1000)(1.5)}{1.5\times10^{-3}}\left(\frac{0.0075}{4.64}\right)^2 = 2.61\text{ m}$$

5-17 $T_f = \frac{90+30}{2} = 60°C = 333\ K \quad \nu = 19.09 \times 10^{-6} \quad k = 0.0288$

$Pr = 0.7 \quad Re = \frac{(20)(0.6)}{19.09 \times 10^{-6}} = 6.29 \times 10^5$

$\bar{h} = \frac{0.0288}{0.6}(0.7)^{1/3}\left[(0.037)(6.29 \times 10^5)^{0.8} - 871\right] = 31.5\ W/m^2 \cdot °C$

$q = (31.5)(0.6)^2(90-30) = 681.3\ W$

5-18

$p = 7\ kN/m^2 \quad T_\infty = 35°C \quad L = 0.3\ m \quad u_\infty = 7.5\ m/sec$

$T_w = 65°C \quad T_f = \frac{65+35}{2} = 50°C = 323°K$

$\rho = \frac{7000}{(287)(323)} = 0.0755\ kg/m^3 \quad \mu = 2.025 \times 10^{-5}\ kg/m\text{-}sec$

$k = 0.02798\ W/m\text{-}°C \quad Pr = 0.71$

$Re = \frac{(0.0755)(7.5)(0.3)}{2.025 \times 10^{-5}} = 8390$

$\bar{h} = \frac{0.02798}{0.3}(0.71)^{1/3}(8390)^{1/2}(0.664) = 5.04\ W/m^2\text{-}°C$

$q = (5.04)(0.3)^2(65-35) = 13.6\ W$

5-19

$T_\infty = 90°C \quad u_\infty = 60\ m/sec \quad L = 60\ cm \quad T_\infty = 10°C$

$T_f = 50°C = 323°K \quad \rho = \frac{1.0132 \times 10^5}{(287)(323)} = 1.093$

$\mu = 1.716 \times 10^{-5} \quad k = 0.0241 \quad Pr = 0.71$

$$Re = \frac{(1.093)(60)(0.6)}{1.716 \times 10^{-5}} = 2.292 \times 10^6$$

$$\bar{h} = \frac{0.0241}{0.6}(0.71)^{1/3}\left[(0.037)(2.29 \times 10^6)^{0.8} - 871\right]$$

$\bar{h} = 131.1\ w/m^2 - °C \quad q = (131.1)(0.6)^2(10-90) = 3776$

5-21

$L = 0.3 \quad T_\infty = 35°C \quad p = 14000\ N/m^2 \quad T_w = 250°C$

$u_\infty = 6\ m/sec \quad T_f = \frac{250+35}{2} = 142.5 = 416\ K$

$\rho = \frac{14000}{(287)(416)} = 0.117 \quad \mu = 2.349 \times 10^{-5}$

$k = 0.03474 \quad Pr = 0.685 \quad Re = \frac{(0.117)(6)(0.3)}{2.349 \times 10^{-5}} = 8966$

$$h = \frac{(0.03474)(0.664)}{0.3}(8966)^{1/2}(0.685)^{1/3} = 6.42\ w/m^2 \cdot °C$$

$q = (6.42)(0.3)^2(250-35) = 124\ W$

5-22 $T_w = 55°C \quad T_\infty = 20°C \quad p = 20000\ N/m^2 \quad x = 0.3\ m$

$u_\infty = 30\ m/sec \quad u = 22.5\ m/sec \quad T_f = \frac{55+20}{2} = 37.5°C = 310\ K$

$\rho = \frac{20000}{(287)(310)} = 0.225\ kg/m^3$

$\mu = 2.001 \times 10^{-5}$

$Re = \frac{(0.225)(30)(0.3)}{2.001 \times 10^{-5}} = 1.01 \times 10^5 \qquad \delta = \frac{(0.3)(4.64)}{(1.01 \times 10^5)^{1/2}} = 4.38 \times 10^{-3}\ m$

$\frac{u}{u_\infty} = \frac{22.5}{30} = \frac{3}{2}\left(\frac{y}{\delta}\right) - \frac{1}{2}\left(\frac{y}{\delta}\right)^3 \qquad \frac{y}{\delta} = 0.56 \qquad y = 2.45 \times 10^{-3}\ m$

5-23 $\frac{C_{fx}}{2} = 0.332\ Re_x^{-\frac{1}{2}}$ @ $x = 15\ cm$ $\quad Re_x = 0.5005 \times 10^5 \quad C_{fx} = 2.968 \times 10^{-3}$

5-24 $T_f = 308\ K \quad \rho = \frac{70000}{(287)(308)} = 0.792 \quad \mu = 2 \times 10^{-5} \quad k = 0.0268 \quad Pr = 0.7$

$Re_{x_0} = \frac{(0.792)(6)(0.15)}{2 \times 10^{-5}} = 35640$

Use Eq. (5-43) and numerically integrate:

$Re_{16.5} = \frac{16.5}{15}(35640) = 38610$

$h_{16.25} = \frac{0.0268}{0.1625}(0.332)(38610)^{1/2}(0.7)^{1/3}\left[1 - \left(\frac{15}{16.25}\right)^{3/4}\right]^{-1/3} = 24.64$

$h_{17.5} = 19.98\left(\frac{16.25}{17.5}\right)^{1/2} = 19.25$

$h_x = (0.332)\, k\, Pr^{1/3}\left(\frac{400}{2}\right)^{1/2} \frac{1}{x^{\frac{1}{2}}} \frac{1}{\left[1 - \left(\frac{x_0}{x}\right)^{3/4}\right]^{1/3}}$

$h_{15.5} = 33.76$

$h_{15.25} = 42.68$

$\bar{h} = \frac{\int h_x\, dx}{\Delta x} = 27.85\ W/m^2 \cdot °C$

$q = (27.85)(0.025)(65-5) = 41.8\ W$

5-25 $T_f = \frac{55+27}{2} = 41°C = 314\ K \quad \nu = 17.94 \times 10^{-6} \quad k = 0.0273$

$Pr = 0.7 \quad Re_L = \frac{(4.5)(15)}{17.94 \times 10^{-6}} = 3.76 \times 10^6$

$$\bar{h} = \frac{0.0273}{15}(0.7)^{1/3}\left[(0.037)(3.76 \times 10^6)^{0.8} - 871\right] = 9.48$$

$$q = (9.48)(15)(55-27) = 3980\ W/m$$

5-26 $T_f = \frac{300+400}{2} = 350\ K \quad \rho = \frac{75000}{(287)(350)} = 0.747\ kg/m^3$

$\mu = 2.075 \times 10^{-5} \quad k = 0.03003 \quad Pr = 0.697$

$$Re = \frac{(0.747)(45)(1.0)}{2.075 \times 10^{-5}} = 1.62 \times 10^6$$

$$\bar{h} = \frac{0.03003}{1.0}(0.697)^{1/3}\left[(0.037)(1.62 \times 10^6)^{0.8} - 871\right]$$

$$= 68.25\ W/m^2 \cdot °C$$

$$q = hA(T_w - T_\infty) = (68.25)(1)^2(400-300)$$

$$= 6825\ W$$

5-27 $T_f = \frac{50+10}{2} = 30°C = 303\ K \quad \rho = \frac{50000}{(287)(303)} = 0.575\ \frac{kg}{m^3}$

$\mu = 1.85 \times 10^{-5} \quad k = 0.0263 \quad Pr = 0.7$

$$Re = \frac{(0.575)(0.5)(20)}{1.85 \times 10^{-5}} = 3.11 \times 10^5$$

$$\bar{h} = \frac{0.0263}{0.5}(0.664)(3.11 \times 10^5)^{1/2}(0.7)^{1/3} = 17.3\ W/m^2 \cdot °C$$

$$q = hA(T_w - T_\infty) = (17.3)(0.5)^2(50-10) = 172.9\ W$$

5-28 $T_f = \frac{100+10}{2} = 55°C = 328\ K$ $\rho = \frac{1.01 \times 10^5}{(2)(287)(328)} = 0.538$

$\mu = 1.974 \times 10^{-5}$ $k = 0.0284$ $Pr = 0.7$

$Re_L = \frac{(0.538)(5)(0.2)}{1.974 \times 10^{-5}} = 27254$ Re @ $x = 10\text{cm} = 13627$

$$\overline{T_w - T_\infty} = \frac{q_w L/k}{0.6795\ Re_L^{1/2}\ Pr^{1/3}}$$

$$q_w = (100)(0.6795)(27254)^{1/2}(0.7)^{1/3}(0.0284)/0.2$$

$$= 1414\ W/m^2$$

$$\frac{hx}{k} = 0.453\ Re_x^{1/2}\ Pr^{1/3}$$

@ $x = 10$ cm

$$h = \frac{0.0284}{0.1}(0.453)(13627)^{1/2}(0.7)^{1/3} = 13.33\ W/m^2 \cdot °C$$

5-29

	ν	u (m/s)
H_2O	9.8×10^{-7}	9.8
air	1.53×10^{-5}	153
F-12	0.198×10^{-6}	1.98
NH_3	0.359×10^{-6}	3.59
He	122.2×10^{-6}	1222

$$u = \frac{10^7 \nu}{L} = 10^7 \nu$$

5-32 Evaluate Properties at 350 K $\nu = 20.76 \times 10^{-6}$

$k = 0.03003$ $Pr = 0.697$ $Re_L = \frac{(3)(0.25)}{20.76 \times 10^{-6}} = 36127$

$$\overline{T_w - T_\infty} = \frac{q_w L/k}{0.6795\, Re_L^{1/2}\, Pr^{1/3}} = \frac{(800)(0.25)/0.03003}{(36127)^{1/2}(0.697)^{1/3}(0.6795)}$$

$$= 58.16\ °C$$

$$\overline{T_w} = 25 + 58.16 = 83.16\ °C$$

$$h\ @\ x = 25\text{ cm} = \frac{k}{x}\, 0.453\, Re_x^{1/2}\, Pr^{1/3}$$

$$= \frac{0.03003}{0.25}(0.453)(36127)^{1/2}(0.697)^{1/3} = 9.17\ \text{W/m}^2\cdot °C$$

or $Nu_x = 76.34$

$$T_w - T_\infty = \frac{q_w x}{k\, Nu_x} = \frac{(800)(0.25)}{(0.03003)(76.34)} = 87.24\ °C$$

$$T_w = 87.24 + 25 = 112.24\ °C$$

5-33 $T_f = 350\text{ K}$ $\nu = 20.76 \times 10^{-6}$ $k = 0.03003$ $Pr = 0.697$

$$u_\infty = \frac{1.1 \times 10^5\, \nu}{L} = 4.57\ \text{m/sec} \qquad x_0 = 25\text{ cm}$$

$$h_x = \frac{k}{x}\, 0.332\, Pr^{1/3}\, Re^{1/2}\left[1 - \left(\frac{x_0}{x}\right)^{3/4}\right]^{-1/3}$$

x (cm)	h_x (W/m²·°C)
26	36.7
35	13.81
44	9.50
50	7.92

$$\int_{x_0}^{L} h_x\, dx = 3.844$$

$$\bar{h} = \frac{3.844}{0.25} = 15.38\ \text{W/m}^2\cdot °C$$

$$q = \bar{h}(L - x_0)L(T_w - T_\infty)$$

$$= (15.38)(0.5 - 0.25)(0.5)(400 - 300) = 192.2\ \text{W}$$

5-34 $u_\infty = 150$ m/sec $T_w = 150°C$ $L = 1$ m $T_\infty = 20°C$ $\mu = 2.11 \times 10^{-5}$

$T_f = \frac{150+20}{2} = 85°C = 358$ K $\rho = \frac{p}{RT} = \frac{14000}{(287)(358)} = 0.136$ kg/m³

$k = 0.03060$ $Pr = 0.695$ $Re = \frac{(0.136)(150)(1)}{2.11 \times 10^{-5}} = 9.67 \times 10^5$

$\bar{h} = \frac{0.03060}{1}(0.695)^{1/3}[(0.037)(9.67 \times 10^5)^{0.8} - 871] = 38.00$ W/m²·°C

$q/A = (38.00)(150-20) = 4939$ W/m²

5-36 $h_x = Cx^{-1/5}$

$$\bar{h} = \frac{1}{L}\int_0^L Cx^{-1/5}\,dx = 1.25\,CL^{-1/5}$$

$$\bar{St}\,Pr^{2/3} = (1.25)(0.0296)\,Re_L^{-1/5} = 0.037\,Re_L^{-1/5}$$

5-37 $T_f = \frac{80+10}{2} = 45°C$ $\rho = 990$ $\mu = 6 \times 10^{-4}$ $k = 0.64$

$Pr = 4.81$ **Max Temp. @ x = L.** $Re_L = \frac{(990)(3)(0.1)}{6 \times 10^{-4}} = 4.95 \times 10^5$

$Nu_x = 0.453\,Re_x^{1/2}\,Pr^{1/3}$

$= (0.453)(4.95 \times 10^5)^{1/2}(4.81)^{1/3} = 538$

$q_w = \frac{Nu_x k}{x}(T_w - T_\infty)$

$= \frac{(538)(0.64)}{0.1}(80-10) = 2.41 \times 10^5$ W/m²

$q = q_w A = (2.41 \times 10^5)(0.1)^2 = 2410$ W

5-38 Take Properties at 350K. $\nu = 20.76 \times 10^{-6}$ $k = 0.03003$

$Pr = 0.697$ $Re_L = \frac{(3)(0.1)}{20.76 \times 10^{-6}} = 14451$

$Nu_x = (0.453)(14451)^{1/2}(0.697)^{1/3} = 48.28$

$q_w = \frac{Nu_x}{x}(T_w - T_\infty)k = \frac{(48.28)(0.03003)}{0.1}(80-10) = 1015$ W/m²

$q = q_w A = (1015)(0.1)^2 = 10.15$ W

5-39 $T_f = \frac{300 + 500}{2} = 400 \text{ K} \quad \nu = 202.9 \times 10^{-6} \quad k = 0.178 \quad Pr = 0.72$

$Re = \frac{(50)(1)}{202.9 \times 10^{-6}} = 2.46 \times 10^5$

$h = \frac{k}{L} 0.664 Re^{1/2} Pr^{1/3} = \frac{0.178}{1}(0.664)(2.46 \times 10^5)^{1/2}(0.72)^{1/3}$

$= 52.59 \text{ W/m}^2 \cdot {}^\circ\text{C}$

$q = hA(T_w - T_\infty) = (52.59)(1)^2(500 - 300) = 10.52 \text{ kW}$

$\frac{\delta}{L} = \frac{4.64}{Re_L^{1/2}} \quad \delta = \frac{(1)(4.64)}{(2.46 \times 10^5)^{1/2}} = 0.0094 \text{ m}$

5-40 $\frac{u}{u_\infty} = \frac{25}{50} = \frac{3}{2}\left(\frac{y}{\delta}\right) - \frac{1}{2}\left(\frac{y}{\delta}\right)^3 \quad \frac{y}{\delta} = 0.334$

$y = (0.334)(9.45 \times 10^{-3}) = 3.16 \times 10^{-3} \text{ m}$

5-41 $\nu = 15.7 \times 10^{-6} \quad Re = 10^7 = \frac{u_\infty x}{\nu} \quad x = \frac{(10^7)(15.7 \times 10^{-6})}{30}$

$= 5.23 \text{ m}$

at $u = 7 \text{ m/s} \quad x = 22.4 \text{ m}$

$u = 12 \text{ m/s} \quad x = 13.08 \text{ m}$

5-42 $u_\infty = 10$ mi/hr $= 4.47$ m/sec $T_\infty = 80°F = 26.67°C = 299.67$ K

$L = 20$ ft $= 6.096$ m $q/A = 110$ Btu/hr·ft² $= 346.9$ W/m² $\rho = 1.177$

$\mu = 1.983 \times 10^{-5}$ $k = 0.0262$ $Pr = 0.71$

$$Re = \frac{(1.177)(4.47)(6.096)}{1.983 \times 10^{-5}} = 1.617 \times 10^6$$

$$\bar{h} = \frac{0.0262}{6.096}(0.71)^{1/3}\left[(0.037)(1.617 \times 10^6)^{0.8} - 871\right] = 9.81 \text{ W/m}^2\cdot°C$$

$$T_w - T_\infty = \frac{346.9}{9.81} = 35.4°C \qquad T_w = 62.0°C$$

5-43 $T_\infty = 400°F$ $T_w = 420°F$ $u_\infty = 1$ ft/sec $= 0.3048$ m/sec

$L = 10$ ft $= 3.048$ m $Re = \frac{(0.3048)(3.048)}{2 \times 10^{-6}} = 4.645 \times 10^5$

$$h = \frac{0.12}{3.048}(0.664)(4.645 \times 10^5)^{1/2}(40)^{1/3} = 60.93 \text{ W/m}^2\cdot°C$$

$$q = (60.93)(3.048)(3)(0.3048)(420-400)(5/9) = 1887 \text{ W}$$

$$\delta = \frac{(3.048)(4.64)}{(4.645 \times 10^5)^{1/2}} = 0.0208 \text{ m}$$

5-44 $T_f = \frac{27+77}{2} = 57°C = 325$ K $\nu = 18.23 \times 10^{-6}$ $k = 0.0281$

$Pr = 0.7$ $Re_L = \frac{(40)(4)}{18.23 \times 10^{-6}} = 8.78 \times 10^6$

$$Nu_L = Pr^{1/3}\left[0.037\, Re_L^{0.8} - 850\right]$$

$$\bar{h} = \frac{0.0281}{4}(0.7)^{1/3}\left[(0.037)(8.78 \times 10^6)^{0.8} - 871\right] = 77.4 \text{ W/m}^2\cdot°C$$

$$q = hA(T_w - T_\infty) = (77.4)(4)^2(77-27)$$

$$= 6.19 \times 10^4 \text{ W}$$

5-45 $T_f = \frac{300 + 273}{2} = 287\ K \quad \nu = 14.51 \times 10^{-6} \quad k = 0.0252$

$Pr = 0.71 \quad u_\infty = 5\ mi/hr. = 7.33\ ft/sec = 2.235\ m/sec.$

$$Re_L = \frac{(2.235)(30)}{14.51 \times 10^{-6}} = 4.62 \times 10^6$$

$$Nu_L = Pr^{1/3}\left[0.037\ Re_L^{0.8} - 850\right]$$

$$\bar{h} = \frac{0.0252}{30}(0.71)^{1/3}\left[(0.037)(4.62 \times 10^6)^{0.8} - 871\right] = 5.30\ W/m^2 \cdot {}^\circ C$$

$$q = \bar{h} A (T_w - T_\infty) = (5.30)(30)(60)(300 - 273)$$

$$= 2.58 \times 10^5\ W$$

5-46 ν @ 300 K $= 15.69 \times 10^{-6}$

$$Re_L = \frac{(10)(0.15)}{15.69 \times 10^{-6}} = 9.56 \times 10^4$$

$$\frac{\delta}{L} = \frac{4.64}{Re^{1/2}} \qquad \delta = (0.15)(9.56 \times 10^4)^{1/2}(4.64)$$

$$= 0.00225\ m$$

5-47 $T_f = \frac{25 + 150}{2} = 87.5°C = 360.5\ K \quad \mu = 2.119 \times 10^{-5}$

$k = 0.0308 \quad Pr = 0.693 \quad \rho = \frac{0.2 \times 10^6}{(287)(360.5)} = 1.933\ kg/m^3$

$$Re_L = \frac{(1.933)(60)(0.5)}{2.119 \times 10^{-5}} = 2.74 \times 10^6$$

$$Nu = Pr^{1/3}\left[0.037\ Re^{0.8} - 850\right]$$

$$\bar{h} = \frac{(0.693)^{1/3}}{0.5}(0.0308)\left[(0.037)(2.74 \times 10^6)^{0.8} - 871\right] = 238.6\ W/m^2 \cdot {}^\circ C$$

$$q = \bar{h} A (T_w - T_\infty) = (238.6)(0.5)^2(150 - 25)$$

$$= 7424\ W$$

5-48 $T_f = \frac{20+166}{2} = 93\,°C = 366\ K$ $\mu = 230.5 \times 10^{-7}$

$k = 0.1691$ $Pr = 0.71$ $\rho = \frac{150 \times 10^3}{(2078)(366)} = 0.197\ kg/m^3$

$$Re_L = \frac{(50)(1)(0.197)}{230.5 \times 10^{-7}} = 4.28 \times 10^5$$

$$\bar{h} = \frac{k}{L}\, 0.664\, Re_L^{1/2}\, Pr^{1/3}$$

$$= \frac{0.1691}{1}(0.664)(4.28 \times 10^5)^{1/2}(0.71)^{1/3} = 98.67\ W/m^2 \cdot °C$$

$$q = h A (T_w - T_\infty) = (98.67)(1)^2(166 - 20) = 14400\ W$$

5-49 $T_f = 300\ K$ $\mu = 1.8462 \times 10^{-5}$ $k = 0.02624$ $Pr = 0.71$

$\rho = \frac{50000}{(287)(300)} = 0.581\ kg/m^3$ $Re_L = \frac{(0.581)(20)(2)}{1.8462 \times 10^{-5}} = 1.26 \times 10^6$

$$\bar{h} = \frac{k}{L} Pr^{1/3}\left[0.037\, Re_L^{0.8} - 850\right]$$

$$= \frac{0.02624}{2}(0.71)^{1/3}\left[(0.037)(1.26 \times 10^6)^{0.8} - 871\right] = 22.7\ W/m^2 \cdot °C$$

$$q = h A (T_w - T_\infty) = (22.7)(2)^2(350 - 250) = 907\ W$$

5-51 $T_f = \frac{50+10}{2} = 30\,°C = 86\,°F$ $\mu = 8.03 \times 10^{-4}$ $k = 0.619$

$Pr = 5.41$ $\rho = 995$

@ $Re = 10^4$ $u_\infty = \frac{(10^4)(8.03 \times 10^{-4})}{(995)(0.3)} = 0.027\ m/sec$

@ $Re = 10^7$ $u_\infty = 27\ m/sec$

@ $Re = 10^4$ $\bar{h} = \frac{0.619}{0.3}(0.664)(10^4)^{1/2}(5.41)^{1/3} = 240.5\ W/m^2 \cdot °C$

$q = (240.5)(0.3)(50 - 10) = 866\ W$

@ $Re = 10^7$ $\bar{h} = \frac{0.619}{0.3}(5.41)^{1/3}\left[(0.037)(10^7)^{0.8} - 871\right]$

$= 49956\ W/m^2 \cdot °C$

$q = (49956)(0.3)^2(50 - 10) = 1.80 \times 10^5\ W$

5-52 $T_f = 350\text{ K}$ $\mu = 19.91 \times 10^{-6}$ $k = 0.0298$ $Pr = 0.7$

$$\rho = \frac{50000}{(287)(350)} = 0.481 \text{ kg/m}^3 \qquad Re_L = \frac{(0.481)(1.2)(100)}{19.91 \times 10^{-6}} = 2.9 \times 10^6$$

$$\bar{h} = \frac{0.0298}{1.2}(0.7)^{1/3}\left[(0.037)(2.9 \times 10^6)^{0.8} - 871\right] = 101.4 \text{ W/m}^2 \cdot {}^\circ\text{C}$$

$$q = \bar{h} A (T_w - T_\infty) = (101.4)(1.2)(400 - 300)$$

$$= 1.22 \times 10^4 \text{ W/m}$$

5-53 $T_f = \frac{15 + 139}{2} = 77\,^\circ\text{C} = 350\text{ K}$ $\mu = 9.954 \times 10^{-6}$ $k = 0.206$

$Pr = 0.697$ $\rho = \frac{(2)(1.01 \times 10^5)}{(4157)(350)} = 0.139 \text{ kg/m}^3$

$$Re_L = \frac{(0.139)(6)(1)}{9.954 \times 10^{-6}} = 83952 \qquad \bar{h} = \frac{k}{L} 0.664\, Re_L^{1/2}\, Pr^{1/3}$$

$$\bar{h} = \frac{0.206}{1}(0.664)(83592)^{1/2}(0.697)^{1/3} = 35.14 \text{ W/m}^2 \cdot {}^\circ\text{C}$$

$$q = \bar{h} A (T_w - T_\infty) = (35.14)(1)^2(139 - 15) = 4357 \text{ W}$$

5-54 $T_f = \frac{10 + 50}{2} = 30\,^\circ\text{C}$ $\nu = 0.349 \times 10^{-6}$ $k = 0.507$ $Pr = 2.01$

$$Re_L = \frac{(5)(0.45)}{0.349 \times 10^{-6}} = 6.45 \times 10^6$$

$$\bar{h} = \frac{k}{L} Pr^{1/3}\left[0.037\, Re_L^{0.8} - 850\right]$$

$$= \frac{0.507}{0.45}(2.01)^{1/3}\left[(0.037)(6.45 \times 10^6)^{0.8} - 871\right] = 13508 \text{ W/m}^2 \cdot {}^\circ\text{C}$$

$$q = \bar{h} A (T_w - T_\infty)$$

$$= (13508)(0.45)^2(50 - 10) = 1.094 \times 10^5 \text{ W}$$

5-55 $T_f = \frac{50+136}{2} = 93\ °C = 366\ K$ $\rho = \frac{45 \times 10^3}{(2078)(366)} = 0.0592\ kg/m^3$

$\mu = 230.5 \times 10^{-7}$ $k = 0.1691$ $Pr = 0.71$

$$Re_L = \frac{(0.0592)(1)(50)}{230.5 \times 10^{-7}} = 1.28 \times 10^5$$

$$\bar{h} = \frac{k}{L} 0.664\ Re_L^{1/2} Pr^{1/3}$$

$$= \frac{0.1691}{1}(0.664)(1.28 \times 10^5)^{1/2}(0.71)^{1/3} = 35.88\ W/m^2 \cdot °C$$

$$q = \bar{h} A (T_w - T_\infty) = (35.88)(1)^2(136-50) = 3086\ W$$

5-56 $T_f = \frac{100+10}{2} = 55°C = 328\ K$ $\nu = (18.53 \times 10^{-6})(10)$
$= 18.53 \times 10^{-5}$

$k = 0.0284$ $Pr = 0.7$

$$Re_L = \frac{(300)(0.8)}{18.53 \times 10^{-5}} = 1.295 \times 10^6$$

$$\bar{h} = \frac{k}{L} Pr^{1/3} [0.037 Re_L^{0.8} - 850]$$

$$= \frac{0.0284}{0.8} (0.7)^{1/3} [(0.037)(1.295 \times 10^6)^{0.8} - 871] = 63.04\ W/m^2 \cdot °C$$

$$q = \bar{h} A (T_w - T_\infty) = (63.04)(0.8)^2(100-10) = 3631\ W$$

5-57 $T_f = \frac{70+130}{2} = 100°F$ $\rho = 993$ $\mu = 6.82 \times 10^{-4}$ $k = 0.63$

$Pr = 4.53$ $Re_L = \frac{(993)(20)(0.3048)^2(1)}{6.82 \times 10^{-4}} = 2.705 \times 10^6$

$$\bar{h} = \frac{k}{L} Pr^{1/3} [0.037 Re_L^{0.8} - 850]$$

$$= \frac{0.63}{(1)(0.3048)} (4.53)^{1/3} [(0.037)(2.705 \times 10^6)^{0.8} - 871]$$

$$= 14720\ W/m^2 \cdot °C$$

$$q = \bar{h} A (T_w - T_\infty) = (14720)(1)^2(0.3048)^2(130-70)(5/9)$$

$$= 45584\ W$$

5-59 $\frac{u}{u_\infty} = \left(\frac{y}{\delta}\right)^{1/7}$ $\dot{m} = \int_0^\delta \rho u\, dy = \int_0^\delta \rho u_\infty \left(\frac{y}{\delta}\right)^{1/7} dy$

$$= \rho \frac{u_\infty}{\delta^{1/7}} \left.\frac{y^{8/7}}{8/7}\right]_0^\delta = 7/8\, \rho u_\infty \delta$$

$\nu = 20.76 \times 10^{-6}$ m²/s @ 350 K $\rho = 0.998$ kg/m³

At $Re_x = 10^6$ $x = \frac{(10^6)(20.76 \times 10^{-6})}{30} = 0.692$ m

$\delta = x\left[(0.381) Re_x^{-1/5} - 10256\, Re_x^{-1}\right] = 9.54 \times 10^{-3}$ m

$\dot{m} = 7/8(0.998)(6)(9.54 \times 10^{-3}) = 4.998 \times 10^{-2}$ kg/s

At $Re_x = 10^7$ $x = 6.92$ m $\delta = 0.0979$ m

$m = 7/8\,(0.998)(6)(0.0979) = 0.513$ kg/s

5-60 $T_f = \frac{600 + 300}{2} = 450$ K $\mu = 2.484 \times 10^{-5}$ $k = 0.03707$

$Pr = 0.683$ $\rho = \frac{50000}{(287)(450)} = 0.387$ kg/m³

$Re_x = \frac{(0.387)(6)(0.2)}{2.484 \times 10^{-5}} = 18703$

$Nu_x = 0.453\, Re_x^{1/2} Pr^{1/3} = (0.453)(18703)^{1/2}(0.683)^{1/3} = 54.56$

$q_w = k\, Nu_x (T_w - T_\infty)/x = (0.03707)(54.56)(600 - 300)/0.2$

$= 3034$ W/m²

$q = q_w A = (3034)(0.2)^2 = 121.3$ W

5-61 $\frac{1}{ur}\frac{\partial}{\partial r}\left(r\frac{\partial T}{\partial r}\right) = \frac{1}{\alpha}\frac{\partial T}{\partial x}$ $\qquad \frac{\partial T}{\partial r} = \frac{u_0}{\alpha}\frac{\partial T}{\partial x}\frac{r}{2} + \frac{C_1}{r}$

$T = \frac{u_0}{\alpha}\frac{\partial T}{\partial x}\frac{r^2}{4} + C_1 \ln r + C_2 \qquad \frac{C_1}{r} = f(x) = C \qquad C_1 = 0$

at $r=0$ $\quad T = T_c \quad T_c = C_2$ $\quad$ then $T = \frac{u_0}{\alpha}\frac{\partial T}{\partial x}\frac{r^2}{4} + T_c$

$$T_b = \frac{\int_0^{r_0} \rho(2\pi r\,dr)\,u\,c_p T}{\int_0^{r_0} \rho(2\pi r\,dr)\,u\,c_p} = \frac{u_0}{8\alpha} r_0^2 \frac{\partial T}{\partial x} + T_c \qquad T = T_w \text{ at } r = r_0$$

$$T_w = \frac{u_0}{4\alpha}\frac{\partial T}{\partial x} r_0^2 + T_c \qquad h = \frac{-k\left(\frac{\partial T}{\partial r}\right)_{r=r_0}}{T_w - T_b}$$

$$\left.\frac{\partial T}{\partial r}\right)_{r=r_0} = \frac{u_0}{\alpha}\frac{\partial T}{\partial x}\cdot\left.\frac{2r}{4}\right)_{r=r_0} = \frac{u_0 r_0}{2\alpha}\frac{\partial T}{\partial x}$$

$$h = \frac{-k\frac{u_0 r_0}{2\alpha}\frac{\partial T}{\partial x}}{\frac{u_0}{4\alpha}\frac{\partial T}{\partial x}r_0^2 + T_c - \left[\frac{u_0}{8\alpha}\frac{\partial T}{\partial x}r_0^2 + T_c\right]} = \frac{-k\frac{u_0 r_0}{2\alpha}\frac{\partial T}{\partial x}}{\frac{u_0}{8\alpha}r_0^2\frac{\partial T}{\partial x}} = -\frac{4k}{r_0}$$

$h = \frac{8k}{d_0}$ or $Nu_d = 8.0$

5-62

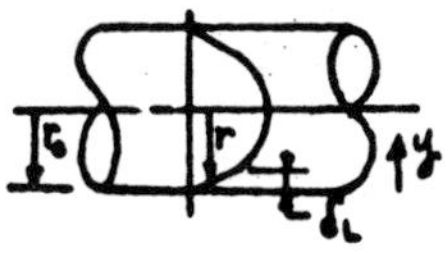

$\frac{u}{u_c} = \left(1 - \frac{r}{r_0}\right)^{1/7} \quad f = 0.316/Re_r^{1/4} \quad u_m = \frac{49}{60}u_c = 0.816\,u_c$

$\frac{dp}{dx} = f\frac{1}{d}\rho\frac{u_m^2}{2} \qquad \pi r_0^2\,dp = \tau_w(2\pi r_0\,dx)$

$\frac{dp}{dx} = 2\frac{\tau_w}{r_0} \quad \tau_w = \mu\left.\frac{du}{dy}\right)_{r=r_0} \quad \tau = \mu b = \frac{\mu u_c}{\delta_L^{6/7} r_0^{1/7}}$

Assume Linear profile in sublayer $u = a + br$ at $r = r_0$ $u = 0$

$a = -br \quad u_c = \left(1 - \frac{r_0 - \delta_L}{y_0}\right)^{1/7} = -b\delta_L \quad b = \frac{-u_c}{\delta_L^{6/7} r_0^{1/7}}$

$$\frac{dp}{dx} = \frac{0.316}{[\rho(0.816)u_c(2r_0)]^{1/4}}\cdot\left(\frac{1}{2r_0}\right)\rho\frac{(0.816)^2 u_c}{2}$$

$$\delta^{6/7} = \frac{2\mu u_c}{r_0^{8/7}(0.316)}\left[\frac{(0.816)\rho u_c(2r_0)}{\mu}\right]^{1/4}\frac{(2r_0)(2)}{(0.816)^2\rho u_c^2} \qquad \delta_L = \left[43\left(\frac{\mu}{u_c\rho}\right)^{3/4} r_0^{3/28}\right]^{7/6}$$

$\delta_L = 81\left(\frac{\mu}{u_c\rho}\right)^{7/8} r_0^{1/8}$ $\underline{\text{or}}$ $\frac{\delta_L}{r_0} = 124\,Re_d^{-7/8}$

5-63 $\Delta p \pi r^2 = \tau (2\pi r)\Delta x \quad \tau = \frac{\Delta p}{\Delta x}\frac{r}{2} = \rho(\nu + \epsilon_m)\frac{du}{dy} \quad \frac{u}{u_0} = \left(1 - \frac{r}{r_0}\right)^{1/7}$

$$\frac{du}{dy} = -\frac{du}{dr} = -u_c\left(\frac{1}{7}\right)\left(1 - \frac{r}{r_0}\right)^{-6/7}\left(-\frac{1}{r_0}\right) \quad \frac{du}{dy} = \frac{u_c}{7r_0}\left(1 - \frac{r}{r_0}\right)^{-6/7}$$

$$\epsilon_m = \frac{\tau}{\rho \frac{du}{dy}} - \nu = \frac{\frac{dp}{dx}\frac{r}{2}}{\rho \frac{u_c}{7r_0}\left(1 - \frac{r}{r_0}\right)^{-6/7}} - \nu \quad u_m = \frac{1}{\pi r_0^2}\int_0^{r_0} 2\pi r u \, dr$$

$$u_m = \frac{1}{\pi r_0^2}\int_0^{r_0} 2\pi r u_c\left(1 - \frac{r}{r_0}\right)^{1/7} dr = \frac{1}{\pi r_0^2}(2\pi u_c)\left(\frac{7}{8}\right)\left(\frac{7}{15}\right) r_0^2 = 0.816\, u_c$$

$$\frac{dp}{dx} = f\frac{1}{2r_0}\rho\frac{u_m^2}{2g_c} \quad f = 0.316\left[\frac{2u_m r_0}{\nu}\right]^{-1/4} = 0.316\left[\frac{2(0.816)u_c r_0}{\nu}\right]^{-1/4}$$

$$\frac{dp}{dx} = 0.316\left[\frac{\nu}{2(0.816)u_c r_0}\right]^{1/4}\frac{1}{2r_0}\rho\frac{(0.816\, u_c)^2}{2g_c}$$

$$\epsilon_m = 0.316\left[\frac{\nu}{2(0.816)u_c r_0}\right]^{1/4}\frac{\rho}{2r_0}\frac{(0.816\, u_c)^2}{2g_c}\frac{r}{2}\frac{\left(1 - \frac{r}{r_0}\right)^{6/7} 7r_0}{\rho u_c} - \nu$$

$$\epsilon_m = 0.162\left(\frac{r}{g_c}\right)u_c^{3/4}\left(\frac{\nu}{r_0}\right)^{1/4}\left(1 - \frac{r}{r_0}\right)^{6/7} - \nu$$

5-65 $Re = 1500 = \frac{\rho u_m d}{\mu} \quad T = 35°C \quad d = 0.025\ m \quad \rho = 993\ \frac{kg}{m^3}$

$\mu = 7.24 \times 10^{-4} \quad k = 0.627$

$$u_m = \frac{(1500)(7.24\times 10^{-4})}{(993)(0.025)} = 0.0437\ m/sec$$

$$u_0 = 2u_m = 0.0875\ m/sec$$

$$h = \frac{(4.364)(0.627)}{0.025} = 109.4\ W/m^2 \cdot °C$$

5-67 Air at $M_\infty = 4$ $p = 3$psia $T_\infty = 0$ °F Plate: 18 in long $T_w = 200$°F $\gamma = 1.402$

$a = \sqrt{\gamma g_c R T} = 1052$ ft/sec $u_\infty = M_\infty = 1.51 \times 10^7$ ft/hr $\rho_\infty = 0.0176$ $\mu_\infty = 0.0394$

$Re_{L\infty} = \frac{u_\infty x}{\nu_\infty} = 9.98 \times 10^6$ $T_0 = T_\infty(1 + \frac{\gamma - 1}{2} M_\infty^2)$ (Laminar portion) $= 1940$ °R

$Pr = 0.681$ (Assume) $r = Pr^{1/2} = 0.825$ $r = \frac{T_{aw} - T_\infty}{T_0 - T_\infty}$ $T_{aw} = 1680$ °R

$T^* = T_\infty + 0.5(T_w - T_\infty) + 0.22(T_{aw} - T_\infty) = 829$ °R $\rho^* = 0.00977$

$\mu^* = 0.061$ $k^* = 0.0218$ $c_p = 0.2444$ Turbulent portion: $Pr = 0.682$

$r = Pr^{1/3} = 0.882$ $T_{aw} = 1800$ °R $T^* = 855$ °R $\therefore Pr = 0.681$ (close enough)

$\rho^* = 0.00947$ $\mu^* = 0.0626$ $c_p^* = 0.245$ $u_\infty = 1.51 \times 10^7$ $x = 1.5$ ft

Laminar Heat Transfer: $x_c = \frac{Re^*_{crit} \mu^*}{\rho^* u_\infty} = 0.206$ ft $\overline{Nu}^* = \frac{\bar{h} x_c}{k^*}$

$\overline{Nu}^* = 0.664 (Re^*_{crit})^{1/2} (Pr^*)^{1/3} = 416$ $q = \bar{h} A (T_w - T_\infty) = -9250$ Btu/hr.

Turbulent Heat Transfer: $h_x = (Pr^*)^{-2/3} (\rho^* c_p^* u_\infty)(0.0288)\left(\frac{\rho^* u_\infty x}{\mu^*}\right)^{-1/5} = 69 x^{-1/5}$

$\bar{h} = \int_{0.206}^{1.5} h_x \, dx \Big/ \int_{0.206}^{1.5} dx = 73.3$ Btu/hr ft² °F $q = \bar{h} A (T_w - T_\infty) = -108000 \frac{\text{Btu}}{\text{hr}}$

Total Cooling $= -117250 \frac{\text{Btu}}{\text{hr.}}$

5-68 $T_w = 65$°C $u_\infty = 600$ m/s $T_\infty = 15$°C $= 288$ K $p = 7000$ N/m² $L = 1$m

$a = [(1.4)(287)(288)]^{1/2} = 340.2$ m/s $M = 600/340.2 = 1.764$ $T_0 = 288[1 + (0.2)(1.764)^2] = 467$ K

Assume $Pr = 0.7$. LAMINAR: $r = (0.7)^{1/2} = \frac{T_{aw} - 288}{467 - 288}$ $T_{aw} = 438$ K

$T^* = 288 + 0.5(467 - 288) + 0.22(438 - 288) = 346$ K $\rho = \frac{7000}{(287)(346)} = 0.0705 \frac{\text{kg}}{\text{m}^3}$

$\mu = 2.07 \times 10^{-5}$ $k = 0.02973$ $Pr = 0.7$ $x_c = \frac{(5 \times 10^5)(2.07 \times 10^{-5})}{(0.0705)(600)} = 0.245$ m

TURBULENT: $r = Pr^{1/3} = (0.7)^{1/3} = \frac{T_{aw} - 288}{467 - 288}$ $T_{aw} = 347$ K $\rho = \frac{7000}{(287)(347)} = 0.0701$

$\mu = 2.07 \times 10^{-5}$ $c_p = 1009$ $k = 0.0298$ $Pr = 0.7$

$h_x = (0.0701)(600)(1009)(0.7)^{2/3}(0.0296)\left[\frac{2.07 \times 10^{-5}}{(0.0701)(600)}\right]^{1/5} x^{-1/5} = 87.24 x^{-1/5}$

$\int_{x_c}^{L} h_x dx = (87.24)(5/4)[(1)^{4/5} - (0.245)^{4/5}] = 73.66$

Laminar heat transfer coeff. $\bar{h} = \frac{(0.664)(0.02973)}{0.245}(5 \times 10^5)^{1/2}(0.7)^{1/3} = 50.59$

For entire plate $\bar{h} = \int \frac{h_x dx}{L} = \frac{(50.59)(0.245) + 73.66}{1} = 86.05$ W/m²·°C

5-69 $Pr = 0.69$ $T_0 = 233[1 + (0.2)(4)^2] = 979\ K$

$r = (0.69)^{1/2} = \dfrac{T_{aw} - 233}{979 - 233}$ $T_{aw} = 853\ K$

5-70 $T_\infty = -40°C = 233\ K$ $a = [(1.4)(1)(287)(233)]^{1/2} = 306\ m/s$

$u_\infty = (2.8)(306) = 856.7$ $T_0 = (233)[1 + (0.2)(2.8)^2] = 598\ K$

$T^* \approx 450\ K$ $Pr = 0.69$

Laminar $r = Pr^{1/2} = 0.83$

$T_{aw} = (0.83)(598 - 233) + 233 = 536\ K$

Turbulent $r = Pr^{1/3} = 0.88$

$T_{aw} = (0.88)(598 - 233) + 233 = 555\ K$

5-72 $T_\infty = 30°C = 303\ K$ $\rho = 1258$ $L = 0.3\ m$ $c_p = 2445$

$u_\infty = 1.5\ m/sec$ $Pr = 5380$ $D = 8.9\ N = \tau_w A$ (both sides)

$\tau_w = \dfrac{8.9}{2(0.3)^2} = 49.44\ N/m^2$ $c_f = \dfrac{(2)(49.44)}{(1258)(1.5)^2} = 0.0349$

$St\ Pr^{2/3} = c_f/2$ $St = \dfrac{0.0349}{2}(5380)^{-2/3} = 5.689 \times 10^{-5}$

$\bar{h} = (5.689 \times 10^{-5})(1258)(1.5)(2445) = 262\ W/m^2 \cdot °C$

5-74 $T_f = 60°C = 333\ K$ $\rho = 1.046$ $c_p = 1042$ $\mu = 19.22 \times 10^{-6}$

$k = 0.02858$ $Re = \dfrac{(1.046)(3.0)(1.3)}{19.22 \times 10^{-6}} = 2.122 \times 10^{-5}$

$h = \dfrac{0.02858}{1.3}(0.664)(2.122 \times 10^5)^{1/2}(0.7)^{1/3} = 5.97$

$q = (5.97)(1.3)^2(100 - 20) = 807.2\ W$

$\bar{c}_f = (2)(0.664)(2.122 \times 10^5)^{-1/2} = 2.883 \times 10^{-3}$

5-75

$$\frac{u}{u_0} = 1 - \frac{r^2}{r_0^2} \qquad \Delta P = f \frac{L}{D} \rho u_m^2$$

$$f = \frac{\Delta P d \, 2g_c}{L \rho u_m^2} = \left(\frac{\Delta P}{L}\right) \frac{d}{\rho u_m^2 / 2g_c}$$

$$\frac{dp}{dx} = \frac{2\mu}{r} \frac{du}{dr} = \frac{\Delta P}{L}$$

$$f = \frac{2\mu}{r} \left(\frac{du}{dr}\right) \frac{d}{\rho u_m^2 / 2g_c} \qquad du = -2r\,dr \left(\frac{u_0}{r_0^2}\right)$$

$$\frac{du}{dr} = -2r \qquad f = \left(\frac{2\mu}{r}\right)\left(\frac{-2r u_0}{r_0^2}\right) \frac{d}{\rho u_m^2 / 2g_c}$$

$$u_m = \frac{\int_0^{r_0} 2\pi r u\,dr}{\pi r_0^2} = \frac{2u_0 \int_0^{r_0} \left(1 - \frac{r^2}{r_0^2}\right) r\,dr}{r_0^2}$$

$$u_m = \frac{u_0}{2} \qquad \therefore f = \frac{64 g_c}{\frac{\rho u_m d}{\mu}} = \frac{64}{Re_d}$$

5-78

$\nu = 15.51 \times 10^{-6}\ m^2/s$

$Re_1 = 5 \times 10^5 = \frac{45x}{15.51 \times 10^{-6}}$ $\quad x = 0.172\ m$

$Re_2 = 10^8 = \frac{45x}{15.51 \times 10^{-6}}$ $\quad x = 34.467\ m$

$\delta_1 = \frac{(5.0)(0.172)}{(5 \times 10^5)^{1/2}} = 1.22 \times 10^{-3}\ m$

$\delta_2 \cong (34.467)[0.381(10^8)^{-1/5} - 10256/10^8] =$
$0.326\ m$

5-79

$x = \frac{5 \times 10^5 \nu}{u_\infty}$ $\quad u_\infty = 20\ m/s$

$\delta = 5x(5 \times 10^5)^{-1/2}$

$= \frac{5\nu}{u_\infty}(5 \times 10^5)^{1/2} = \frac{3536\ \nu}{u_\infty} = 1768\ \nu$

a) $\nu = 14.2 \times 10^{-6}\ m^2/s$ $\quad \delta = 0.025\ m$

b) $\nu = 1.31 \times 10^{-6}$ $\quad \delta = 0.0023\ m$

c) $\nu = 99.69 \times 10^{-6}$ $\quad \delta = 0.176\ m$

d) $\nu = 0.368 \times 10^{-6}$ $\quad \delta = 6.5 \times 10^{-4}\ m$

e) $\nu = 0.203 \times 10^{-6}$ $\quad \delta = 3.59 \times 10^{-4}\ m$

5-80

$$\bar{h}_L = \frac{1}{L}\int_0^L h_x \, dx = \frac{1}{L}\int_0^L kC\left(\frac{\rho u_\infty}{\mu}\right)^n x^{n-1} f(P_r)\, dx$$

$$= \frac{1}{L}\, C k f(P_r)\left(\frac{\rho u_\infty}{\mu}\right)^n \frac{L^n}{n}$$

$$= \frac{h_{x=L}}{n}$$

$$\frac{\bar{h}_L}{h_{x=L}} = \frac{1}{n}$$

5-81 $\nu = 0.0009\ m^2/s$

$P_r = 10400$

Eq (5-44) $Nu_x = (0.332)(10{,}000)^{1/2}(10400)^{1/3}$
$= 724.7$

Eq. (5-51) $Nu_x = \dfrac{(0.3387)(10\,000)^{1/2}(10400)^{1/3}}{\left[1 + (0.0468/10400)^{2/3}\right]^{1/4}}$
$= 739.3$

5-82 $T_f = 325\,K \quad \nu = 18.23 \times 10^{-6}$

$k = 0.02814, \quad Pr = 0.7, \quad \rho = 1.086$

$$Re = \frac{(45)(0.75)}{18.23 \times 10^{-6}} = 1.85 \times 10^6$$

$$\bar{h} = \frac{(0.02814)}{0.75}(0.7)^{1/3}\left[(0.037)(1.85 \times 10^6)^{0.8} - 871\right]$$

$$= 98.3 \text{ W/m}^2\text{-°C}$$

$$q = (98.3)(0.75)^2(350 - 300) = 2764 \text{ W}$$

$$\bar{C}_f = \frac{0.074}{(1.85 \times 10^6)^{0.2}} - \frac{1055}{1.85 \times 10^6} = 0.00356$$

$$D = (0.00356)(0.75)^2(1.086)(45)^2/2$$
$$= 2.2 \text{ N}$$

Laminar Portion $$L = \frac{(5 \times 10^5)(18.23 \times 10^{-6})}{45} = 0.203 \text{ m}$$

$$h = \frac{(0.02814)}{0.203}(0.7)^{1/3}(5 \times 10^5)^{1/2}(0.332) = 28.96 \text{ W/m}^2\text{-°C}$$

$$q = (28.96)(0.203)(0.75)(350 - 300) = 220 \text{ W}$$

5-83 $x_c = 0.203\ m$

$$\delta_c = (5)(0.203)/(5\times10^5)^{1/2} = 0.0014 m$$

$$\delta_L = (0.75)\left[(0.381)(1.85\times10^6)^{-1/5} - 10256/1.85\times10^6\right]$$

$$= 0.0118\ m$$

All turbulent $\delta = \dfrac{(0.75)(0.381)}{(1.85\times10^6)^{0.2}} = 0.0159\ m$

5-84 $T_W = 500\ K \qquad T_f = 400\ K$

(a) Properties @ $T_\infty = 300K$, $\nu = 15.69\times10^{-6}$, $k = 0.02624$, $Pr = 0.708$

(b) Properties @ $T_f = 400K$, $\nu = 25.9\times10^{-6}$, $k = 0.03365$, $Pr = 0.689$

(c) Properties @ $T_w = 500K$, $\nu = 37.9\times10^{-6}$, $k = 0.04038$, $Pr = 0.68$

(a) $Re = (45)(0.75)/15.69\times10^{-6} = 215\times10^{4}$

$$h = \frac{0.02624}{0.78}(0.708)^{1/3}\left[(0.037)(2.15\times10^6)^{0.8} - 871\right] = 107.2$$

$$q = (107.2)(0.75)^2(500-300) = 12053\ W$$

(b) $Re = (45)(0.75)/25.9\times10^{-6} = 1.303\times10^6$

$$h = \frac{0.03365}{0.75}(0.689)^{1/3}\left[(0.037)(1.303\times10^6)^{0.8} - 871\right]$$

$$= 79.82$$

$$q = (79.82)(0.75)^2(500-300) = 8979\ W$$

(c) $Re = (45)(0.75)/37.9\times10^{-6} = 8.9\times10^5$

$$h = \frac{0.04038}{0.75}\left[(0.037)(8.9\times10^5)^{0.8} - 871\right](0.68)^{1/3} = 59.5$$

$$q = (59.5)(0.75)^2(500-300) = 6694\ W$$

Properties rather strongly dependent on temperature.

5-85 $q_w = 700\ W/m^2$ $L = 0.3\ m$

Properties at 300K

$\nu = 15.69 \times 10^{-6}$ $k = 0.02624$ $Pr = 0.708$

at $x = 15$ cm $Re = \dfrac{(10)(0.15)}{15.69 \times 10^{-6}} = 95600$

$Nu_x = 0.453\,(95600)^{1/2}\,(0.708)^{1/3} = 124.8$

$T_w - T_\infty = \dfrac{(700)(0.15)}{(124.8)(0.02624)} = 32°C$

300 K is close to average film temperature.

@ $x = 1$ cm $Re = 6373$

$Nu_x = 32.2$ $T_w - T_\infty = 8.3°C$ $T_w = 258.3°K$

@ $x = 5$ cm $Re = 31865$ $Nu_x = 72.1$

$T_w - T_\infty = 18.5°C$ $T_w = 268.5$ K

@ $x = 10$ cm $Re = 63730$ $Nu_x = 101.5$

$T_w - T_\infty = 26.3°C$ $T_w = 276$ K

@ $x = 20$ cm $Re = 128670$ $Nu_x = 144.8$

$T_w = T_\infty = 36.8°C$ $T_w = 286.8$ K

@ $x = 30$ cm $Re = 191200$ $Nu_x = 176.5$

$T_w - T_\infty = 45.3°C$ $T_w = 295.3$ K

5-86 at 30°C = T_f $\nu = 0.00057$ $k = 0.144$
$Pr = 6635$

$$Re = \frac{(10)(0.2)}{0.00057} = 3509$$

$$h = \frac{(0.144)}{0.2}(0.664)(3509)^{1/2}(6635)^{1/3} = 532\ W/m^2\text{-}°C$$

$$q = (532)(0.2)^2(40-20) = 426\ W$$

at 20°C $\nu = 0.0009$ $\rho = 888.2$

$$Re = \frac{(10)(0.2)}{0.0009} = 2228$$

$$\overline{C}_{f/2} = 0.664(2228)^{-1/2} = 0.141$$

$$D = (0.0141)(0.2)^2(888.2)(10)^2 = 50.1\ N$$

5-87 $T_f = \frac{27+77}{2} = 52°C = 325K$

$\nu = 18.23 \times 10^{-6}$ $k = 0.02814$ $Pr = 0.7$

$u_\infty = 44\ ft/s = 13.4\ m/s$

$$Re_L = \frac{(13.4)(4)}{18.23 \times 10^{-6}} = 2.94 \times 10^6$$

$$\bar{h} = \frac{0.02814}{4}(0.7)^{1/3}\left[(0.037)(2.94 \times 10^6)^{0.8} - 871\right]$$
$$= 29.1\ W/m^2\text{-}°C$$

$$q = (29.1)(4)(1)(77-27) = 5820\ W$$

$$h_{x=L} = (0.0296)(2.94 \times 10^6)^{0.8}(0.7)^{1/3}(0.02814)/4$$
$$= 27.64\ W/m^2\text{-}°C$$

5-87 (contd)

$q/A = (27.64)(50) = 1383 \text{ W/m}^2$

at $x = 50$ cm $Re_L = 3.68 \times 10^5$

$$h_x = \frac{0.02814}{0.5}(0.332)(3.68 \times 10^5)^{1/2}(0.7)^{1/3} = 10.06$$

$q/A = (10.06)(50) = 500 \text{ W/m}^2$

5-88 at $T_f = 350$ K

$\nu = 20.76 \times 10^{-6}$ $k = 0.03003$ $Pr = 0.697$

$$Re_L = \frac{(30)(0.1)}{20.76 \times 10^{-6}} = 1.445 \times 10^5$$

$x_0 = 0.05$ m

x	Re_x	$[1-(\frac{x_0}{x})^{3/4}]^{-1/3}$	Nu_x	h_x
0.06	0.867×10^5	1.985	171.8	85.98
0.075	1.084×10^5	1.562	151.2	60.54
0.1	1.445×10^5	1.351	151	45.34

$$\bar{h} = \left[\frac{85.98 + 60.54}{2}(0.015) + \frac{60.54 + 45.34}{2}(0.025)\right]\frac{1}{0.04}$$

$$= 60.56 \text{ W/m}^2\text{-°C}$$

$q = (60.56)(0.1 - 0.05)(0.1)(400 - 300) = 30.3$ W

5-89

at $x = 5$ cm $\quad Re_x = 72250$

$$h_x = \frac{0.03003}{0.05}(0.332)(72250)^{1/2}(0.697)^{1/3}$$

$$= 47.52 \ \text{W/m}^2\text{-°C}$$

$$q = (47.52)(0.1)(0.005)(400-300) = 2.38 \ \text{W}$$

5-90

$\nu = 15.69 \times 10^{-6}$

$$Re_L = \frac{(15)(0.2)}{15.69 \times 10^{-6}} = 1.91 \times 10^5$$

$$\delta = \frac{(5)(0.2)}{(1.91 \times 10^5)^{0.8}} = 0.00229 \ \text{m}$$

$$= 0.23 \ \text{cm}$$

No interference

6-1 For $L = 20$ cm $L/d = 40$ @ 120°C $Pr = 175$ $k = 0.135$ $c_p = 2.307$

$\mu = (0.124 \times 10^{-4})(829)$

1st iteration

$$h = \frac{0.135}{0.005}(1.86)(175000)^{1/3}(1/40)^{1/3} = 821.3$$

$$Re = 1000 = \frac{\dot{m}(0.005)(4)}{\pi(0.005)^2(0.124 \times 10^{-4})(829)}$$

$\dot{m} = 0.0404$ kg/sec

$$(0.0404)(2307)(120 - T_2) = (821.3)\pi(0.005)(2)\left[60 + \frac{T_2}{2} - 50\right] \quad (a)$$

$T_2 = 118.1$°C Small change in temp.

@ $T_w = 50$°C $\nu = 0.00057$

$$\bar{h} = (821.3)\left(\frac{0.124 \times 10^{-4}}{0.00057}\right)^{0.14} = 480.6$$

Recalculating T_2 from eq. (a) gives $T_2 = 118.9$ °C

$$q = (0.0404)(2307)(120 - 118.9) = 104.8 \text{ W}$$

6-3 $Nu_T = 2.47$ $k = 0.521$

$$D_H = \frac{4A}{P} = \frac{(4)(1)(1/2)(0.866)}{3} = 0.5744 \text{ cm}$$

$$h = \frac{(2.47)(0.521)}{0.005744} = 222.9 \text{ W/m}^2\text{ °C}$$

$$q/L = (222.9)(0.03)(50 - 20) = 200.6 \text{ W/m}$$

6-4 $D_H = \dfrac{(4)(5)(10)}{30} = 6.667$ mm $k = 0.6$ $Nu_T = 3.657$

$$h = \frac{(3.657)(0.6)}{6.667 \times 10^{-3}} = 329.1$$

$$q/L = (329.1)(30 \times 10^{-3})(60 - 20) = 394.9 \text{ W/m}$$

6-5 $q = (3)(4175)(15-5) = 125850\ W$ @ $10°C$ $\mu = 1.31 \times 10^{-3}$

$k = 0.585$ $Pr = 9.40$ $Re = \dfrac{(0.05)(3)(4)}{\pi(0.05)^2(1.31\times10^{-3})} = 58316$

$$h = \frac{0.585}{0.05}(0.023)(58316)^{0.8}(9.4)^{0.4} = 4283\ W/m^2\cdot°C$$

$q = 125850 = (4283)\pi(0.05)L(90-10)$ $L = 2.338\ m$

6-6 $q = (0.8)(4221)(40-35) = 16884\ W$ $\mu = 6.82\times10^{-4}$

$\rho = 993$ $k = 0.63$ $Pr = 4.53$ $Re = \dfrac{(0.025)(0.8)(4)}{\pi(0.025)^2(6.82\times10^{-4})} = 59741$

$$h = \frac{(0.023)(0.63)}{0.025}(59741)^{0.8}(4.53)^{0.4} = 7024\ W/m^2\cdot°C$$

$q = 16884 = (7024)\pi(0.025)L(90-37.5)$ $L = 0.583\ m$

6-7 $L/D = \dfrac{1.5}{0.025} = 60$ $T_f \approx \dfrac{50+20}{2} = 35°C$ @ $20°C$ $\rho = 998$

$c = 4180$ @ $35°C$ $Pr = 5.45$ $1.0 = (998)\dfrac{\pi(0.025)^2 u_m}{4}$

$u_m = 2.04\ m/sec$ $7000 = f(60)\dfrac{(998)(2.04)^2}{2}$ $f = 0.0562$

$$St = \frac{(0.0562)(5.45)^{-2/3}}{8} = 2.268\times10^{-3}$$

$h = (2.268\times10^{-3})(998)(4180)(2.04) = 19297\ W/m^2\cdot°C$

$$q = (19297)\pi(0.025)(1.5)\left[50 - \frac{T_{b_{out}}+20}{2}\right] = 1.0(4180)\left[T_{b_{out}} - 20\right]$$

$T_{b_{out}} = 32.83°C$

6-8 $T_{b_{avg}} = \dfrac{15+50}{2} = 32.5°C$ $q = 1.0(4170)(50-15) = 145950\ W$

$Pr = 5.1$ $\mu = 7.7\times10^{-4}$ $k = 0.623$ $Re = \dfrac{(0.025)(1.0)(4)}{\pi(0.025)^2(7.7\times10^{-4})} = 66142$

$$h = \frac{(0.023)(0.623)}{0.025}(66142)^{0.8}(5.1)^{0.4} = 7901\ W/m^2$$

$q = 145950 = (7901)\pi(0.025)L(14)$ $L = 16.8\ m$

6-9 Assume $T_{b\,avg}$ about 50°C $\rho = 870$ $c_p = 2000$ $u_m = 30$ cm/s

$k = 0.139$ $\nu = 1.24 \times 10^{-4}$ m²/sec $Pr = 1960$

$$Re = \frac{(0.30)(0.0125)}{1.24 \times 10^{-4}} = 30.24$$

$$Nu = (1.86)\left[(30.24)(1960)\left(\frac{0.0125}{3}\right)\right]^{1/3}\left(\frac{1.24}{0.723}\right)^{0.14} = 12.59$$

$$h = \frac{(12.59)(0.139)}{0.0125} = 139.9 \text{ W/m}^2\cdot°\text{C}$$

$$q = (139.9)\pi(0.0125)(3)\left[65 - \frac{T_c}{2} - \frac{38}{2}\right] = (870)\frac{\pi(0.0125)^2}{4}(0.3)(2000)(T_c - 38)$$

$49.839 = 1.287\,T_c$ $T_c = 44.16$°C $q = 394.6$ W

6-11 Gas @ 700 K $\rho = 0.503$ kg/m³ $\mu = 3.332 \times 10^{-5}$ $Pr = 0.684$

$k = 0.0523$ $$Re = \frac{(0.025)(0.8)(4)}{\pi(0.025)^2(3.332 \times 10^{-5})} = 1.22 \times 10^6$$

$$h_g = \frac{(0.0523)(0.023)}{0.025}(1.22 \times 10^6)^{0.8}(0.684)^{0.3} = 3176 \text{ W/m}^2\cdot°\text{C}$$

Water @ 150°C = 423 K $\mu = 1.86 \times 10^{-4}$ $\rho = 918$ $k = 0.684$

$Pr = 1.17$ $D_H = 5 - 2.5 - 2(0.16) = 2.18$ cm

$$Re = \frac{(4)(1.5)}{\pi(0.05 + 0.0282)(1.86 \times 10^{-4})} = 1.313 \times 10^5$$

$$h_w = \frac{(0.684)(0.023)}{0.0218}(1.313 \times 10^5)^{0.8}(1.17)^{0.4} = 9555 \text{ W/m}^2\cdot°\text{C}$$

Neglect conduction resistance, $q = 17.5$ kW

$$175000 = \frac{700 - 423}{\frac{1}{h_g A_i} + \frac{1}{h_w A_o}}$$

$$= \frac{(700 - 423)L}{\frac{1}{(3176)\pi(0.025)} + \frac{1}{(9555)\pi(0.0282)}}$$

$L = 0.328$ m

6-12 $\mu = 1.31 \times 10^{-3}$ $k = 0.585$ $Pr = 9.4$

$$Re = \frac{(0.5)(0.025)(4)}{\pi(0.025)^2(1.31\times10^{-3})} = 19439$$

$$h = \frac{(0.023)(0.585)}{0.025}(19439)^{0.8}(9.4)^{0.4} = 3557 \text{ W/m}^2\cdot{}^\circ\text{C}$$

$$q = (3557)\pi(0.025)(15)(15) = (0.5)(4190)\Delta T_b \quad \Delta T_b = 30.01^\circ\text{C}$$

$$T_{out} = 40.01\ ^\circ\text{C}$$

6-13 $\Delta p = f\frac{L}{D}\rho\frac{u_m^2}{2g_c}$ @ $T_f = \frac{27+55}{2} = 41^\circ\text{C}$ $Pr_f = 4.23$

$\rho = 996$ $\mu = 8.6\times10^{-4}$ $k = 0.614$ $Pr = 5.85$ $c_p = 4179$

$$u_m = \frac{0.7}{(996)\pi(0.0125)^2} = 1.43 \text{ m/s} \quad f = \frac{(2000)(0.025)(2)}{(6)(996)(1.43)^2} = 8.16\times10^{-3}$$

$$St_b\, Pr_f^{2/3} = f/8 \quad h = \frac{(8.16\times10^{-3})(996)(1.43)(4179)}{8}(4.23)^{-2/3} = 2321 \frac{\text{W}}{\text{m}^2\cdot{}^\circ\text{C}}$$

$$q = hA(\overline{T_w - T_b}) = \dot{m}c_p\Delta T_b$$

$$\Delta T_b = \frac{(2321)(\pi)(0.025)(6)(55-27)}{(0.7)(4179)} = 10.47\ ^\circ\text{C}$$

$$T_{exit} = 27 + \frac{10.47}{2} = 32.24\ ^\circ\text{C}$$

6-14 $Re = \frac{(0.3)(0.0025)}{1.6\times10^{-4}} = 4.69$ $Re\,Pr\frac{d}{L} = (4.69)(1960)\frac{0.25}{60} = 38.28$

$$h = \frac{0.14}{0.0025}(1.86)(38.28)^{1/3} = 351 \text{ W/m}^2\cdot{}^\circ\text{C} \quad c_p = \frac{Pr\,k}{\mu} = \frac{(1960)(0.14)}{(1.6\times10^{-4})(860)}$$

$$c_p = 1994 \text{ J/kg}\ ^\circ\text{C} \quad \dot{m} = \frac{(860)\pi(0.0025)^2(0.3)}{4} = 1.266\times10^{-3} \text{ kg/sec}$$

$$q = (1.266\times10^{-3})(1994)(T_e - 20) = (351)\pi(0.0025)(0.6)\left(120 - \frac{T_e - 20}{2}\right)$$

$$T_e = 69.36\ ^\circ\text{C} \quad q = 120.67 \text{ W}$$

6-15 $m = 1\ \text{lbm/sec} = 0.454\ \frac{\text{kg}}{\text{sec}}$ $T_b\,\text{avg} = \frac{10+38}{2} = 24\,°\text{C}$

$\rho = 605.6$ $\nu = 0.355 \times 10^{-6}$ $k = 0.515$ $c_p = 4840$ $Pr = 2.02$

$$Re = \frac{(0.025)(0.454)(4)}{\pi(0.025)^2(0.355\times10^{-6})(605.6)} = 1.14\times10^5$$

$$h = \frac{0.515}{0.025}(0.023)(1.14\times10^5)^{0.8}(2.02)^{0.4} = 6969\ \text{W/m}^2\cdot°\text{C}$$

$q = (0.454)(4840)(38-10) = 61526\ \text{W} = (6969)\pi(0.025)(2.5)(T_w - 24)$

$T_w = 69\,°\text{C}$

6-16

Freon 12 $\rho = 1364$ $k = 0.073$ $\nu = 0.203\times10^{-6}$ $Pr = 3.6$

$$Re = \frac{(3)(0.0125)}{0.203\times10^{-6}} = 184700$$

$$h = \frac{0.073}{0.0125}(0.023)(184700)^{0.8}(3.6)^{0.4} = 3663\ \text{W/m}^2\cdot°\text{C}$$

water $\rho = 999$ $\mu = 1.31\times10^{-3}$ $k = 0.585$ $Pr = 9.4$

$$Re = \frac{(999)(3)(0.0125)}{1.31\times10^{-3}} = 57195$$

$$h = \frac{0.585}{0.0125}(0.023)(57195)^{0.8}(9.4)^{0.4} = 16870\ \text{W/m}^2\cdot°\text{C}$$

6-17 $T_f = \frac{50+10}{2} = 30°\text{C}$ $L = 6\ \text{m}$ $d = 2.5\ \text{cm}$ $m = 0.4\ \text{kg/s}$ $\Delta p = 3000\ \text{N/m}^2$

@ 10°C $\rho = 999$ $c_p = 4195$ @ 30°C $Pr = 5.22$ $0.4 = \frac{(999)\pi(0.025)^2 u_m}{4}$

$u_m = 0.816\ \text{m/s}$ $\Delta p = f\frac{L}{D}\rho\frac{u_m^2}{2}$ $f = \frac{(3000)(2)(0.025)}{(6)(0.816)^2(999)} = 0.0376$

$St_b Pr_f^{2/3} = f/8$ $St_b = \frac{(0.0376)(5.22)^{-2/3}}{8} = 1.56\times10^{-3}$

$h = (1.56\times10^{-3})(999)(4195)(0.816) = 5338\ \text{W/m}^2\ °\text{C}$

$q = (5338)\pi(0.025)(6)(50-10) = (0.4)(4195)\Delta T_b$

$q = 100600\ \text{W}$ $\Delta T_b = 60\,°\text{C}$ $T_{out} \simeq \frac{60}{2} + 10 = 40\,°\text{C}$

6-18 $q = (0.5)(4175)(71-32) = 81412\ W$ $\quad T_b\ avg = 51.5°C = 125°F$

$\mu = 5.38 \times 10^{-4}$ $\quad k = 0.647$ $\quad Pr = 3.47$ $\quad \rho = 987$

(a) $12.5\ mm = d$ $\quad Re = \dfrac{(0.5)(0.0125)(4)}{\pi(0.0125)^2(5.38 \times 10^{-4})} = 94664$

$$h = \frac{0.647}{0.0125}(0.023)(94664)^{0.8}(3.47)^{0.3} = 16549\ W/m^2 \cdot °C$$

$q = 81412 = (16549)\pi(0.0125)L(51.5-4)$ $\quad L = 2.64\ m$

$u_m = \dfrac{(0.5)(4)}{\pi(0.0125)^2(987)} = 4.13\ m/s$ $\quad f = 0.0185$

$$\Delta p = (0.0185)\frac{(2.64)}{0.0125}\frac{(987)}{2}(4.13)^2 = 32.9\ kPa$$

(b) $d = 25\ mm$ $\quad T_w = 20°C$ $\quad Re = 47332$

$$h = \frac{(0.647)(0.023)(47332)^{0.8}(3.47)^{0.3}}{0.025} = 4701$$

$81412 = (4701)\pi(0.025)L(51.5-20)$ $\quad L = 7.0\ m$

$f = 0.021$ $\quad u = 4.13/4 = 1.033$

$$\Delta p = \frac{(0.021)(7)}{0.025}\frac{987}{2}(1.033)^2 = 3.09\ kPa$$

6-19 $T_b = 550\ K$ $\quad \mu = 2.848 \times 10^{-5}$ $\quad k = 0.0436$ $\quad Pr = 0.68$ $\quad c_p = 1.039$

$\rho = \dfrac{1.4 \times 10^6}{(287)(550)} = 8.87\ kg/m^3$ $\quad Re = \dfrac{(0.075)(4)(0.5)}{\pi(0.075)^2(2.848 \times 10^{-5})} = 298\,000$

$$Nu = 0.036\ Re^{0.8} Pr^{1/3}(d/L)^{0.0555}$$

$$h = \frac{0.0436}{0.075}(0.036)(298000)^{0.8}(0.68)^{1/3}\left(\frac{0.075}{6}\right)^{0.0555} = 346.5\ W/m^2 \cdot °C$$

$q = (346.5)\pi(0.075)(6)(550-500) = (0.5)(1039)\Delta T_b = 24493\ W$

$\Delta T_b = 47.15°C$

6-20 $\bar{T}_b = 30°C \quad \nu_b = 13.94 \times 10^{-6} \quad k = 0.252 \quad Pr = 148 \quad c_p = 2.428$

$\nu_w = 2.98 \times 10^{-6} \quad \rho = 1109 \quad D_H = 5 - 4 = 1\text{ cm} = 0.01\text{ m}$

$\dot{m} = \rho A u_m = (1109)\frac{\pi}{4}(0.05^2 - 0.04^2)(6.9) = 5.409\text{ kg/s}$

$q = \dot{m} c_p \Delta T_b = (5.409)(2428)(40-20) = 2.627 \times 10^5\text{ W}$

$$Re_{DH} = \frac{(6.9)(0.01)}{13.94 \times 10^{-6}} = 4950$$

$$h = \frac{0.252}{0.01}(0.027)(4950)^{0.8}(148)^{1/3}\left(\frac{13.94}{2.98}\right)^{0.14} = 4033\text{ W/m}^2\cdot°C$$

$q = h \pi d_i L (T_w - \bar{T}_b)$

$2.627 \times 10^{-5} = (4033)\pi(0.04)(L)(80-30) \quad L = 10.37\text{ m}$

6-21 At 300K & 1 atm. $\nu = 15.69 \times 10^{-6} \quad \rho = 1.1774 \quad k = 0.02624$

$Pr = 0.708 \quad D_H = \frac{(4)(45)(90)}{(2)(45+90)} = 60\text{ cm} = 0.6\text{ m}$

$Re = \frac{(0.6)(7.5)}{15.69 \times 10^{-6}} = 2.87 \times 10^5 \quad h = \frac{0.02624}{0.6}(0.023)(2.87 \times 10^5)^{0.8}(0.708)^{0.4}$

$h = 20.35\text{ W/m}^2\cdot°C \quad \Delta p = f \frac{L}{d} \rho \frac{u_m^2}{2}, \quad f = 0.0145$

$$\Delta p = (0.0145)\frac{(1)}{0.6}\frac{(1.1774)(7.5)^2}{2} = 0.8\text{ Pa}$$

6-22 Assume $\bar{T}_b$ about $38°C = 100°F \quad \rho = 993 \quad c_p = 4.174$

$\mu = 6.82 \times 10^{-4} \quad k = 0.63 \quad Pr_f = 2.9$

$Re = \frac{\rho u_m d}{\mu} = 10^5 = \frac{(993)u(0.03)}{6.82 \times 10^{-4}} \quad u = 2.29\text{ m/sec} \quad \rho u = 2274$

$f = 0.0255$

$St_b Pr_f^{2/3} = f/8 \quad h = \frac{0.0255}{8}(2274)(4174)(2.9)^{-2/3} = 14878\text{ W/m}^2\cdot°C$

6-23 $\bar{T}_b = 10°C \quad T_w = 40°C \quad c_p = 0.9345 \quad @\ 0°C \quad \nu = 0.214 \times 10^{-6}$

$\rho = 1397$

$$700 = \frac{(u)(0.0035)}{0.214 \times 10^{-6}} \qquad u = 0.0428 \text{ m/sec}$$

$$q = (1397)(934.5)(0.0428)\ \frac{\pi}{4}(0.0035)^2(20-0) = 10.75 \text{ W}$$

$@\ 10°C \quad k = 0.073 \quad Pr = 3.6$

$$q = hA(T_w - \bar{T}_b) = h\pi(0.0035)L(40-10) = 10.75$$

$$\bar{h}L = 32.59 \qquad Gz = Re\,Pr\frac{d}{L} = (700)(3.6)\frac{d}{L} = 2520\frac{d}{L}$$

$$\overline{Nu}_d = \frac{hL}{k} \times \frac{d}{L} = \frac{32.59}{0.073} \cdot \frac{d}{L} = 446.4\frac{d}{L}$$

From Figure:

$\frac{1}{Gz}$	$\frac{d}{L}$	$\overline{Nu}_d$	$446.4\frac{d}{L}$
0.1	0.00397	4.3	1.771
0.01	0.00397	7.5	17.72
0.03	0.0132	5.4	5.89
0.035	0.0134	5.2	5.06

$$\frac{d}{L} \approx 0.035 \qquad L = \frac{0.0035}{0.035} = 0.1 \text{ m}$$

6-24 $\bar{T}_b = \frac{27+77}{2} = 52°C = 325\text{ K} \quad \mu = 1.96 \times 10^{-5} \quad k = 0.0282$

$Pr = 0.7 \quad c_p = 1.007 \quad L = 30 \text{ cm} \quad \rho = 1.088$

$$D_H = \frac{(4)(0.003)^2/2}{(3)(0.003)} = 0.002$$

$$Re = \frac{(0.002)(5\times10^{-5})(2)}{(0.003)^2(1.96\times10^{-5})} = 1134$$

$$q = \dot{m}c_p\Delta T_b = (5\times10^{-5})(1007)(77-27) = 2.518 \text{ W}$$

$$Re\,Pr\frac{d}{L} = (1134)(0.7)\left(\frac{0.002}{0.3}\right) = 5.292$$

From Figure for $1/Gz = \frac{1}{5.292} = 0.189$

$\overline{Nu}_d \cong 4.0$ and flow is nearly developed.

6-24 cont Assuming ratio for Nu_T from table-

$$\overline{Nu}_{DH} = (2.47)\frac{(4.0)}{3.66} = 2.7$$

$$h = \frac{(2.7)(0.0282)}{0.002} = 38.1 \ W/m^2 \cdot {}^\circ C$$

$$q = hA(T_w - T_b) \cdot \qquad T_w - T_b = \frac{2.518}{(38.1)(3)(0.3)(0.003)} = 24.5\ {}^\circ C$$

$$T_w = 24.5 + 52 = 76.5\ {}^\circ C$$

6-25 At 27°C = 300 K $\mu = 1.8462 \times 10^{-5}$ $k = 0.02624$ $Pr = 0.708$

$$\rho = \frac{9000}{(287)(300)} = 1.045\ kg/m^3 \qquad Re = \frac{(0.004)(7\times10^{-5})(4)}{\pi(0.004)^2(1.8462\times10^{-5})} = 1207$$

$$Gz = (1207)(0.708)(0.004/0.12) = 28.48 \quad > 10$$

At $T_w = 27 + 70 = 97\ {}^\circ C = 390\ K \approx 400\ K$ $\mu_w = 2.286 \times 10^{-5}$

$$h = \frac{k}{d}(1.86)\,Gz^{1/3}\left(\frac{\mu}{\mu_w}\right)^{0.14} = \frac{0.02624}{0.004}(1.86)(28.48)^{1/3}\left(\frac{1.8462}{2.286}\right)^{0.14}$$

$$= 36.16 \ W/m^2 \cdot {}^\circ C$$

$$q = hA(T_w - T_b) = \dot{m} c_p \Delta T_b$$

$$\Delta T_b = \frac{(36.16)\pi(0.004)(0.12)(70)}{(7\times10^{-5})(1007)} = 54.15\ {}^\circ C$$

$$T_{be} = 54.15 + 27 = 81.15\ {}^\circ C$$

6-26 At $40°C = 313\ K$ $\mu = 1.906 \times 10^{-5}$ $k = 0.0272$ $Pr = 0.7$

$$\rho = \frac{110 \times 10^3}{(287)(313)} = 1.225\ kg/m^3$$

$$Re = \frac{(0.006)(8 \times 10^{-5})(4)}{\pi(0.006)^2(1.906 \times 10^{-5})} = 891 \quad Gz = (891)(0.7)(0.006/0.14) = 26.73$$

$$1/Gz = 0.0374 \quad \overline{Nu}_d = 5.2 \quad h = \frac{(5.2)(0.0272)}{0.006} = 23.57\ W/m^2 \cdot °C$$

$$q = \bar{h} A (T_w - \bar{T}_b) = \dot{m} c_p (T_{be} - T_{bi})$$

$$(23.57)\pi(0.006)(0.14)(140 - 20 - \frac{T_{be}}{2}) = (8 \times 10^{-5})(1005)(T_{be} - 40)$$

$$T_{be} = 95.8\ °C$$

6-27 At $40°C$ $\rho = 876\ kg/m^3$ $c_p = 1.964\ kJ/kg\,°C$

$\nu = 0.00024$ $k = 0.144$ $Pr = 2870$

$$Re = 50 = \frac{(u)(0.01)}{0.00024} \quad u = 1.2\ m/s$$

$$\dot{m} = (876)(1.2)\pi(0.01)^2/4 = 0.0826\ kg/sec$$

$$Gz = (50)(2870)(0.01/0.08) = 1.794 \times 10^4$$

at $T_w = 80°C$ $\nu_w = 0.375 \times 10^{-4}$

$$Nu_d = (1.86)(1.794 \times 10^4)^{1/3}\left(\frac{2.4}{0.375}\right)^{0.15} = 64.3$$

$$h = \frac{(64.3)(0.144)}{0.01} = 926 \quad q = (926)\pi(0.01)(0.08)(80 - 20 - \frac{T_e}{2})$$

$$= (0.0826)(1964)(T_e - 40)$$

$$T_e = 40.57°C$$

6-28 $\bar{T}_b = 25°C = 77°F$ $\rho = 996$ $\mu = 8.96 \times 10^{-4}$ $k = 0.611$

$Pr = 6.13$ $c_p = 4180$ $Re = \frac{(996)(8)(0.02)}{8.96 \times 10^{-4}} = 1.78 \times 10^5$

$$h = (0.023)\frac{(0.611)}{0.02}(1.78 \times 10^5)^{0.8}(6.13)^{0.4} = 23003 \ W/m^2 \cdot °C$$

$$q = hA(\bar{T}_w - \bar{T}_b) = \dot{m} c_p \Delta T_b$$

$$(23003)\pi(0.02)(10)(\bar{T}_w - 25) = (996)(8)\pi\frac{(0.02)^2}{4}(4180)(10)$$

$$T_w = 32.2 °C$$

6-29 At 20°C $\rho = 888$ $c_p = 1880$ J/kg°C $\nu = 0.0009$ $k = 0.145$

$Pr = 10400$ $Re = \frac{(1.2)(0.002)}{0.0009} = 2.67$

$Gz = (2.666)(10400)(0.002/1) = 55.47 > 10$ $1/Gz = 0.018$

$\overline{Nu}_d = 6.1$ $h = \frac{(6.1)(0.145)}{0.002} = 442$

$$q = (442)\pi(0.002)(1.0)(60 - 10 - \frac{T_e}{2}) = (888)(1.2)\frac{\pi}{4}(0.002)^2(1880)(T_e - 20)$$

$$T_e = 34.46 °C$$

6-30 $\bar{T}_b = 40°C = 104°F$ $\rho = 993$ $\mu = 6.55 \times 10^{-4}$ $k = 0.633$

$Pr = 4.33$ $c_p = 4175$ $Re = \frac{(993)(1)(0.0254)(10)(0.3048)}{6.55 \times 10^{-4}} = 1.17 \times 10^5$

$$h = \frac{0.633}{0.0254}(0.023)(1.17 \times 10^5)^{0.8}(4.33)^{0.4} = 11709 \ W/m^2 \cdot °C$$

$$q = \dot{m} c_p \Delta T_b = h A (T_w - T_b)$$

$$(993)\pi\frac{(0.0254)^2}{4}(3.048)(4175)(50-30) = (11709)\pi(0.0254)(L)(20)$$

$$L = 6.85 \ m$$

6-31 $\bar{T}_b = \frac{21+32}{2} = 26.5\,°C \quad \rho = 996 \quad \mu = 8.6\times10^{-4}$

$Re = 600 = \frac{\rho u d}{\mu} \qquad u = \frac{(600)(8.6\times10^{-4})}{(996)(0.003)} = 0.173\ m/s$

$\dot{m} = \rho A u = (996)\,\pi\frac{(0.003)^2}{4}(0.173) = 0.00122\ kg/s$

6-32 $\bar{T}_b = 80°F \quad \rho = 996 \quad \mu = 8.6\times10^{-4} \quad k = 0.614 \quad Pr = 5.85$

$c_p = 4180 \qquad Re = \frac{(0.03)(1.0)(4)}{\pi(0.03)^2(8.6\times10^{-4})} = 49350$

$h = \frac{(0.023)(0.614)}{0.03}(49350)^{0.8}(5.85)^{0.4} = 5423\ W/m^2\cdot°C$

$q = \dot{m} c_p \Delta T_b = hA(T_w - \bar{T}_b)$

$(1.0)(4180)(100-60)(5/9) = (5423)\pi(0.03)L(140-80)(5/9)$

$L = 5.45\ m$

6-33 At $\bar{T}_b = 20°C \quad \rho = 1264 \quad c_p = 2386 \quad k = 0.286 \quad Pr = 12.5\times10^3$

$\nu = 0.003 \qquad Re = 10 = \frac{ud}{\nu} \qquad u = \frac{(10)(0.003)}{0.005} = 6\ m/sec$

$\dot{m} = \rho A u = (1264)\,\pi\frac{(0.005)^2}{4}(6) = 0.149\ kg/sec$

$q = \dot{m} c_p \Delta T_b = (0.149)(2386)(30-10) = 7106\ W$

$= hA(T_w - \bar{T}_b) = h\pi(0.005)L(40-20) \qquad hL = 22619 \qquad h = \frac{22619}{L}$

$\frac{hd}{k} = 1.86\left(Re\,Pr\,\frac{d}{L}\right)^{1/3}\left(\frac{\mu}{\mu_w}\right)^{0.14}$

@ $T_w = 40°C \qquad \nu_w = 0.00022$

$\frac{(22619)(0.005)}{(0.286)L} = 1.86\left[(10)(12500)(0.005)/L\right]^{1/3}\left(\frac{0.003}{0.00022}\right)^{0.14}$

$L^{2/3} = 17.25$

$L = 71.66\ m \qquad Gz = (10)(12500)(0.005/71.66) = 8.72 < 10$

Below range of equation:

$1/Gz \approx 1/8.72 = 0.115$

6-33 cont'd From Figure: $\overline{Nu}_d \simeq 4.2$

$h = \frac{(4.2)(0.286)}{0.005} = 240.2 \quad L = 22619/240.2 = 94.2\ m$

New $1/Gz = (0.115)(94.2/71.66) = 0.15$

New iteration: $\overline{Nu}_d \approx 4.0 \quad h = 228.8 \quad L = 98.8\ m$

Close enough to fully developed tube, could take $\overline{Nu}_d = 3.66$.

6-34 $T_f = \frac{100+10}{2} = 55°C = 328\ K \quad \mu = 19 \times 10^{-6} \quad k = 0.0282$

$Pr = 0.7 \quad \rho = \frac{(2)(1.01 \times 10^5)}{(297)(328)} = 2.08\ kg/m^3 \quad Re = \frac{(2.08)(5)(0.05)}{19.6 \times 10^{-6}} = 27377$

$C = 0.193 \quad n = 0.618 \quad h = \frac{k}{d} C\, Re^n\, Pr^{1/3} = \frac{0.0282}{0.05}(0.193)(27377)^{0.618}(0.7)^{1/3}$

$= 53.4\ W/m^2 \cdot °C$

$\frac{q}{L} = h \pi d (T_w - T_\infty) = (53.4)\pi(0.05)(100-10) = 755\ W/m$

6-35 $T_f = \frac{54+0}{2} = 27°C = 300\ K \quad \nu = 15.69 \times 10^{-6} \quad k = 0.02624$

$Pr = 0.708 \quad Re = \frac{\rho u d}{\mu} = \frac{(25)(0.04)}{15.69 \times 10^{-6}} = 63735$

$C = 0.0266 \quad n = 0.805 \quad h = \frac{k}{d} C\, Re^n\, Pr^{1/3}$

$h = \frac{0.02624}{0.04}(0.0266)(63735)^{0.805}(0.708)^{1/3} = 114.6\ W/m^2 \cdot °C$

$q/L = h \pi d (T_w - T_\infty) = (114.6)\pi(0.04)(54-0) = 778\ W/m$

6-36 $T_f = \frac{80+10}{2} = 45°C = 318\ K \quad \mu = 1.929 \times 10^{-5} \quad k = 0.0276$

$Pr = 0.7 \quad \rho = \frac{200 \times 10^3}{(287)(318)} = 2.191\ kg/m^3 \quad Re = \frac{\rho u d}{\mu} = \frac{(2.191)(20)(0.2)}{1.929 \times 10^{-5}}$

$Re = 5.68 \times 10^5$

Churchill Equation:

$$Nu_d = 0.3 + \frac{(0.62)(5.68 \times 10^5)^{1/2}(0.7)^{1/3}}{\left[1 + \left(\frac{0.4}{0.7}\right)^{2/3}\right]^{1/4}}\left[1 + \left(\frac{5.68 \times 10^5}{282000}\right)^{5/8}\right]^{4/5}$$

$= 769.7$

$$h = \frac{k}{d} Nu = \frac{(0.0276)(769.7)}{0.2} = 106.2\ W/m^2 \cdot °C$$

$q/L = h \pi d (T_w - T_\infty) = (106.2)\pi(0.2)(80-10) = 4672\ W/m$

From Figure $C_D = 0.3$

$$F_D = C_D A \frac{\rho u_\infty^2}{2 g_c} = \frac{(0.3)(0.2)(2.191)(25)^2}{2} = 41.1\ N/m \text{ Length}$$

6-37 $T_f = \frac{71+43}{2} = 57°C = 135°F \quad Pr_f = 3.15 \quad \rho_b = 991 \quad c = 4174$

$\mu = 6.16 \times 10^{-4} \quad Re = \frac{(0.05)(6)(4)}{\pi(0.05)^2(6.16 \times 10^{-4})} = 2.48 \times 10^5 \quad f = 0.014$

$\dot{m} = \rho A u_m \quad u_m = \frac{(6)(4)}{(991)\pi(0.05)^2} = 3.08\ m/sec$

$St_b Pr_f^{2/3} = f/8$

$$h = \frac{(0.014)(991)(3.08)(4174)}{(8)(3.15)^{2/3}} = 10388\ W/m^2 \cdot °C$$

$q = hA(T_w - \bar{T}_b) = \dot{m} c_p \Delta T_b$

$(10388)\pi(0.05)(9)(71 - 21.5 - T_e/2) = (6)(4174)(T_e - 43)$

$T_e = 55.7°C$

6-38 @ 38°C $\rho = 993$ $\mu = 6.82 \times 10^{-4}$ $k = 0.63$ $Pr = 4.53$

$c_p = 4180$ $Re = \frac{(993)(1.5)(0.0064)}{6.82 \times 10^{-4}} = 13978$

$$h = \frac{0.63}{0.0064}(0.036)(13978)^{0.8}(453)^{1/3}\left(\frac{0.0064}{0.15}\right)^{0.055} = 10213 \text{ W/m}^2\cdot°C$$

$$q = (10213)\pi(0.0064)(0.15)(28) = (993)\frac{\pi}{4}(0.0064)^2(1.5)(4180)(T_e - 38)$$

$= 862.5 \text{ W}$ $T_e = 42.31°C$

6-39 $\rho = 1094$ $c_p = 2518$ $k = 0.258$ $Pr = 72$ $\nu = 6.72 \times 10^{-6}$

$\dot{m} = (1094)(10)\pi(0.03)^2/4 = 7.733$ kg/sec

$q = (7.733)(2518)(60-40) = 3.894 \times 10^5$ W

$Re = \frac{(10)(0.03)}{6.72 \times 10^{-6}} = 4.464 \times 10^4$ $\nu_w = 19.18 \times 10^{-6}$

$$h = \frac{0.258}{0.03}(0.027)(4.464 \times 10^4)^{0.8}(72)^{1/3}\left(\frac{6.72}{19.18}\right)^{0.14} = 4375 \text{ W/m}^2\cdot°C$$

$(4375)\pi(0.03)L(50-20) = 3.894 \times 10^5$ $L = 31.5$ m

6-41 $T = 20°C = 293$ K $\mu = 1.91 \times 10^{-5}$ $\rho = \frac{7000}{(287)(293)} = 0.832$

$Re = \frac{(0.832)(20)(0.05)}{1.91 \times 10^{-5}} = 43582$ $C_D = 1.2$

$F_D = (1.2)(0.05)(0.832)(20)^2 = 9.98$ N/m length

6-42 $T_f = \frac{450 + 325}{2} = 387.5$ K $\nu = 24.62 \times 10^{-6}$ $k = 0.0327$

$Pr = 0.69$ $Re = \frac{(30)(0.025)}{24.62 \times 10^{-6}} = 30463$ $C = 0.193$ $n = 0.618$

$$h = \frac{0.0327}{0.025}(0.193)(30463)^{0.618}(0.69)^{1/3} = 131.6 \text{ W/m}^2\cdot°C$$

$q/L = h\pi d(T_w - T_\infty) = (131.6)\pi(0.025)(450-325) = 1293$ W/m

6-43 $T_f = 52.5°F$ $\mu_f = 0.0428$ $\rho = P/RT = 0.0778$ $k_f = 0.0144$

$Re_{d_f} = \frac{\rho u_\infty d}{\mu_f} = 288{,}000$ $h = \frac{(0.0239) k_f}{d}\left(\frac{\rho u_\infty d}{\mu_f}\right)^{0.805} = 8.5$ Btu/hr ft² °F

$q = hA(T_w - T_\infty) = 7200$ Btu/hr.

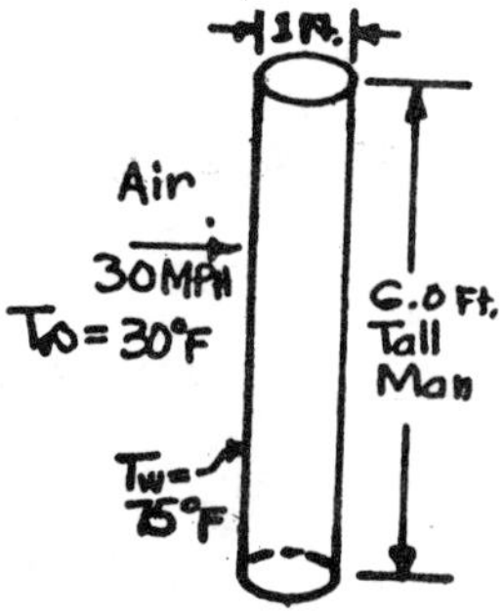

6-45 $T_f = \frac{175 + (-30)}{2} = 72.5°C = 345.5\ K$ $\rho = \frac{54000}{(287)(345.5)} = 0.544$

$\mu = 2.07 \times 10^{-5}$ $k = 0.03$ $Pr = 0.7$ $Re = \frac{(0.544)(0.00013)(230)}{2.07 \times 10^{-5}} = 786.6$

$h = \frac{0.03}{0.00013}(0.683)(786.6)^{0.466}(0.7)^{1/3} = 3129\ W/m^2 \cdot °C$

$q = (3129)\pi(0.00013)(0.0125)(175 + 30) = 3.275\ W$

6-46 $T_f = \frac{90 + 150}{2} = 120°C = 393\ K$ $\rho = 0.899$ $\mu = 2.256 \times 10^{-5}$

$k = 0.0331$ $Pr = 0.69$ $Re = \frac{(0.899)(1/16)(0.0254)(6)}{2.256 \times 10^{-5}} = 379.6$

$h = \frac{0.0331}{(1/16)(0.0254)}(0.683)(379.6)^{0.466}(0.69)^{1/3} = 200.3\ W/m^2 \cdot °C$

$q/L = (200.3)(\pi)(1/16)(0.0254)(150 - 90) = 59.95\ W/m$

6-48 $T_f = \frac{425 + 325}{2} = 375\ K$ $\nu = 181.4 \times 10^{-6}$ $k = 0.192$

$Pr = 0.71$ $Re = \frac{(0.003)(9)}{181.4 \times 10^{-6}} = 148.8$ $C = 0.683$ $n = 0.618$

$h = \frac{0.192}{0.003}(0.683)(148.8)^{0.618}(0.71)^{1/3} = 85.86\ W/m^2 \cdot °C$

$q/L = h\pi d(T_w - T_\infty) = (85.86)\pi(0.003)(425 - 325) = 809\ W/m$

6-49 $T_f = \frac{65+20}{2} = 42.5\ °C = 315.5\ K$

Air: $\rho = \frac{1.0132\times10^5}{(287)(315.5)} = 1.119$ $\mu = 2.012\times10^{-5}$ $k = 0.0274$ $Pr = 0.7$

$C = 0.911$ $n = 0.385$ $Re = \frac{(1.119)(6)(0.025\times10^{-3})}{2.012\times10^{-5}} = 8.342$

$h = \frac{0.0274}{0.025\times10^{-3}}(0.911)(8.342)^{0.385}(0.7)^{1/3} = 2006\ W/m^2\cdot°C$

$q/L = (2006)\pi(0.025\times10^{-3})(65-20) = 7.091\ W/m$

Water: $\rho = 991$ $\mu = 6.2\times10^{-4}$ $k = 0.635$ $Pr = 4.1$ $C = 0.683$

$n = 0.466$ $Re = \frac{(991)(6)(0.025\times10^{-3})}{6.2\times10^{-4}} = 239.8$

$h = \frac{0.635}{0.025\times10^{-3}}(0.683)(239.8)^{0.466}(4.1)^{1/3} = 3.57\times10^5\ W/m^2\cdot°C$

$q/L = (3.57\times10^5)\pi(0.025\times10^{-3})(65-20) = 1261\ W/m$

6-52 $T_f = \frac{50-35}{2} = 7.5\ °C = 280.5\ K$ $Pr = 0.71$ $\mu = 1.79\times10^{-5}$

$\rho = \frac{1.0132\times10^5}{(287)(280.5)} = 1.259$ $k = 0.0247$ $Re = \frac{(1.259)(13)(0.5)}{1.79\times10^{-5}} = 457000$

$C = 0.266$ $n = 0.805$

$h = \frac{0.0247}{0.5}(0.0266)(457000)^{0.805}(0.71)^{1/3} = 42.21\ W/m^2\cdot°C$

$q/L = (42.21)\pi(0.5)(50+35) = 5636\ W/m$

6-53 $T_f = \frac{50+27}{2} = 38.5\ °C = 311.5\ K$ $\nu = 17.74\times10^{-6}$

$k = 0.02711$ $Pr = 0.7$ $Re = \frac{(20)(0.04)}{17.74\times10^{-6}} = 4.51\times10^4$

circ. tube: $h = \frac{0.02711}{0.04}(0.0266)(4.51\times10^4)^{0.805}(0.7)^{1/3} = 89.32\ W/m^2\cdot°C$

Sq tube: $h = \frac{0.02711}{0.04}(0.102)(4.51\times10^4)^{0.675}(0.7)^{1/3} = 85.04\ W/m^2\cdot°C$

$q/L\ (circ) = (89.32)\pi(0.04)(50-27) = 258.1\ W/m$

$q/L\ (sq.) = (85.04)(4)(0.04)(50-27) = 312.9\ W/m$

6-54 $T_f = \frac{200+50}{2} = 125°C = 398\ K$ $\nu = 14.35 \times 10^{-6}$ $k = 0.0246$

$Pr = 0.74$ $Re = \frac{u_\infty d}{\nu} = \frac{(40)(0.03)}{14.35 \times 10^{-6}} = 83624$ $C = 0.0266$ $n = 0.805$

$$h = \frac{0.0246}{0.03}(0.0266)(83624)^{0.805}(0.74)^{1/3} = 181\ W/m^2 \cdot °C$$

$$q/L = h\pi d(T_w - T_\infty) = (181)\pi(0.03)(200-50) = 814\ W/m$$

Compare Churchill Equation

$$Nu_d = 0.3 + \frac{(0.62)(83624)^{1/2}(0.74)^{1/3}}{[1 + (0.4/0.74)^{2/3}]^{1/4}}\left[1 + \left(\frac{83624}{282000}\right)^{1/2}\right] = 220.8$$

$$h = \frac{(220.8)(0.0246)}{0.03} = 181\ W/m^2 \cdot °C$$

Very close check. No need to use more complicated relation.

6-55 $T_f = \frac{100+20}{2} = 60°C = 333\ K$ $\mu = 216 \times 10^{-7}$ $k = 0.159$

$Pr = 0.7$ $\rho = \frac{150 \times 10^3}{(2078)(333)} = 0.217$ $Re = \frac{(0.217)(50)(0.3)}{216 \times 10^{-7}} = 1.5 \times 10^5$

$C = 0.0266$ $n = 0.805$

$$h = \frac{0.159}{0.3}(0.0266)(1.5 \times 10^5)^{0.805}(0.7)^{1/3} = 184.3\ W/m^2 \cdot °C$$

$$q = h\pi dL(T_w - T_\infty) = (184.2)\pi(0.3)(6)(100-20) = 83360\ W$$

<u>Churchill Equation</u>

$$Nu_d = 0.3 + \frac{(0.62)(1.5 \times 10^5)^{1/2}(0.7)^{1/3}}{[1 + (0.4/0.7)^{2/3}]^{1/4}}\left[1 + \left(\frac{1.5 \times 10^5}{282000}\right)^{5/8}\right]^{4/5}$$

$$= 323.3$$

$h = (323.3)(0.159)/0.3 = 171.4\ W/m^2 \cdot °C$ or about 7% less.

6-56 $T_f = \frac{300+30}{2} = 115°C = 388\ K$ $\nu = 13.62\times10^{-6}$ $k = 0.0236$

$Pr = 0.742$ $Re = \frac{(0.25)(0.0254)(35)}{13.62\times10^{-6}} = 16318$ $c = 0.193$ $n = 0.618$

$$h = \frac{0.0236}{(0.25)(0.0254)}(0.193)(16318)^{0.618}(0.742)^{1/3} = 260.6\ W/m^2\cdot°C$$

$$q = h\pi dL(T_w - T_\infty) = (260.5)\pi(0.0254)(0.25)(4.5)(300-30)$$

$$= 6316\ W$$

6-57 $T_f = \frac{300+400}{2} = 350\ K$ $\nu = 11.19\times10^{-6}$ $k = 0.02047$

$Pr = 0.755$ $Re = \frac{(50)(0.2)}{11.19\times10^{-6}} = 8.94\times10^{-5}$

Churchill Equation:

$$Nu_d = 0.3 + \frac{(0.62)(8.94\times10^{-5})^{1/2}(0.755)^{1/3}}{\left[1+(0.4/0.755)^{2/3}\right]^{1/4}}\left[1+\left(\frac{8.94\times10^{5}}{282000}\right)^{5/8}\right]^{4/5}$$

$$= 1151$$

$$h = \frac{(1151)(0.0247)}{0.2} = 117.8\ W/m^2\cdot°C$$

$$q/L = h\pi d(T_w - T_\infty) = (117.8)\pi(0.2)(400-300) = 7401\ W/m$$

6-58 $T_f = \frac{20+85}{2} = 52.5°C = 325.5\ K$ $\mu = 1.96\times10^{-5}$ $k = 0.0281$

$Pr = 0.7$ $\rho = \frac{(0.6)(1.01\times10^5)}{(287)(325.5)} = 0.651\ kg/m^3$ $c = 0.102$ $n = 0.675$

$$Re = \frac{(0.651)(10)(0.04)}{1.96\times10^{-5}} = 13281$$

$$h = \frac{0.0281}{0.04}(0.102)(13281)^{0.675}(0.7)^{1/3} = 38.62\ W/m^2\cdot°C$$

$$q/L = h\pi d(T_w - T_\infty) = (38.62)\pi(0.04)(85-20) = 315.4\ W/m$$

6-59 @ 38°C $\rho = 993$ $\mu = 6.82\times10^{-4}$ $k = 0.63$ $Pr = 4.53$

@ 93°C $\mu = 3.06\times10^{-4}$ $Re_\infty = \dfrac{(993)(0.003)(6)}{6.82\times10^{-4}} = 26208$

Eq. 6-20 $h = \dfrac{0.63}{0.003}\left[1.2 + 0.53(26208)^{0.54}\right]\left(\dfrac{6.82}{3.06}\right)^{0.25}(4.53)^{0.3}$

$h = 52520\ \text{W/m}^2\cdot{}^\circ\text{C}$

$q = (52520)\,4\pi(0.0015)^2(93-38) = 81.67\ \text{W}$

6-61 $T_\infty = 293\ \text{K}$ $\nu = 15.96$ $k = 0.026$ $Pr = 0.71$ $T_w = 313\ \text{K}$

$\nu = 17.86$ $Re = \dfrac{(6)(4)}{15.96\times10^{-6}} = 1.5\times10^6$

$h = \dfrac{0.026}{4.0}\left\{2 + \left[(0.4)(1.5\times10^6)^{1/2} + (0.06)(1.5\times10^6)^{2/3}\right]\times(0.71)^{0.4}\left(\dfrac{15.96}{17.86}\right)^{\frac{1}{4}}\right\}$

$= 7.045\ \text{W/m}^2\cdot{}^\circ\text{C}$

$q = (7.045)\,4\pi(2)^2(40-20) = 7082\ \text{W}$

6-62 At $T_\infty = 20^\circ\text{C}$ $\mu = 9.75\times10^{-4}$ $k = 0.6$ $Pr = 6.7$ $\rho = 997$

At $T_w = 90^\circ\text{C}$ $\mu_w = 3.2\times10^{-4}$ $Nu\,Pr^{-0.3}\left(\dfrac{\mu_w}{\mu}\right)^{0.25} = 1.2 + 0.53\,Re_d^{0.54}$

$Re = \dfrac{(997)(3.5)(0.03)}{9.75\times10^{-4}} = 1.073\times10^5$

$h = \dfrac{0.6}{0.03}\left[1.2 + 0.53(1.073\times10^5)^{0.54}\right]\left(\dfrac{9.75}{3.2}\right)^{0.25}(6.7)^{0.3}$

$= 12957\ \text{W/m}^2\cdot{}^\circ\text{C}$

$q = h\,4\pi r^2(T_w - T_\infty) = (12957)(4)\pi(0.015)^2(90-20) = 2564\ \text{W}$

6-63 $T_f = \dfrac{220+20}{2} = 120°C = 393 K \quad \nu = 2.256 \times 10^{-5}$

$k = 0.0331 \quad Pr = 0.69 \quad Re = \dfrac{u_\infty d}{\nu} = \dfrac{(20)(0.006)}{2.256 \times 10^{-5}} = 5319$

$$\frac{hd}{k} = 0.37\, Re^{0.6}$$

$$h = \frac{0.0331}{0.006}(0.37)(5319)^{0.6} = 351.1 \text{ W/m}^2\cdot°C$$

$$q = h\,4\pi r^2 (T_w - T_\infty) = (351.1)(4)\pi(0.003)^2(220-20) = 7.94 \text{ W}$$

6-64 $T_f = \dfrac{200+30}{2} = 115°C = 388 K \quad \mu = 2.235 \times 10^{-5}$

$k = 0.0328 \quad Pr = 0.69 \quad \rho = \dfrac{(3)(1.01\times 10^5)}{(287)(388)} = 2.73 \text{ kg/m}^3$

$$Re = \frac{(2.73)(100)(1)}{2.235\times 10^{-5}} = 1.22 \times 10^7$$

$$\bar{h} = \frac{k}{L} Pr^{1/3}\left[0.037\, Re^{0.8} - 850\right]$$

$$= \frac{0.0328}{1}(0.69)^{1/3}\left[(0.037)(1.22\times 10^7)^{0.8} - 850\right] = 476 \text{ W/m}^2\cdot°C$$

$$q = \bar{h} A (T_w - T_\infty) = (476)(1)^2(200-30) = 8.1 \times 10^4 \text{ W}$$

6-65 @ 38°C $\rho = \dfrac{3.5\times 10^6}{(287)(311)} = 39.2 \text{ kg/m}^3$

$\dot{m} = (39.2)(9)(1.5)(20)(0.025) = 264.7 \text{ kg/sec} \quad T_f \approx \dfrac{200+38}{2} = 119°C = 392 K$

$\rho_f = \dfrac{3.5\times 10^6}{(287)(392)} = 31.1 \text{ kg/m}^3 \quad \mu_f = 2.25\times 10^{-5} \quad k_f = 0.0331 \quad Pr_f = 0.69$

$c_p \approx 1010 \text{ J/kg}\cdot°C \quad u_{max} = u_\infty\left(\dfrac{S_n}{S_n - d}\right) = 9\left(\dfrac{2.5}{2.5-1.25}\right) = 18 \text{ m/sec}$

$\dfrac{S_n}{d} = 2 \quad S_p/d = 3 \quad Re_{max} = \dfrac{(31.1)(18)(0.0125)}{2.25\times 10^{-5}} = 311\,000$

$C = 0.488 \quad n = 0.562 \quad h = \dfrac{0.331}{0.0125}(0.488)(311000)^{0.562}(0.69)^{1/3} = 1395 \text{ W/m}^2\cdot°C$

$$q = \dot{m} c_p \Delta T = h A (T_w - T_{avg})$$

$$q = (264.7)(1010)(T_e - 38) = 1395(400)\pi(0.0125)(1.5)\left(200 - \frac{T_e - 38}{2}\right)$$

$T_e = 56.76 °C \quad q = 5.016 \text{ MW}$

6-66 $\frac{S_n}{d} = \frac{1.9}{0.633} = 3$ $S_p/d = 3$ $u_\infty = 4.5$ m/sec $T_\infty = 293$ K

$p = 1$ atm $\rho_\infty = 1.18$ kg/m³ $T_f = \frac{90+20}{2} = 55$ °C $= 328$ K $\rho = 1.077$

$\mu = 2.034 \times 10^{-5}$ $k = 0.0284$ $Pr = 0.7$ $c_p = 1007$ $u_{max} = u_\infty \left(\frac{S_n}{S_n - d}\right)$

$u_{max} = (4.5)(15) = 6.75$ m/sec $Re = \frac{(1.077)(6.75)(0.00633)}{2.034 \times 10^{-5}} = 2262$

$C = 0.317$ $n = 0.608$ $h = \frac{0.0284}{0.00633}(0.317)(2262)^{0.608}(0.7)^{1/3}(0.94) = 131.4\ \frac{W}{m^2 \cdot °C}$

$q/L = (131.4)(6)(50)\pi(0.00633)\left(90 - \frac{T_e}{2} - \frac{20}{2}\right) = (1.18)(50)(0.019)(4.5)(1007)(T_e - 20)$

$T_e = 30.03$ °C $q/L = 50944$ W/m

$Re_{max} = \frac{(7.27)(0.00633)}{2.034 \times 10^{-5}} = 2262$ $G_{max} = (1.077)(6.75) = 7.27$ kg/m²·sec

$= 5361$ lbm/ft²-hr

$\rho = 0.0737$ lbm/ft³

$f' = \left\{0.044 + \frac{(0.08)(3)}{(2)^{0.807}}\right\}(2262)^{-0.15} = 0.0569$ @90°C $\mu_w = 2.17 \times 10^{-5}$

@ 20°C $\mu = 1.98 \times 10^{-5}$ $\Delta p = \frac{(0.0569)(5361)^2(6)}{(0.0737)(2.09 \times 10^8)}\left(\frac{2.17}{1.98}\right)^{0.14} = 0.645$ lbf/ft²

$\Delta p = 30.89$ N/m²

6-69 $S_n/d = S_p/d = 1.5$ $T_f = \frac{150+35}{2} = 92.5$ °C $= 365.5$ K $\rho = 5.226$

$\rho_f = \frac{(3)(1.0132 \times 10^5)}{(189)(365.5)} = 4.404$ @ 35°C = 308 K $c_p = 921$ $\mu_f = 17.82 \times 10^{-6}$

$Pr_f = 0.75$ $k_f = 0.0218$ $u_{max} = (5)\left(\frac{1.875}{1.875 - 1.25}\right) = 15$ m/s $C = 0.278$

$Re_{max} = \frac{(4.404)(15)(0.0125)}{17.82 \times 10^{-6}} = 46339$ $n = 0.620$

$h = \frac{0.0218}{0.0125}(0.278)(46339)^{0.620}(0.75)^{1/3} = 344.2$ W/m²·°C

$q = (344.2)(100)\pi(0.0125)(0.6)\left(150 - \frac{T_e - 35}{2}\right) = (5.226)(10)(0.6)(0.01875)(5)(921) \cdot (T_e - 35)$

$T_e = 62.35$ °C $q = 74059$ W

6-70

$T_f = 55°C = 328\text{ K}$ $\quad u_\infty = 12\text{ m/s}$ $\quad Pr = 0.7$

$k = 0.0284$ $\quad \nu = 19.04 \times 10^{-6}$ $\quad d = 2.5\text{ cm}$

$u_{max} = (12)\dfrac{5}{5-2.5} = 24\text{ m/s}$ $\quad Re_{max} = \dfrac{(24)(0.025)}{19.04 \times 10^{-6}} = 3.15 \times 10^4$

$c = 0.112$ $\quad n = 0.702$ $\quad \bar{h} = \dfrac{0.0284}{0.025}(0.112)(3.15 \times 10^4)^{0.702}(0.7)^{1/3}(0.96)$

$h = 156\text{ W/m}^2\cdot°C$ $\quad q/L = (156)\pi(0.025)(15)(7)(90-20) = 9 \times 10^4\text{ W/m}$

$$f' = \left\{0.044 + \frac{(0.08)(1.5)}{1}\right\}(3.15 \times 10^4)^{-0.15} = 3.468 \times 10^{-2}$$

$p = \dfrac{1.013 \times 10^5}{(287)(328)} = 1.076$ $\quad G_{max} = (1.076)(24) = 25.83$ $\quad \rho_\infty = 1.205$

$$\Delta P = \frac{(2)(3.468 \times 10^{-2})(25.83)^2(7)}{1.205}\left(\frac{2.12}{1.96}\right)^{0.14} = 272\text{ N/m}^2 = 0.033\text{ psi}$$

6-71

$T_f = \dfrac{350+300}{2} = 325\text{ K}$ $\quad \nu = 18.23 \times 10^{-6}$ $\quad k = 0.0281$

$Pr = 0.7$ $\quad u_{max} = (10)(5/2) = 20\text{ m/s}$ $\quad Re_{max} = \dfrac{(20)(0.025)}{18.23 \times 10^{-6}} = 27427$

$S_P/d - S_n/d = 5/2.5 = 2.0$ $\quad C = 0.254$ $\quad n = 0.632$

$$h = \frac{0.0281}{0.025}(0.254)(27427)^{0.632}(0.7)^{1/3} = 161.8\text{ W/m}^2\cdot°C$$

correction for 5 tubes = 0.92

$h = (161.8)(0.92) = 148.8\text{ W/m}^2\cdot°C$

Zukauskas $\quad c = 0.27$ $\quad n = 0.63$ and $h = 166.9$

correction for 5 tubes = 0.92

$h = (166.9)(0.92) = 153.5\text{ W/m}^2\cdot°C$ only 3% difference

6-72

$$T_f = \frac{90+20}{2} = 55°C = 328\ K \qquad \rho = \frac{1.01 \times 10^5}{(287)(328)} = 1.076$$

$$\mu = 2.035 \times 10^{-5} \qquad k = 0.0284 \qquad Pr = 0.7$$

$$Re = \frac{(1.076)(0.05)(15)}{2.035 \times 10^{-5}} = 39656 \qquad C = 0.102 \qquad n = 0.675$$

$$h = \frac{0.0284}{0.05}(0.102)(39656)^{0.675}(0.7)^{1/3} = 65.34\ W/m^2 \cdot °C$$

$$q/L = (65.34)(4)(0.05)(90-20) = 915\ W/m$$

6-73 $T_f = \frac{870+20}{2} = 445°C = 718\ K$ $Pr = 0.685$ $\mu = 3.386 \times 10^{-5}$

$\rho = \frac{1.01 \times 10^5}{(287)(718)} = 0.492$ $k = 0.053$ $Re = \frac{(0.492)(2)(0.006)}{3.386 \times 10^{-5}} = 174.37$

$C = 0.228$ $n = 0.731$ (Approximate values)

$$h = \frac{0.053}{0.006}(0.228)(174.37)^{0.731}(0.685)^{1/3} = 77.2\ W/m^2 \cdot °C$$

$q = (77.2)(7)(2)(0.006)(0.35)(870-20) = 1929\ W$ this value is probably low because of the range of Table 6-2

6-74 $T_f = \frac{50+30}{2} = 40°C = 313\ K$ $\mu = 2.007 \times 10^{-5}$ $k = 0.0272$

$Pr = 0.7$ $\rho = \frac{1.01 \times 10^5}{(287)(313)} = 1.128$ $Re = \frac{(1.128)(6)(0.3)}{2.007 \times 10^{-5}} = 101165$

$C = 0.102$ $n = 0.675$

$$h = \frac{0.0272}{0.3}(0.102)(101165)^{0.675}(0.7)^{1/3} = 19.63\ W/m^2 \cdot °C$$

$q/L = (19.63)(4)(0.3)(50-30) = 471\ W/m$

If velocity halved:

$h_{1/2} = (19.63)\left(\frac{1}{2}\right)^{0.675} = 12.29$ q reduced by 37.3%

6-77 Bismuth @ 400°C $\rho = 9950\ kg/m^3$ $\mu = 1.47 \times 10^{-3}\ kg/m\ sec$

$c_p = 0.15\ kJ/kg\ °C$ $k = 16.3$ $Pr = 0.013$ $Re\,Pr = 450.4$

$$Re = \frac{(0.025)(1)(4)}{\pi(0.025)^2(1.47 \times 10^{-3})} = 34645$$

$$h = \frac{16.3}{0.025}\left[5.0 + 0.025(450.4)^{0.8}\right] = 5423\ W/m^2 \cdot °C$$

$$q = (1)(150)(T_e - 400) = (5423)\pi(0.025)(0.6)\left(450 - \frac{T_e - 400}{2}\right)$$

$T_e = 445.98\ °C$ $q = 6898\ W$

6-78 Sodium at 134.5 °C $\rho = 900$ $\mu = 0.45 \times 10^{-3}$ $c_p = 1345$

$k = 82$ $Pr = 0.08$ $q = (2.3)(1345)(149-120) = 151.6$ kW

$$Re = \frac{(0.025)(2.3)(4)}{(0.025)^2(0.45\times10^{-3})} = 260,300 \qquad Re\,Pr = 2082$$

$$h = \frac{8.2}{0.025}\left[4.82 + 0.01825\,(2082)^{0.8277}\right] = 49500 \ \mathrm{W/m^2 \cdot °C}$$

$$151600 = (49500)\pi(0.025)L(200-149) \qquad L = 0.765 \text{ m}$$

6-79 $h = \frac{k}{x}\,0.53\,Re_x^{1/2}\,Pr^{1/2} = Cx^{-1/2}$

$$\bar{h} = 2h_{x=L} \qquad \frac{\bar{h}L}{k} = \overline{Nu}_L = 1.06\,Re_L^{1/2}\,Pr^{1/2}$$

6-80

Air $T_f = \frac{93+15}{2} = 54°C = 327$ K $\mu = 2.03\times10^{-5}$ $k = 0.028$ $Pr = 0.7$

$$\rho = \frac{1.01\times10^5}{(287)(327)} = 1.08 \ \mathrm{kg/m^3} \qquad Re = \frac{(1.08)(15)(0.0516)}{2.03\times10^{-5}} = 41163$$

$$h_a = \frac{0.028}{0.0516}(0.0266)(41163)^{0.805}(0.7)^{1/3} = 66.44 \ \mathrm{W/m^2 \cdot °C}$$

Water @ 93°C $\mu = 3.06\times10^{-4}$ $k = 0.678$ $Pr = 1.90$

$$Re = \frac{(0.05)(0.8)(4)}{\pi(0.05)^2(3.06\times10^{-4})} = 66574$$

$$h_w = \frac{(0.678)(0.023)(66574)^{0.8}(1.9)^{0.4}}{0.5} = 2911 \ \mathrm{W/m^2 \cdot °C}$$

$$\frac{q}{L} = \frac{(93-15)\pi(1)}{\frac{1}{(66.44)(0.0516)} + \frac{1}{(2911)(0.05)}} = 821 \ \mathrm{W/m}$$

6-81 $G = \dot{m}/A = \frac{0.035}{\pi(0.0625)^2} = 2.852$

Properties at 325 K

$\mu = 1.96 \times 10^{-5}$ kg/m·s

$k = 0.02813$ W/m·°C

$Pr = 0.7$

$$Re = \frac{dG}{\mu} = \frac{(0.0125)(2.852)}{1.96 \times 10^{-5}} = 1819 \text{ Laminar}$$

$$Re\,Pr\,\frac{d}{L} = (1819)(0.7)(0.0125/12) = 1.33$$

$$h = \frac{0.02813}{0.0125}\left[3.66 + \frac{(0.0668)(1.33)}{1 + (0.04)(1.33)^{2/3}}\right]$$

$$= 8.32 \text{ W/m}^2 \cdot °\text{C}$$

$$q = \dot{m} c_p \Delta T_b = hA(T_w - \bar{T}_b)$$

$$= (0.035)(1005)(T_e - 300) = (8.32)\pi(0.0125)(12)(350 - 150 - T_e/2)$$

$$T_e = 305.3\text{ K}$$

$$q = (0.035)(1005)(305.3 - 300) = 186 \text{ W}$$

6-82

$$T_f = \frac{77+20}{2} = 48.5°C = 321.5\ K$$

Properties $\nu = 17.87 \times 10^{-6}$ $k = 0.028$ $Pr = 0.7$

$$Re = \frac{u_\infty d}{\nu} = \frac{(25)(0.05)}{17.87\times10^{-6}} = 69950$$

$C = 0.0266$ $n = 0.805$

$$h = \frac{0.028}{0.05}(0.0266)(69950)^{0.805}(0.7)^{1/3}$$

$$= 105.1\ W/m^2 \cdot °C$$

$$q/L = h\pi d(T_w - T_\infty) = (105.1)\pi(0.05)(77-20)$$

$$= 941\ W/m$$

6-83

Properties at 20°C,

$\nu = 0.0009\ m^2/s$ $Pr = 10400$

$k = 0.145\ W/m \cdot °C$, $\rho = 888\ kg/m^3$ $c_p = 1880\ J/kg \cdot °C$

$$G = \frac{(0.4)(4)}{\pi(0.0254)^2} = 789$$

$$Re\ \frac{dG}{\mu} = \frac{(0.0254)(789)}{(888)(0.0009)} = 25.1$$

$$Re\,Pr\,d/L = (25.1)(10400)(0.0254)/8 = 828$$

From Fig. 6-5 $\overline{Nu} = 16$, $h = (16)(0.145)/0.0254 = 913\ W/m^2 \cdot °C$

$$q = \pi d L h(T_w - \bar{T}_b) = \dot{m} c_p \Delta T_b$$

$$= \pi(0.0254)(8)(91.3)(80 - 10 - Te/2) = (0.4)(1880)(Te - 20)$$

$$19120 = 781\ Te$$

$$Te = 24.4\ °C$$

6-84 at 325 °K, properties, $k = 0.02814$ W/m·°C

$\nu = 18.23 \times 10^{-6}$ m²/s, $Pr = 0.7$, $\mu = 1.96 \times 10^{-5}$

$\rho = 1.09$ kg/m³ $\quad G = \frac{0.2}{(0.1)(0.2)} = 10$

$$D_H = \frac{(4)(0.1)(0.2)}{(2)(0.1+0.2)} = 0.133$$

$$Re = \frac{D_H G}{\mu} = \frac{(0.133)(10)}{1.96 \times 10^{-5}} = 68030$$

$$h = \frac{0.028}{0.133}(0.023)(68030)^{0.8}(0.7)^{0.4} = 30.8 \text{ W/m}^2\cdot{}^\circ\text{C}$$

$$q = \dot{m} c_p \Delta T_b = hA(T_w - \bar{T}_b)$$

$$= (0.2)(1005)(Te - 300) = 30.8\,(2)(0.1+0.2)(2.5)(400 - 150 - Te/2)$$

$$224\,T_e = 71850$$

$$Te = 320.6 \text{ K}$$

$q = 4144$ W

6-85 constant temperature tube

$Pr = 0.7 \quad d = 1.5$ mm $\quad k = 0.02624$ w/m·°C

$Re\,Pr = (1200)(0.7) = 840 \quad Re\,Pr\,d = 1.26$

Use Fig. 6-5

x, m	$Gz^{-1} = x/d\,(Re\,Pr)$	$\overline{Nu}_d$	$\bar{h}$, W/m²·°C
0.01	7.94×10^{-3}	7.9	138
0.1	7.94×10^{-2}	4.5	78.8
0.2	0.159	4.0	70
1.0	0.793	3.66	64.1

6-86

at 15°C $C_p = 4180$, $\rho = 999$, $\mu = 1.12 \times 10^{-3}$

$k = 0.595$ W/m·°C $Pr = 7.88$

$Re = 50{,}000$

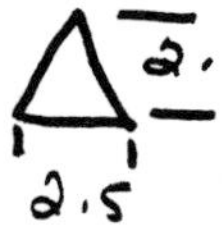

$2.5 \sin 60° = 2.165$

$$A_c = (\tfrac{1}{2})(2.165)(2.5) = 2.71\ cm^2$$

$$D_H = \frac{(4)(2.71)}{(3)(2.5)} = 1.443\ cm = 0.01443\ m$$

$$h = \frac{0.595}{0.01443}(0.023)(50{,}000)^{0.8}(7.88)^{0.4} = 12438\ W/m^2\text{-}°C$$

$$q = \dot{m}\, c_p \Delta T_b = hA(T_w - \bar{T}_b), \quad Re = \frac{D_H \frac{\dot{m}}{A_c}}{\mu}$$

$$\dot{m} = \frac{(50{,}000)(1.12 \times 10^{-3})(2.71)(10^{-4})}{0.01443} = 1.052\ kg/sec$$

$$q = (1.052)(4180)(10) = (12438)(3)(0.025)L(15) = 43961\ W$$

$L = 3.14$ m

Checking Fig. 6-6 fully developed flow in present

6-87 $Re = 10^4$ $k = 0.026\ W/m\text{-}°C$, $Pr = 0.7$

Geom	C	n	Nu_d	h	A	$q \sim hA$
○	0.193	0.618	50.8	$50.8 \frac{k}{d}$	πd	$159.6 k$
□	0.102	0.675	45.4	$45.4 \frac{k}{d}$	$4d$	$181.6 k$

6-88 $Pr = 0.7$ $k = 0.026$

Eq. (6-25) $Nu = (0.37)(50{,}000)^{0.6} = 244$

Eq. (6-26) $Nu = 2 + [0.25 + 3\times10^{-4}(50000)^{1.6}]^{1/2} = 101.5$

Eq. (6-30) $Nu = 2 + [0.4(50000)^{1/2} + 0.06(50000)^{2/3}](0.7)^{0.4} = 150.1$

6-89

at 10°C, $Pr = 9.4$ $\mu = 1.31\times10^{-3}$ $k = 0.585$ $\rho = 999$

at 60°C, $\mu_w = 1.12\times10^{-3}$

$$Re = \frac{(999)(4)(0.025)}{1.31\times10^{-3}} = 76260$$

Use Eq. (6-29)

$$Nu = (9.4)^{0.3}\left(\frac{1.31}{1.12}\right)^{0.25}\left[1.2 + (0.53)(76260)^{0.54}\right] = 470$$

$$h = \frac{(470)(0.585)}{0.025} = 10990\ W/m^2\text{-}°C$$

$$q = hA(T_w - T_\infty) = (10990)\,4\pi\left(\frac{0.025}{2}\right)^2(60-10) = 1079\ W$$

6-90 $U_{max} = (6)\left(\frac{2}{2-1.5}\right) = 24$ m/s

Take $T_f = 325$ K $\rho = 1.09$ $\mu = 1.96 \times 10^{-5}$ $\rho_{inlet} = 1.18$

$k = 0.02814$ $Pr = 0.7$ $c_p = 1005$

$$Re = \frac{\rho u_\infty d}{\mu} = \frac{(1.09)(24)(0.015)}{1.96 \times 10^{-5}} = 20020$$

$$\frac{S_n}{d} = \frac{S_p}{d} = \frac{2}{1.5} = 1.33$$

From Table 6.4, by interpolation

$n = 0.597$, $C = 0.364$

$$h = \frac{0.02814}{0.015}(0.364)(20020)^{0.597}(0.7)^{1/3} = 224 \text{ W/m}^2\text{-°C}$$

$$q = hA(T_w - \bar{T}_\infty) = \dot{m} c_p \Delta T_\infty$$

$A = (144)\pi(0.015)(1.0) = 6.786$ m^2

$\dot{m} = (1.18)(6)(12)(1.0)(0.02) = 1.69$ kg/s

$(224)(6.786)(350-150-Te/2) = (1.69)(1005)(Te-300)$

$8.14 \times 105 = 2458\, Te$

$Te = 330.9$ °K

$q = 52500$ W

6-91 $k = 0.026\ W/m \cdot °C \quad Pr = 0.7$

a) $\bar{h} = \frac{0.026}{0.1}(0.0266)(50\,000)^{0.805}(0.7)^{1/3} = 37.2\ W/m^2\text{-}°C$

b) $\bar{h} = \frac{0.026}{0.1}(0.023)(50000)^{0.8}(0.7)^{0.4} = 29.8\ W/m^2\text{-}°C$

c) $\bar{h} = \frac{0.026}{0.1}(0.664)(50000)^{1/2}(0.7)^{1/3} = 34.3\ W/m^2\text{-}°C$

6-92 $h = \frac{k}{d}(0.023)\left[\frac{(m)(x)\,d}{\pi d^2 \mu}\right]^{0.8} Pr^{0.4}$

$Pr \sim$ Const $= 0.7$

$h = c \quad k\mu^{-0.8}$

T (K)	k	$\mu \times 10^5$	$k\mu^{-0.8}$
300	0.02624	1.8462	161
400	0.03365	2.286	174
500	0.04038	2.671	184
800	0.05779	3.625	206

Comment:

h varies approximately as $[T(K)]^{0.25}$ for constant mass flow.

6-93 Helium – Same relation as Prob. 6-92

T(K)	k	$\mu \times 10^5$	$k\mu^{-0.8}$
255	0.1357	1.817	842
477	0.197	2.75	877
700	0.251	3.475	927

Comment:

h varies approximately as $[T(K)]^{0.1}$ for constant mass flow, i.e., not strongly dependent on temperature

7-3 Show that $\beta = 1/T$ for an ideal Gas.

$$\beta = \frac{1}{V}\left(\frac{\partial V}{\partial T}\right)_p \qquad pV = nRT \qquad V = \frac{nRT}{p} \qquad \left(\frac{\partial V}{\partial T}\right)_p = \frac{nR}{p}$$

$$\therefore \quad \beta = \frac{1}{\frac{nRT}{p}}\left(\frac{nR}{p}\right) = \frac{1}{T}$$

7-4 $T_f = \frac{65+15}{2} = 40°C = 313 K \qquad \rho = \frac{1.01 \times 10^5}{(287)(313)} = 1.128$

$\mu = 2.006 \times 10^{-5} \qquad \beta = \frac{1}{313} = 3.195 \times 10^{-3} \qquad k = 0.0272 \qquad Pr = 0.7$

$x = 1\ ft = 0.3048\ m \qquad c_p = 1006$

$$Gr\,Pr = \frac{(9.806)(1006)(3.195\times10^{-3})(65-15)(1.128)^2(0.3048)^3}{(2.006\times10^{-5})(0.0272)} = 1.041 \times 10^8$$

$$h = \frac{0.0272}{0.3048}(0.59)(1.041\times10^8)^{1/4} = 5.318\ W/m^2\cdot°C$$

$$q = (5.318)(0.3048)^2(65-15) = 24.7\ W$$

$$u_{max} = \left(\frac{4}{27}\right)(5.17)\left(\frac{2.006\times10^{-5}}{1.128}\right)\left(\frac{20}{21}+0.7\right)^{-\frac{1}{2}} \times$$
$$\left[\frac{(9.806)(3.195\times10^{-3})(50)(0.3048)(1.128)^2}{(2.006\times10^{-5})^2}\right]^{1/2} = 0.412\ m/sec$$

7-6 $\frac{u}{u_x} = \left(\frac{y}{\delta}\right)\left(1-\frac{y}{\delta}\right)^2$

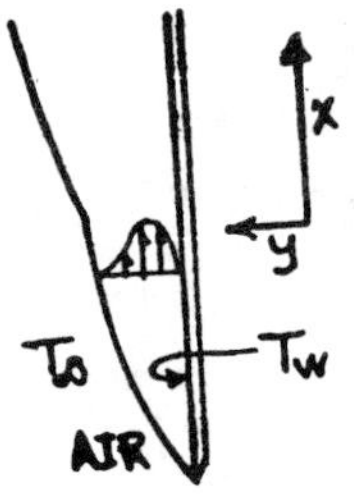

$$\frac{du}{dy} = u_x\left[\frac{y}{\delta}\left(-\frac{2}{\delta}\right)\left(1-\frac{y}{\delta}\right) + \frac{1}{\delta}\left(1-\frac{y}{\delta}\right)^2\right] = 0$$

$$y = \frac{\frac{4}{\delta} \pm \sqrt{(16/\delta^2) - 4(3/\delta^2)}}{6/\delta^2}$$

$y = \delta \qquad y = \delta/3 \qquad$ Max u occurs at $\delta/3$

$$u_{max} = u_x\left(\frac{y}{3y}\right)\left(1-\frac{y}{3y}\right)^2 = \frac{4}{27}u_x = \frac{4}{27}C_1 x^m$$

7-8 $L = 30\text{ cm}$ $T_f = 57.5°C = 330.5\text{ K}$ $\nu = 19.23 \times 10^{-6}$

$$Gr_L = \frac{(9.8)(1/330.5)(100-15)(0.3)^3}{19.23 \times 10^{-6}} = 1.84 \times 10^8$$

$$D = \frac{(0.3)(35)}{(1.84 \times 10^8)^{1/4}} = 9.01 \times 10^{-2}\text{ m} = 9.01\text{ cm}$$

7-9 $T_f = \frac{400+25}{2} = 212.5°C = 485.5\text{ K}$ $\nu = 36.1 \times 10^{-6}$ $k = 0.0394$

$Pr = 0.68$ $$GrPr = \frac{(9.8)(1/485.5)(400-25)(1)^3(0.68)}{(36.1 \times 10^{-6})^2} = 3.95 \times 10^9$$

$$h = \frac{0.0394}{1}(0.1)(3.95 \times 10^9)^{1/3} = 6.22\text{ W/m}^2\cdot°C$$

$$q = hA(T_w - T_\infty) = (6.22)(1)^2(400-25) = 2334\text{ W}$$

7-10 $\delta \sim x^{1/4}$ $$\frac{\delta_{24}}{\delta_{14}} = \left(\frac{24}{14}\right)^{1/4}$$

$$\delta_{24} = (1.0)\left(\frac{24}{14}\right)^{1/4} = 1.14\text{ in}$$

7-11 $T_f = \frac{93+30}{2} = 61.5°C = 334.5\text{ K}$ $\beta = \frac{1}{334.5} = 2.99 \times 10^{-3}$

$\nu = 19.19 \times 10^{-6}$ $k = 0.0289$ $Pr = 0.7$

$$GrPr = \frac{(9.806)(2.99\times10^{-3})(93-30)(1.8)^3}{(19.19 \times 10^{-6})^2}(0.7) = 2.05 \times 10^{10}$$

$$h = \frac{0.0289}{1.8}(0.1)(2.05 \times 10^{10})^{1/3} = 4.394\text{ W/m}^2\cdot°C$$

$$q = (4.394)\pi(0.075)(1.8)(93-30) = 117.4\text{ W}$$

7-12 $q/A = 1100 - 95 = 1005\text{ W/m}^2$ @ 300 K $\beta = 3.33 \times 10^{-3}$

$\nu = 15.68 \times 10^{-6}$ $Pr = 0.7$ $k = 0.02624$

$$Gr^* = \frac{(9.806)(3.33\times10^{-3})(1005)(6)^4}{(15.68 \times 10^{-6})^2(0.02624)} = 6.59 \times 10^{15}$$

$$h = \frac{0.02624}{6}(0.17)(6.59 \times 10^{15})^{1/3} = 6.7\text{ W/m}^2\cdot°C$$

$$\Delta T = \frac{1005}{6.7} = 150°C$$ $T_{wall} \approx 150 + 20 = 170°C$

This does not take into account radiation which would lower the temperature substantially.

7-13 $\Delta T = 75 - 68 = 7°F = 3.888\ °C$

$h = (0.95)(3.888)^{1/3} = 1.494\ W/m^2 \cdot °C$

$q = (1.494)\pi(0.3048)(1.8288)(3.888) = 10.17\ W$

$q = 34.71$ Btu/hr.

7-14 $q_w = \frac{20}{(0.3)^2} = 222.2\ W/m^2$ Take properties @ 300 K,

$k = 0.02624$ $\beta = 3.33 \times 10^{-3}$ $\nu = 15.68 \times 10^{-6}$ $Pr = 0.71$

$$Gr^* = \frac{(9.806)(3.33 \times 10^{-3})(222.2)(0.15)^4}{(0.02624)(15.68 \times 10^{-6})^2} \ @\ 15\text{ cm} = 5.694 \times 10^8$$

$Gr^* = 9.11 \times 10^9$ @ 30 cm.

@ 15 cm $h = \frac{0.02624}{0.15}(0.6)\left[(5.694 \times 10^8)(0.7)\right]^{1/5} = 5.51\ W/m^2 \cdot °C$

@ 30 cm $h = \frac{0.02624}{0.3}(0.6)\left[(9.11 \times 10^9)(0.7)\right]^{1/5} = 4.796\ W/m^2 \cdot °C$

$\bar{h} = 5/4\ h_{x=L} = (1.25)(4.796) = 6.0\ W/m^2 \cdot °C$

7-15 $T_f = \frac{120 + 65}{2} = 92.5\ °F = 307\ K$ $\nu = 16.4 \times 10^{-6}$

$k = 0.0268$ $Pr = 0.7$

$$Gr\,Pr = \frac{(9.8)(1/307)(120 - 65)(5/9)(0.3048)^3(0.7)}{(16.4 \times 10^{-6})^2} = 7.19 \times 10^7$$

$C = 0.59$ $m = 1/4$

$h = \frac{0.0268}{0.3048}(0.59)(7.19 \times 10^7)^{1/4} = 4.78\ W/m^2 \cdot °C$

$q = 2hA(T_w - T_\infty) = (2)(4.78)(0.3048)^2(120 - 65)(5/9) = 27.1\ W$

7-16 $T_f = \frac{100 + 20}{2} = 60\ °C = 333K$ $\mu = 2.16 \times 10^{-7}$ $\rho = \frac{(2)(1.01 \times 10^5)}{(333)(278)} = 0.293\ \frac{kg}{m^3}$

$k = 0.159$ $Pr = 0.7$

$$Gr\,Pr = \frac{(9.8)(1/333)(0.293)^2(100 - 20)(2)^3(0.3048)^3(0.7)}{(2.16 \times 10^{-7})^2} = 6.87 \times 10^7$$

$$h = \frac{(0.159)(0.59)(6.87 \times 10^7)^{1/4}}{2(0.3048)} = 14\ W/m^2 \cdot °C$$

$q = hA(T_w - T_\infty) = (14)(2)^2(0.3048)^2(100 - 20) = 416.5\ W$

7-17 $T_f = \frac{135+40}{2} = 87.5°F = 304\ K$ $\nu = 16.1 \times 10^{-6}$ $k = 0.0265$

$Pr = 0.7$ $Gr\,Pr = \frac{(9.8)(1/304)(135-40)(5/9)(20)^3(0.3048)^3(0.7)}{(16.1\times10^{-6})^2} = 1.04 \times 10^{12}$

$$\overline{Nu}^{1/2} = 0.825 + \frac{0.387\ Ra^{1/6}}{\left[1 + (0.492/Pr)^{9/16}\right]^{8/27}} = 33.44$$

$Nu = 1118$ $h = \frac{(1118)(0.0265)}{(20)(0.3048)} = 4.86\ W/m^2 \cdot °C$

$q = hA(T_w - T_\infty) = (4.86)(20)(4)(0.3048)^2(135-40)(5/9) = 1907\ W$

using $C = 0.1$ and $m = 1/3$

$Nu = (0.1)(1.04 \times 10^{12})^{1/3} = 1012$ ⇒ 9.5% lower

7-18 $T_f = \frac{120+70}{2} = 95°F = 308\ K$ $\nu = 16.5 \times 10^{-6}$ $k = 0.0268$ $Pr = 0.7$

$Gr\,Pr = \frac{(9.8)(1/308)(120-70)(5/9)(1)^3(0.7)}{(16.5\times10^{-6})^2} = 2.27 \times 10^9$

$$\overline{Nu}^{1/2} = 0.825 + \frac{0.387\ Ra^{1/6}}{\left[1 + (0.492/Pr)^{9/16}\right]^{8/27}} = 12.57 \qquad Nu = 158.1$$

$h = \frac{(158.1)(0.0268)}{1} = 4.24\ W/m^2 \cdot °C$

$q = hA(T_w - T_\infty) = (4.24)(1)^2(120-70)(5/9) = 117.7\ W$

using $C = 0.1$ and $m = 1/3$

$Nu = (0.1)(2.27 \times 10^9)^{1/3} = 131.3$ ⇒ 17% lower

7-19 Take Properties @ 300 K

$\nu = 15.69 \times 10^{-6}$ $k = 0.02624$ $Pr = 0.7$

$10^{12} = \frac{(9.8)(1/300)(10)\ L^3(0.7)}{(15.69\times10^{-6})^2}$ $L = 10.25\ m$

7-20 Take Properties @ 15.56 °C $k = 0.595$ $\frac{g\beta\rho^2 c_p}{\mu k} = 1.08 \times 10^{10}$

$$Gr^* Pr = \frac{(1.08 \times 10^{10})(1000)(0.25)^4}{0.595} = 7.09 \times 10^{10}$$

$$h_x = \frac{0.595}{0.25}(0.6)(7.09 \times 10^{10})^{1/5} = 211$$

$$\bar{h} = 1.25\, h_x = 264 \text{ W/m}^2 \cdot °C$$

$$\overline{\Delta T} = q_w / \bar{h} = 1000/264 = 3.79 °C$$

$$\bar{T}_w = 3.79 + 15 = 18.79 °C$$ it is very nearly isothermal.

7-21 cylinder: $h = 1.32\left(\frac{\Delta T}{d}\right)^{1/4}$

Vert. plate: $h = 1.42\left(\frac{\Delta T}{L}\right)^{1/4}$

$\frac{1}{2}$ cyl: $q = \frac{1}{2}(1.32)\left(\frac{\Delta T}{d}\right)^{1/4} \pi d\, \Delta T = 2.073\, d^{3/4} \Delta T^{5/4}$

Plate: $q = (1.42)\left(\frac{\Delta T}{L}\right)^{1/4} \frac{\pi d}{2} \Delta T = 1.992\, d^{3/4} \Delta T^{5/4}$

Very close

7-23 $T_f = \frac{70 + 20}{2} = 45 °C = 318\ K$ $\nu = 0.335 \times 10^{-6}$ $k = 0.485$

$Pr = 2.0$ $\beta = 2.45 \times 10^{-3}$

$$Gr\, Pr = \frac{(9.8)(2.45 \times 10^{-3})(70 - 20)(0.03)^3 (2)}{(0.335 \times 10^{-6})^2} = 5.78 \times 10^8$$

$C = 0.53$ $m = 1/4$

$$h = \frac{0.485}{0.3}(0.53)(5.78 \times 10^8)^{1/4} = 1328 \text{ W/m}^2 \cdot °C$$

$$q = h \pi d L (T_w - T_\infty) = (1328)\pi(0.03)(1)(70 - 20) = 6260 \text{ W}$$

7-24 $h = (1.32)\left(\frac{120 - 17}{0.075}\right)^{1/4} = 8.036 \text{ W/m}^2 \cdot °C$

$q = 100000$ Btu/hr. $= 29300$ W $= 8.036\, \pi\, (0.075) L (120 - 17)$

$L = 150.2$ m

7-25 $T_f = \frac{93+38}{2} = 65.5\ °C$ $k = 0.659$

$Gr\,Pr = (7.62 \times 10^{10})(93-38)(0.0004)^3 = 268.2$

from Fig 7-8 $\log Nu = 0.41$ $Nu = 2.57$

$$h = \frac{(2.57)(0.659)}{0.0004} = 4235\ W/m^2 \cdot °C$$

$q = (4235)\pi(0.0004)(0.1)(93-38) = 29.27\ W$

7-26 Laminar:

$$h = 1.32\left(\frac{85-20}{0.03}\right)^{1/4} = 9.006\ W/m^2 \cdot °C$$

$q = (9.006)\pi(0.03)(15)(85-20) = 827.6\ W$

7-27 $T_f = \frac{140+20}{2} = 60°C = 353K$ $\nu = 21.07 \times 10^{-6}$ $k = 0.0303$ $Pr = 0.697$

$$Gr\,Pr = \frac{(9.8)(1/353)(140-20)(0.08)^3(0.697)}{(21.07 \times 10^{-6})^2} = 2.68 \times 10^6 \quad C = 0.53 \quad m = 1/4$$

$$h = \frac{0.0303}{0.08}(0.53)(2.68 \times 10^6)^{1/4} = \mathbf{8.12}\ W/m^2 \cdot °C$$

$q/L = h\pi d(T_w - T_\infty) = (8.12)\pi(0.08)(140-20) = \mathbf{245}\ W/m$

7-28 $h = 1.32\left(\frac{250-20}{0.0125}\right)^{1/4} = 15.38\ W/m^2 \cdot °C$

$q/L = (15.38)\pi(0.0125)(250-20) = 138.9\ W/m$

7-29 $T_f = \frac{150+93}{2} = 121.5\ °C$ $\nu = 0.124 \times 10^{-4}$ $k = 0.135$ $Pr = 175$

$\beta = 0.7 \times 10^{-3}$ $$Gr\,Pr = \frac{(9.8)(0.7 \times 10^{-3})(150-93)(0.025)^3}{(0.124 \times 10^{-4})^2} = 6.96 \times 10^6$$

$$h = \frac{0.135}{0.025}(0.53)(6.96 \times 10^6)^{1/4} = 147\ W/m^2 \cdot °C$$

$q/L = (147)\pi(0.025)(150-93) = 658\ W/m = 200.6\ W/ft$

7-30 $T_f = \frac{80+60}{2} = 70°F = 294\ K$ $\Delta T = 20°F = 11.11°C$

laminar:

Top $h = 1.32\left(\frac{11.11}{0.3048}\right)^{1/4} = 3.243\ W/m^2\cdot°C$

$q = (3.243)(0.3048)^2(11.11) = 3.348\ W$

Sides $h = 1.42\left(\frac{11.11}{0.3048}\right)^{1/4} = 3.489$

$q = (3.489)(2)(0.3048)^2(11.11) = 7.202\ W$

Bottom $h = 0.61\left[\frac{11.11}{(0.3048)^2}\right]^{1/5} = 1.588\ W/m^2\cdot°C$

$q = (1.588)(0.3048)^2(11.11) = 1.639\ W$

$q\,(\text{total}) = 3.348 + 7.202 + 1.639 = 12.19\ W/ft$

7-31 $T_f = \frac{240+10}{2} = 125°C = 398\ K$ $\nu = 6.76 \times 10^{-5}$ $k = 0.178$ $Pr = 0.71$

$$Gr\,Pr = \frac{(9.8)(1/398)(240-10)(2.54\times10^{-5})^3}{(6.76\times10^{-5})^2}(0.71) = 1.42\times10^{-5}$$

$$Nu = 0.36 + \frac{(0.518)(1.42\times10^{-5})^{1/4}}{\left[1 + \left(\frac{0.559}{0.7}\right)^{9/16}\right]^{4/9}} = 0.384$$

$$h = \frac{(0.384)(0.178)}{2.54\times10^{-5}} = 2691$$

$q/L = (2691)\pi(2.54\times10^{-5})(240-10) = 49.4\ W/m$

7-32 $T_f = 135°C = 408\ K$ $\nu = 26.83\times10^{-6}$ $k = 0.0342$ $Pr = 0.688$

$$Gr\,Pr = \frac{(9.8)(1/408)(250-20)(3)^3(0.688)}{(26.83\times10^{-6})^2} = 1.43\times10^{11}$$

$$Nu^{1/2} = 0.6 + 0.387\left\{\frac{1.43\times10^{11}}{\left[1+\left(\frac{0.559}{0.688}\right)^{9/16}\right]^{16/9}}\right\}^{1/6} = 23.78$$

$Nu = 565.3$ $h = \frac{(565.3)(0.0342)}{3.0} = 6.44$

$q/L = (6.44)\pi(3.0)(250-20) = 13.97\ kW/m$

7-33 $T_f = 40°C = 313K$ $\nu = 0.00022$ $k = 0.286$ $\beta = 0.5 \times 10^{-3}$

$Pr = 2.45$ $Gr\,Pr = \frac{(9.8)(0.5\times10^{-3})(60-20)(0.02)^3}{(0.00022)^2}(2.45) = 79.37$

$$Nu = 0.36 + \frac{(0.518)(79.37)^{1/4}}{\left[1 + \left(\frac{0.559}{2.45}\right)^{9/16}\right]^{4/9}} = 1.677$$

$$h = \frac{(1.677)(0.286)}{0.02} = 23.98$$

$q = (23.98)\pi(0.02)(0.6)(60-20) = 36.15$ W

7-34 $q/A = 1500$ W/m² $\beta = 1/293$ $q/L = (1500)\pi(0.035) = 165$ W/m

Properties @ 350 K $\Delta T \approx 100°C$ $\nu = 20.76\times10^{-6}$ $k = 0.03003$ $Pr = 0.7$

$\theta = 65°$ Take $L = 1$ m

$$Gr\,Pr = \frac{(9.8)(1/293)(100)(1.0)^3}{(20.76\times10^{-6})^2}(0.7) = 5.4\times10^9$$

$$\tfrac{1}{4} + \tfrac{1}{12}(\sin\theta)^{1.75} = 0.32$$

$$Nu_L = \left[0.6 - (0.488)(\sin 65)^{1.03}\right](5.4\times10^9)^{0.32} = 206.9$$

$h = \frac{(206.9)(0.03003)}{1} = 6.21$ h is insensitive to L because of 0.32 exponent.

$$\Delta T = \frac{1500}{6.21} = 242°C$$

With Properties at 400 K & $\Delta T = 242$

$\nu = 25.9\times10^{-6}$ $k = 0.03365$ $Nu_L = 237.6$ $Gr\,Pr = 8.32\times10^9$

$h = 8.0$ $\Delta T = \frac{1500}{8} = 187.5°C$

$T_w = 20 + 187.5 = 207.5°C = 480.5$ K

$T_f = 386$ K close enough

7-35 $T_f = \frac{25+20}{2} = 22.5\ °C = 295.5\ K \quad \nu = 15.3 \times 10^{-6} \quad k = 0.0259$

$Pr = 0.7 \quad Gr\,Pr = \frac{(9.8)(1/295.5)(25-20)(0.3)^3(0.7)}{(15.3 \times 10^{-6})^2} = 1.34 \times 10^7$

$C = 0.53 \quad m = 1/4$

$$h = \frac{0.0259}{0.3}(0.53)(1.34 \times 10^7)^{1/4} = 2.77\ W/m^2 \cdot °C$$

$$q/L = h \pi d (T_w - T_\infty) = (2.77)\pi(0.3)(25-20) = 13.04\ W/m$$

7-36 $T_f = \frac{500+60}{2} = 280\ °F = 411\ K \quad \nu = 27.18 \times 10^{-6} \quad k = 0.0344$

$Pr = 0.687 \quad Gr\,Pr = \frac{(9.8)(1/411)(500-60)(5/9)(5)^3(0.0254)^3(0.687)}{(27.18 \times 10^{-6})^2} = 1.11 \times 10^7$

$C = 0.53 \quad m = 1/4$

$$h = \frac{0.0344}{(5)(0.0254)}(0.53)(1.11 \times 10^7)^{1/4} = 8.29\ W/m^2 \cdot °C$$

$$q = (125000)(3.413) = (8.29)\pi(5)(0.0254)L(500-60)(5/9)$$

$L = 527.9\ m = 1732\ ft$

There will also be substantial radiation heating.

7-37 $T_f = \frac{180+60}{2} = 120\ °F \quad k = 0.644 \quad \frac{g\beta\rho^2 c_p}{\mu k} = 4.89 \times 10^{10}$

$$Gr\,Pr = (4.89 \times 10^{10})(0.05)^3(180-60)(5/9) = 4.075 \times 10^8$$

$C = 0.53 \quad m = 1/4$

$$h = \frac{0.644}{0.05}(0.53)(4.078 \times 10^8)^{1/4} = 970\ W/m^2 \cdot °C$$

$$q = h\pi d L (T_w - T_\infty) = (970)\pi(0.05)(3)(180-60)(5/9)$$

$$= 3.047 \times 10^4\ W$$

7-38 $T_f = \frac{77+27}{2} = 52°C = 325\,K$ $\nu = 12.83 \times 10^{-6}$ $k = 0.0281$

$Pr = 0.7$ $Gr\,Pr = \frac{(9.8)(1/325)(77-27)(2)^3(0.7)}{(18.23 \times 10^{-6})^2} = 2.54 \times 10^{10}$

$$Nu^{1/2} = 0.6 + 0.387 \left\{ \frac{Gr\,Pr}{\left[1 + \left(\frac{0.559}{Pr}\right)^{9/16}\right]^{16/9}} \right\}^{1/6}$$

$Nu^{1/2} = 17.998$ $Nu = 323.9$

$h = \frac{(323.9)(0.0281)}{2} = 4.55\ W/m^2 \cdot °C$

$q = h\pi dL(T_w - T_\infty) = (4.55)\pi(2)(20)(77-27) = 2.86 \times 10^4\ W$

Using $C = 0.13$ and $m = 1/3$

$Nu = (0.13)(2.54 \times 10^{10})^{1/3} = 293.7 \Rightarrow 9\%$ Lower

7-39 $T_f = \frac{90+20}{2} = 55°C = 328\,K$ $\beta = 3.049 \times 10^{-3}$ $\nu = 18.52 \times 10^{-6}$

$k = 0.0284$ $Pr = 0.7$

$Gr\,Pr = \frac{(9.806)(3.049 \times 10^{-3})(90-20)(0.3)^3(0.7)}{(18.52 \times 10^{-6})^2} = 1.153 \times 10^8$

$h = \frac{0.0284}{0.3}\left[2 + 0.43(1.153 \times 10^8)^{1/4}\right] = 4.407\ W/m^2 \cdot °C$

$q = (4.407)\pi(4)(0.15)^2(90-20) = 87.23\ W$

7-40 $T_f = \frac{32+10}{2} = 21°C$ $k = 0.604$

$Gr\,Pr = (1.46 \times 10^{10})(20-10)(0.025)^3 = 5.019 \times 10^6$

$h = \frac{0.604}{0.025}\left[2 + 0.43(5.019 \times 10^6)^{1/4}\right] = 540\ W/m^2 \cdot °C$

$q = (540)4\pi(0.0125)^2(32-10) = 23.32$

7-41 $T_f = -25°C = 248\ K$ $\quad \rho = \frac{1400}{(287)(248)} = 1.967 \times 10^{-2}$ $\quad Pr = 0.72$

$\mu = 1.488 \times 10^{-5}$ $\quad \beta = 4.032 \times 10^{-3}$ $\quad k = 0.0223$

$$Gr\,Pr = \frac{(9.8)(4.032 \times 10^{-3})(1.967 \times 10^{-2})(50)(2.4)^3}{(1.488 \times 10^{-5})^2}(0.72) = 3.44 \times 10^7$$

$$h = \frac{0.0223}{2.4}\left[2 + 0.43(3.44 \times 10^7)^{1/4}\right] = 0.325\ W/m^2 \cdot °C$$

$$q = (0.325)\pi(4)(1.2)^2(50) = 294\ W$$

Forced Convection:

$$Re = \frac{(1.967 \times 10^{-2})(0.3)(2.4)}{1.488 \times 10^{-5}} = 952$$

$$h = \frac{0.0223}{2.4}(0.37)(952)^{0.6} = 0.211\ W/m^2 \cdot °C$$

$$q = (0.211)(4\pi)(1.2)^2(50) = 191\ W$$

7-42 $T_f = \frac{38 + 15}{2} = 26.5\ °C$ $\quad k = 0.614$

$$Gr\,Pr = (1.91 \times 10^{10})(38 - 15)(0.025)^3 = 6.86 \times 10^6$$

$$h = \frac{0.614}{0.025}\left[2 + 0.43(6.86 \times 10^6)^{1/4}\right] = 589.7\ W/m^2 \cdot °C$$

$$q = (589.7)(4\pi)(0.0125)^2(38 - 15) = 26.63\ W$$

7-45 $T_f = \frac{150 + 20}{2} = 85°C = 358\ K$ $\quad \beta = 2.793 \times 10^{-3}$ $\quad \nu = 21.58 \times 10^{-6}$

$k = 0.0306$ $\quad Pr = 0.7$

$$Gr\,Pr = \frac{(9.806)(2.793 \times 10^{-3})(150 - 20)(0.15)^3}{(21.58 \times 10^{-6})^2}(0.7) = 1.806 \times 10^7$$

$$h = \frac{0.0306}{0.15}(0.15)(1.806 \times 10^7)^{1/3} = 8.03\ W/m^2 \cdot °C$$

$$q = (8.03)\pi(0.075)^2(150 - 20) = 18.4\ W$$

7-46 $T_f = \frac{60+20}{2} = 40°C$ $\nu = 0.00024$ $k = 0.144$ $Pr = 2870$ $\beta = 0.7\times10^{-3}$

$$Gr\,Pr = \frac{(9.8)(0.7\times10^{-3})(60-20)(0.3)^3(2870)}{(0.00024)^2} = 3.69\times10^8$$

$$h = \frac{0.144}{0.3}(0.15)(3.69\times10^8)^{1/3} = 51.6\ W/m^2\cdot°C$$

$$q = (51.6)(0.3)^2(60-20) = 185.8\ W$$

7-47 laminar: $L = 6mm$ $\Delta T = 500-20$ $q = 2\ kW$

Top: $h = 1.32\left(\frac{480}{0.006}\right)^{1/4} = 22.2\ W/m^2\cdot°C$

Bottom: $h = 0.61\left(\frac{480}{(0.006)^2}\right)^{1/5} = 16.23\ W/m^2\cdot°C$

$q = 2000 = (16.23 + 22.2)(0.006)\,L\,(480)$ $L = 18.07\ m$

7-48 $T_f = \frac{25+28}{2} = 26.5°C = 299.5K$ $\nu = 16.84\times10^{-6}$

$k = 0.02624$ $Pr = 0.71$

$$Gr\,Pr = \frac{(9.8)(1/299.5)(3)(10)^3(0.71)}{(16.84\times10^{-6})^2} = 2.42\times10^{11}$$

$$h = \frac{0.02624}{10}(0.15)(2.42\times10^{11})^{1/3} = 2.454$$

$$q = (2.454)(10)^2(3) = 736\ W$$

7-49 $T_f = \frac{15+50}{2} = 32.5°C = 305.5\ K$ $\nu = 16.25\times10^{-6}$ $k = 0.0267$ $Pr = 0.7$

$$Gr\,Pr = \frac{(9.8)(1/305.5)(50-15)(4)^3(0.7)}{(16.25\times10^{-6})^2} = 1.9\times10^{11}$$

Upper: $h_u = \frac{0.0267}{4}(0.15)(1.9\times10^{11})^{1/3} = 5.75\ W/m^2\cdot°C$

Lower: $h_l = \frac{0.0267}{4}(0.27)(1.9\times10^{11})^{1/4} = 1.19\ W/m^2\cdot°C$

$$q = (5.75 + 1.19)(4)^2(50-15) = 3886\ W$$

7-50 $T_f = \frac{400+300}{2} = 350\ K \quad \nu = 2.075\times10^{-6} \quad k = 0.03003 \quad Pr = 0.697$

$$x = A/p = \frac{(L)(L/2)}{3L} = \frac{L}{6} = 45/6 = 7.5\ cm$$

$$GrPr = \frac{(9.8)(1/350)(400-300)(0.075)^3(0.697)}{(2.075\times10^{-6})^2} = 1.91\times10^8$$

$$h = \frac{0.03003}{0.075}(0.15)(1.91\times10^8)^{1/3} = 34.58\ W/m^2\cdot{}^\circ C$$

$$q = (34.58)\frac{(0.45)^2}{2}(400-300) = 350\ W$$

7-51 $T_w = 40^\circ C \quad T_\infty = 20^\circ C \quad T_e = 40 - (0.25)(40-20) = 35^\circ C$

$k = 0.626 \quad GrPr = (2.89\times10^{10})(0.2)^3(40-20) = 4.624\times10^9$

$$\theta = 30^\circ \quad \bar{h} = \frac{0.626}{0.2}(0.56)\left[(4.624\times10^9)\cos 30^\circ\right]^{1/4} = 440\ W/m^2\cdot{}^\circ C$$

$$q = (440)(0.2)^2(40-20) = 352.7\ W$$

7-52 $\theta = -30^\circ \quad \overline{Nu}_e = (0.14)\left[(4.624\times10^9)^{1/3} - (2\times10^9)^{1/3}\right]$

$$+ 0.56\left[(4.624\times10^9)^{1/3}\cos 30^\circ\right]^{1/4}$$

$$\overline{Nu}_e = 56.85 + 140.87 = 197.72$$

$$h = \frac{(0.626)(197.72)}{0.2} = 618.9\ W/m^2\cdot{}^\circ C$$

$$q = (168.9)(0.2)^2(40-20) = 495.1\ W$$

7-53 Assume $T = 20^\circ C$ inside

$T_m = 20 + \frac{30}{2} = 35^\circ C = 308K \quad \beta = 3.247\times10^{-3}$

$\nu = 16.49\times10^{-6} \quad k = 0.0268 \quad Pr = 0.7$

$$Gr_\delta Pr = \frac{(9.8)(3.247\times10^{-3})(30)(0.0125)^3}{(16.49\times10^{-6})^2}(0.7) = 4803$$

$$k_e \approx k \quad q = \frac{(0.0268)(1.8)(1.2)(30)}{0.0125} = 138.9\ W$$

7-54 $T_m = \frac{160+40}{2} = 100°C = 373\ K$ $\beta = 2.68 \times 10^{-3}$ $k = 0.0317$

$Pr = 0.69$ $\rho = \frac{1.01 \times 10^5}{(287)(373)} = 0.0946$ $\mu = 2.172 \times 10^{-5}$

$$Gr_\delta Pr = \frac{(9.8)(2.68 \times 10^{-3})(0.946)^2(160-40)(0.08)^3(0.69)}{(2.172 \times 10^{-5})^2} = 2.11 \times 10^4$$

$\theta = 20°$ $C = 0.212$ $n = 1/4$

$$k_e/k = (0.212)\left[(2.11 \times 10^4)\cos 20°\right]^{1/4} = 2.516$$

$$q = (2.516)(0.0317)(1)^2(160-40) = 9.57\ W$$

7-55 $q_w = 700\ W/m^2$ $L = 1\ m$ $\theta = 60°$ @ $30°C$ $\nu = 15.98 \times 10^{-6}$

$Pr = 0.7$ $k = 0.027$ $\beta = 3.3 \times 10^{-3}$

$$Gr^* = \frac{(9.8)(3.3 \times 10^{-3})(700)(1)^4}{(0.027)(15.98 \times 10^{-6})^2} = 3.286 \times 10^{12}$$

$$h = \frac{0.027}{1}(0.17)\left[(3.286 \times 10^{12})(0.7)\right]^{1/4} = 5.65\ W/m^2 \cdot °C$$

$$\Delta T = \frac{700}{5.65} = 123.8°C \qquad T_w \approx 123.8 + 30 = 153.8°C$$

7-56 $\sigma \epsilon A(T_w^4 - T_s^4) = hA(T_s - T_a)$

$$(5.669 \times 10^{-8})(0.5)(308^4 - 303^4) = (6.5)(303 - T_a)$$

$$T_a = 300.5\ K = 27.5°C$$

7-57 $T_e = 80 - \frac{1}{4}(80-20) = 65°C = 338\ K$ $\beta = 1/308$

$\nu = 19.54 \times 10^{-6}$ $k = 0.0291$ $Pr = 0.7$

Downward facing $\theta = 45°$

$$Gr\,Pr = \frac{(9.8)(1/308)(80-20)(0.1)^3(0.7)}{(19.54 \times 10^{-6})^2} = 3.5 \times 10^6$$

$$h_d = \frac{0.0291}{0.1}(0.56)(3.5 \times 10^6 \times 0.707)^{1/4} = 6.46\ W/m^2 \cdot °C$$

7-57 cont

Upward facing $Gr_c = 1.05 \times 10^9$ $Gr_c Pr = 7.35 \times 10^8$

$Gr_L < Gr_c$ so that

$$h_u = \frac{0.0291}{0.1}(0.56)(3.5 \times 10^6 \times 0.707)^{1/4} = 6.46 \text{ W/m}^2 \cdot °C$$

$$q = (2)(6.46)(0.1)^2(80-20) = 7.752 \text{ W}$$

7-58 $T_e = 50 - \frac{1}{4}(50-20) = 42.5°C$ $\rho = 990$ $\mu = 6.2 \times 10^{-4}$

$k = 0.635$ $Pr = 4.1$ $\theta = 30°$ $\cos\theta = 0.866$

$$Gr\,Pr = \frac{(9.8)(2.07 \times 10^{-4})(50-20)(0.05)^3(4.1)(990)^2}{(6.2 \times 10^{-4})^2} = 7.95 \times 10^7$$

$$h_d = \frac{0.635}{0.05}(0.56)(7.95 \times 10^7 \times 0.866)^{1/4} = 647.8 \text{ W/m}^2 \cdot °C$$

Since $Gr_L < Gr_c$ $h_u = h_d$

$$q = (2)(647.8)(0.05)^2(50-20) = 97.2 \text{ W}$$

7-60 $T_f = \frac{93+20}{2} = 56.5°C = 329.5 \text{ K}$ $\beta = 3.03 \times 10^{-3}$ $\nu = 18.73 \times 10^{-6}$

$k = 0.0285$ $Pr = 0.7$ $\frac{1}{L} = \frac{1}{2.5} + \frac{1}{5}$ $L = 1.67 \text{ cm}$

$$Gr\,Pr = \frac{(9.8)(3.03 \times 10^{-3})(93-20)(0.0167)^3}{(18.73 \times 10^{-6})^2}(0.7) = 2.016 \times 10^4$$

$$\bar{h} = \frac{0.0285}{0.0167}(0.6)(2.016 \times 10^4)^{1/4} = 12.2 \text{ W/m}^2 \, °C$$

$$A = (2.5)^2(2) + (4)(2.5)(5) = 62.5 \text{ cm}^2$$

$$q = (12.2)(62.5 \times 10^{-4})(93-20) = 5.567 \text{ W}$$

7-61 $A = \frac{1}{2}(40)^2 \sin 60° = 692.8 \text{ cm}^2 \quad P = (3)(40) = 120 \text{ cm}$

$L = A/P = \frac{692.8}{120} = 5.77 \text{ cm} \quad T_f = \frac{55+25}{2} = 40°C = 313 \text{ K}$

$\beta = 3.195 \times 10^{-3} \quad \nu = 17 \times 10^{-6} \quad k = 0.0272 \quad Pr = 0.7$

$$Gr\,Pr = \frac{(9.8)(3.195 \times 10^{-3})(55-25)(0.0577)^3}{(17 \times 10^{-6})^2}(0.7) = 4.37 \times 10^5$$

$$h = \frac{0.0272}{0.0577}(0.54)(4.37 \times 10^5)^{1/4} = 6.546 \text{ W/m}^2 \cdot °C$$

$$q = (6.546)(0.0693)(55-25) = 13.61 \text{ W}$$

7-62 $T_f = 40°C = 313 \text{ K} \quad \beta = 3.195 \times 10^{-3} \quad \nu = 17 \times 10^{-6} \quad k = 0.0272$

$Pr = 0.7 \quad Gr\,Pr = \frac{(9.8)(3.195 \times 10^{-3})(50-30)(0.03)^3(0.7)}{(17 \times 10^{-6})^2} = 4.1 \times 10^4$

$$h = \frac{0.0272}{0.03}(0.54)(4.1 \times 10^4)^{1/4} = 6.97 \text{ W/m}^2 \cdot °C$$

$$q = (6.97)\pi(0.015)^2(50-30) = 0.0985 \text{ W}$$

7-63 $T_f = \frac{400+27}{2} = 213.5°C = 486.5 \text{ K} \quad \nu = 36.23 \times 10^{-6} \quad k = 0.03948$

$Pr = 0.681 \quad \beta = 1/486.5 \quad 1/L = \frac{1}{15} + 1/8 \quad L = 5.22 \text{ cm}$

$$Gr\,Pr = \frac{(9.8)(1/486.5)(400-27)(0.0522)^3(0.681)}{(36.23 \times 10^{-6})^2} = 5.54 \times 10^5$$

$$h = \frac{0.03948}{0.0522}(0.6)(5.54 \times 10^5)^{1/4} = 12.38 \text{ W/m}^2 \cdot °C$$

$$A = (2)(15)^2 + (4)(15)(8) = 930 \text{ cm}^2$$

$$q = (12.38)(930 \times 10^{-4})(400-27) = 429 \text{ W}$$

7-64 $\frac{1}{L} = \frac{1}{6} + \frac{1}{6}$ $L = 3 \text{ in} = 7.62 \text{ cm}$ $\nu = 17 \times 10^{-6}$

$k = 0.027$ $Pr = 0.7$

$$Gr\,Pr = \frac{(9.8)(1/300)\,\Delta T\,(0.0762)(0.7)}{(17 \times 10^{-6})^2} = 5 \times 10^4\ \Delta T$$

$$A = 6^3 = 216 \text{ in}^2 = 0.354 \text{ m}^2$$

$$h = \frac{0.027}{0.03}(0.6)(5 \times 10^4)^{1/4}\,\Delta T^{1/4} = 8.075\,\Delta T^{1/4}$$

$$50 = (8.075)(0.354)(\Delta T)^{5/4} \qquad \Delta T = 9.87°C = 17.8°F$$

$$T_w = 17.8 + 70 = 87.8°F$$

7-65 $\sigma\epsilon(T_t^4 - T_w^4) = h(T_a - T_t)$

$$(5.669 \times 10^{-8})(0.95)[303^4 - 283^4] = 5(T_a - 303)$$

$$T_a = 325 \text{ K} = 52°C$$

7-66 $L = 30 + 15 = 45 \text{ cm}$ $T_f = \frac{120 + 70}{2} = 95°F = 308 \text{ K}$

$\nu = 16.5 \times 10^{-6}$ $k = 0.0268$ $Pr = 0.7$

$$Gr\,Pr = \frac{(9.8)(1/308)(50)(5/9)(0.45)^3(0.7)}{(16.5 \times 10^{-6})^2} = 2.07 \times 10^8$$

$C = 0.52,\ m = 1/4$

$$h = \frac{0.0268}{0.45}(0.52)(2.07 \times 10^8)^{1/4} = 3.72 \text{ W/m}^2 \cdot °C$$

$$q/L = h\frac{A}{L}(T_w - T_\infty) = (3.72)(2)(0.3 + 0.15)(50)(5/9)$$

$$= 92.9 \text{ W/m}$$

7-68 $T_m = \frac{38+60}{2} = 49°C$ $k = 0.644$ $Pr = 3.64$

$Gr_\delta Pr = (4.89 \times 10^{10})(60-38)(0.0125)^3 = 2.1 \times 10^6$

$$\frac{k_e}{k} = (0.42)(2.1 \times 10^6)^{1/4}(3.64)^{0.012}\left(\frac{30}{1.25}\right)^{-0.3} = 6.26$$

$$q = \frac{(6.26)(0.644)(60-38)(0.3)^2}{0.0125} = 638.5 \text{ W}$$

7-69 $T_m = \frac{80+20}{2} = 50°C = 323 \text{ K}$ $k = 0.156$ $\nu = \frac{143.3 \times 10^{-6}}{1.3}$

$\nu = 1.102 \times 10^{-4}$ $Pr = 0.7$

$$Gr_\delta Pr = \frac{(9.8)(1/323)(80-20)(0.02)^3(0.7)}{(1.102 \times 10^{-4})^2} = 839$$

$k_e = k$

$$q = \frac{(0.156)(0.4)^2(80-20)}{0.02} = 74.88 \text{ W}$$

7-70 $T_m = 30°C$ $k = 0.62$ $Gr_\delta Pr = (2.25 \times 10^{10})(40-20)(0.02)^3 = 3.6 \times 10^6$

$C = 0.40$ $n = 0.20$ $k_e/k = (0.40)(3.6 \times 10^6)^{0.2} = 8.191$

$$q = \frac{(2\pi)(8.191)(0.62)(1)(40-20)}{\ln(10/8)} = 2860 \text{ W}$$

7-71 $T_m = \frac{30-10}{2} = 10°C = 283 \text{ K}$ $\rho = \frac{(0.05)(1.01 \times 10^5)}{(287)(283)} = 0.0624$

$k = 0.0249$ $\beta = 3.53 \times 10^{-3}$ $\mu = 1.814 \times 10^{-5}$ $Pr = 0.71$

$$Gr_\delta Pr = \frac{(9.8)(0.0624)^2(3.53 \times 10^{-3})(40)(0.05)^3}{(1.814 \times 10^{-5})^2}(0.71) = 1454$$

$k_e/k = (0.228)(1454)^{0.226} = 1.182$

$$q = \frac{4\pi(1.182)(0.0249)(1)(1.05)(40)}{1.05-1} = 310.7 \text{ W}$$

7-73 $T_m = \frac{400+140}{2} = 270\,°F = 132.2\,°C = 405\ K$ $\quad \nu = 26.2 \times 10^{-6}$

$k = 0.034 \quad \beta = 2.47 \times 10^{-3} \quad Pr = 0.7$

$$Gr_\delta Pr = \frac{(9.8)(2.47\times10^{-3})(400-140)(5/9)(0.03)^3}{(26.2\times10^{-6})^2}(0.7) = 9.63\times10^4$$

$C = 0.197 \quad n = 1/4 \quad m = -1/9$

$$h\delta/k = (0.197)(9.63\times10^4)^{1/4}\left(\frac{35}{3}\right)^{-1/9} = 2.64$$

$$q/A = \frac{(2.64)(0.034)(400-140)(5/9)}{0.03} = 432\ W/m^2$$

7-75 $T_m = \frac{200+90}{2} = 145\,°C = 418\ K \quad \nu = 26.97\times10^{-6} \quad k = 0.0349$

$\beta = 2.39\times10^{-3} \quad Pr = 0.685$

$$Gr_\delta Pr = \frac{(9.8)(2.39\times10^{-3})(200-90)(0.025)^3(0.685)}{(26.97\times10^{-6})^2} = 3.79\times10^4$$

$C = 0.197 \quad n = 1/4 \quad m = -1/9$

$$k_e/k = (0.197)(3.79\times10^4)^{1/4}\left(\frac{30}{2.5}\right)^{-1/9} = 2.086$$

$$q = (2.086)(0.0349)(0.3)^2\left(\frac{200-90}{0.025}\right) = 28.83\ W$$

7-76 $T_m = 90 + \frac{165}{2} = 172.5\,°C = 445.5\ K \quad Pr = 0.684$

$\beta = 2.245\times10^{-3} \quad \nu = 28.59\times10^{-6} \quad k = 0.0368$

$$Gr_\delta Pr = \frac{(9.8)(2.245\times10^{-3})(165)(0.0016)^3}{(28.59\times10^{-6})^2}(0.684) = 12.45$$

$k_e = k \qquad \frac{q}{A} = \frac{(0.0368)(165)}{0.0016} = 3795\ W/m^2$

7-77 @ 172.5°C $k = 0.669$ $Gr_\delta Pr = (10.11 \times 10^{10})(165)(0.0016)^3$

$Gr_\delta Pr = 6.83 \times 10^4$ $C = 0.13$ $n = 0.3$ $m = 0$

$$\frac{k_e}{k} = (0.13)(6.83 \times 10^4)^{0.3} = 3.67$$

$$k_e = (3.67)(0.669) = 2.453 \quad q/A = \frac{(2.453)(165)}{0.0016} = 2.53 \times 10^5 \text{ W/m}^2$$

7-79 $T_m = \frac{50+20}{2} = 35°C$ $\rho = 994$ $\mu = 7.2 \times 10^{-4}$ $k = 0.626$

$Pr = 4.82$ $\frac{g\beta\rho^2 c_p}{\mu k} = 2.89 \times 10^{10}$

$$Gr_\delta Pr = (2.89 \times 10^{10})(0.04)^3(30) = 5.55 \times 10^7$$

$$k_e/k = (0.046)(5.55 \times 10^7)^{1/3} = 17.55$$

$$q = (17.55)(0.626)(0.5)^2 \frac{(50-20)}{0.04} = 2060 \text{ W}$$

7-80 $$\frac{k_e}{k} = (0.057)(5.55 \times 10^7)^{1/3} = 21.74$$

$$q = (21.74)(0.626)(0.5)^2 \frac{(50-20)}{0.04} = 2552 \text{ W}$$

7-83 Assume 20°C inside

$T_m = 20 + \frac{17}{2} = 28.6°C = 301 \text{ K}$ $\beta = 3.32 \times 10^{-3}$ $Pr = 0.71$

$\nu = 15.68 \times 10^{-6}$ $k = 0.0262$

$$Gr\,Pr = \frac{(9.8)(3.32 \times 10^{-3})(17)(0.1)^3(0.71)}{(15.68 \times 10^{-6})^2} = 1.598 \times 10^6$$

$C = 0.073$ $n = 1/3$ $m = -1/9$

$$k_e/k = (0.073)(1.598 \times 10^6)^{1/3}\left(\frac{2}{0.1}\right)^{-1/9} = 6.118$$

$$q/A = \frac{(6.118)(0.0262)(17)}{0.1} = 27.25 \text{ W/m}^2$$

7-85 $T_m = \frac{100+20}{2} = 60°C = 333\ K$ $\nu = 19.04 \times 10^{-6}$ $k = 0.0287$

$Pr = 0.7$ $Gr_\delta Pr = \frac{(9.8)(1/333)((100-20)(0.08)^3(0.7)}{(19.04 \times 10^{-6})^2} = 2.33 \times 10^6$

$C = 0.073$ $n = 1/3$ $m = -1/9$

$$\frac{k_e}{k} = (0.073)(2.33 \times 10^6)^{1/3}(1/0.08)^{-1/9} = 7.304$$

$$q = \frac{(0.0287)(7.304)(1)^2(100-20)}{0.08} = 209.6\ W$$

7-86 Take properties @ 300 K $\nu = 15.69 \times 10^{-6}$ $k = 0.02624$

$Pr = 0.7$ $Gr_\delta Pr = \frac{(9.8)(1/300)(30)(0.01)^3(0.7)}{(15.69 \times 10^{-6})^2} = 2787$

For horizontal Plate $C = 0.059$ $n = 0.4$

$$\frac{k_e}{k} = (0.059)(2787)^{0.4} = 1.41$$

$$q = (1.41)(0.02624)(0.3)^2(30)/0.01 = 9.98\ W$$

For 30 cm Vertical Plate

$$Gr\,Pr = (2787)\left(\frac{30}{1}\right)^3 = 7.52 \times 10^7 \quad C = 0.59 \quad m = 1/4$$

$$h = \frac{0.02624}{0.3}(0.59)(7.52 \times 10^7)^{1/4} = 4.806\ W/m^2 \cdot °C$$

$$q = hA(T_w - T_\infty) = (4.806)(0.3)^2(30) = 129.8\ W$$

The 9.98 W is a reduction of 93%

7-87 $T_m = \frac{120+20}{2} = 70°C = 343 K$ $\mu = 2.04 \times 10^{-5}$ $k = 0.0295$

$Pr = 0.7$ $\delta = 1$ cm

$$Gr_\delta = 1700 = \frac{\rho^2 (9.8)(1/343)(120-20)(0.01)^3 (0.7)}{(2.04 \times 10^{-5})^2}$$

$\rho^2 = 0.3537$ $\rho = 0.595$

At 1atm $\rho = 1.023$

Thus at $\delta = 1$ cm $p = \frac{0.595}{1.023} = 0.582$ atm (58.96 kPa)

For the other spacings:

δ (cm)	ρ^2	ρ	p (atm, kPa)
2	0.0442	0.2103	0.2056, 20.83
5	0.00283	0.0532	0.052, 5.27
10	3.537×10^{-3}	0.0188	0.0184, 1.86

7-89 $q = h\pi d L \Delta T$

$h = 1.32\left(\frac{\Delta T}{d}\right)^{1/4}$

$q \sim \Delta T^{5/4}$

$\frac{q_{83}}{175} = \left(\frac{83-27}{55-27}\right)^{5/4} \qquad q_{83} = 416\ W$

7-90 $T_f = \frac{140+70}{2} = 105°F = 40.6°C = 314K$

$\nu = 17.11 \times 10^{-6}\ m^2/s \quad k = 0.0273 \quad Pr = 0.7$

$Gr\,Pr = 10^9 = \frac{(9.8)(1/314)(140-70)(5/9)\,x^3}{(17.11\times10^{-6})^2}(0.7)$

$x = 0.701\ m$

$h = \frac{0.0273}{0.701}(0.59)(10^9)^{1/4} = 4.09\ W/m^2\text{-}°C$

$q/A = h\Delta T = (4.09)(140-70)(5/9) = 159\ W/m^2$

$u = u_{max}$ at $y = \delta/3 = 0.148\,u_x$

$u_x = c_1 x^{1/2}$

$c_1 = (5.17)(17.11\times10^{-6})\left[\frac{20}{21}+0.7\right]^{-1/2}\left[\frac{(9.8)(1/314)(70)(5/9)}{(17.11\times10^{-6})^2}\right]^{1/2}$

$= 4.43$

$u_x = (4.43)(0.701)^{1/2} = 3.71$

$u_{max} = (0.148)(3.71) = 0.55\ m/s$

7-91

$\delta = 0.025\ m \qquad T_f = 50°C = 323K$

$\nu = 18.02 \times 10^{-6}\ m^2/s \qquad k = 0.02798$

$Pr = 0.7$

$$Gr_\delta Pr = \frac{(9.8)(1/323)(70-30)(0.025)^3(0.7)}{(18.02 \times 10^{-6})^2}$$

$$= 40878$$

$$\frac{k_e}{k} = (0.212)(40878)^{1/4} = 3.014$$

$k_e = 0.0843\ W/m\text{-}°C$

$$q/A = (0.0843)\frac{(70-30)}{0.025} = 135\ W/m^2$$

For $\delta = 0.01\ m \qquad Gr_\delta Pr = 2616$

$$\frac{k_e}{k} = 0.059\ (2616)^{0.4} = 1.374$$

$$q/A = (1.374)(0.02798)\frac{(70-30)}{0.01} = 154\ W/m^2$$

7-92

$$Gr_\delta Pr = 1700 = 40878\left(\frac{p}{1atm}\right)^2 \quad \text{since } \rho = p/RT$$

$p = 0.204\ atm$

7-93 $T_f = \frac{120+80}{2} = 100°F = 37.8°C = 311\ K$

a) Air $\nu = 16.81 \times 10^{-6}$ $k = 0.02707$ $Pr = 0.7$

$$Gr\,Pr = \frac{(9.8)(1/311)(120-80)(5/9)(0.025)^3(0.7)}{(16.81 \times 10^{-6})^2}$$

$$= 27100$$

$$h = \frac{0.02707}{0.025}\left[2 + 0.43(27100)^{1/4}\right] = 8.14\ W/m^2\text{-}°C$$

$$q = hA\Delta T = (8.14)\,4\pi(0.0125)^2(40)(5/9)$$

$$= 0.36\ W$$

b) WATER $\frac{g\beta\rho^2 c_p}{\mu k} = 3.3 \times 10^{10}$ $k = 0.630$

$$Gr\,Pr = (3.3 \times 10^{10})(0.025)^3(120-80)(5/9)$$

$$= 1.146 \times 10^7$$

$$h = \frac{0.63}{0.025}\left[2 + 0.50(1.146 \times 10^7)^{1/4}\right]$$

$$= 783\ W/m^2\text{-}°C$$

$$q = hA\Delta T = (783)\,4\pi(0.0125)^2(40)(5/9)$$

$$= 34.2\ W$$

7-94 $T_f = 325\ K$ $\nu = 18.23 \times 10^{-6}$ $k = 0.0281$

$Pr = 0.7$

$r_i = 0.04$ $r_o = 0.05$ $\delta = 0.01$

$$Gr_\delta Pr = \frac{(9.8)(1/325)(350-300)(0.01)^3(0.7)}{(18.23 \times 10^{-6})^2} = 3176$$

$$\frac{k_e}{k} = (0.228)(3176)^{0.226} = 1.41$$

$$k_e = 0.0396$$

$$q = \frac{4\pi(0.0396)(0.04)(0.05)(50)}{0.05-0.04}$$

$$= 4.98\ W$$

7-95

$$T_f = \frac{15-10}{2} = 2.5°C = 275\ K$$

$\nu = 13.5 \times 10^{-6}\ m^2/s \quad k = 0.0243 \quad Pr = 0.71$

$\delta = 0.025\ m \qquad L = 1.0\ m$

$$Gr_\delta Pr = \frac{(9.8)(1/275)(0.025)^3(0.71)(15+10)}{(13.5 \times 10^{-6})^2}$$

$$= 54231$$

$C = 0.197, \quad n = 1/4, \quad m = -1/9$

$$\frac{k_e}{k} = (0.197)(54231)^{1/4}(1/0.025)^{1/9}$$

$$= 1.995$$

$k_e = 0.0485\ W/m \cdot °C$

$$q/A = \frac{\Delta T}{R} = \frac{\Delta T}{\delta/k_e}$$

$$R = \delta/k_e = \frac{0.025}{0.0485} = 0.516$$

For fiberglass blanket, $k = 0.04\ W/m \cdot °C$

$$R = \frac{0.025}{0.04} = 0.625$$

7-96 $d = L = 0.08\ m$ $C = 0.775$ $m = 0.21$

$T_f = 287.5\ K$ $\nu = 14.6 \times 10^{-6}$ $k = 0.0252$

$Pr = 0.71$

$$Gr\,Pr = \frac{(9.8)(1/287.5)(300-275)(0.08)^3(0.71)}{(14.6 \times 10^{-6})^2}$$

$$= 1.45 \times 10^6$$

$$h = \frac{0.0252}{0.08}(0.775)(1.45 \times 10^6)^{0.21}$$

$$= 4.81\ W/m^2\text{-}^\circ C$$

$$hA = (4.81[2\pi(0.04)^2 + \pi(0.08)(0.08)]$$

$$= 0.1452$$

$$\rho c V = (999)(4180)\pi(0.04)^2(0.08)$$

$$= 1679$$

$$hA/\rho c V = 0.1452/1679 = 8.65 \times 10^{-5}$$

$$\frac{290-275}{300-275} = \exp[-8.65 \times 10^{-5}\tau]$$

$$\tau = 5905\ sec = 1.64\ hr.$$

7-97 $L = \frac{A}{P} = \frac{\pi d^2}{4\pi d} = d/4$

$T_f = \frac{120+80}{2} = 100°F = 37.8 °C$

$\frac{g\beta\rho^2 c_p}{\mu k} = 3.3 \times 10^{10}$ $k = 0.63$

$Gr\,Pr = (3.3 \times 10^{10})(0.05/4)^3(120-80)(5/9)$

$= 1.43 \times 10^6$

Upper surface $C = 0.54$ $m = 1/4$

$h = \frac{0.63}{0.05/4}(0.54)(1.43 \times 10^6)^{1/4} = 942 \text{ W/m}^2\text{-°C}$

$q = (942)\pi(0.025)^2(40)(5/9) = 41.1 \text{ W}$

Lower surface $c = 0.27$ $m = 1/4$

$h = 471 \text{ W/m}^2\text{-°C}$

$q = 20.6 \text{ W}$

7-98

$\theta = 45°$ $\quad L = 0.1\ m$ $\quad T_f = 350\ K$

$T_e = 400 - 0.25(400-300)\ 375\ K$

$\nu = 23.33 \times 10^{-6}$ $\quad k = 0.0318$ $\quad Pr = 0.69$

$$Gr_e Pr_e \cos\theta = \frac{(9.8)(1/350)(400-300)(0.1)^3(0.7)(0.707)}{(23.33 \times 10^{-6})^2}$$

$$= 2.54 \times 10^6$$

Eq (7-43)

$$h = \frac{0.0318}{0.1}(0.56)(2.54 \times 10^6)^{1/4} = 7.11\ W/m^2 \cdot °C$$

$$q = hA\Delta T = (7.11)(0.1)^2(400-300) = 7.11\ W$$

7-99

Upward facing surface

Because $Gr_e < Gr_c$ 1st term of Eq. (7-46) drops out and the result is the same as in Prob. (7-98).

7-100

$$\frac{D}{L} \geq \frac{35}{Gr_L^{1/4}} \qquad L = 50\ cm = 0.5\ m$$

$T_f = 350\ K$

$\nu = 20.76 \times 10^{-6}$ $\quad k = 0.03003$ $\quad Pr = 0.7$

$$Gr_L = \frac{(9.8)(1/350)(400-300)(0.5)^3}{(20.76 \times 10^{-6})^2} = 8.12 \times 10^8$$

$$d_{min} = (0.5)(35)(8.12 \times 10^8)^{-1/4} = 0.104\ m$$

8-1

(a) $T = 800°C = 1073\ K$ $\quad \lambda_1 T = (0.2)(1073) = 214.6$ $\quad \lambda_2 T = (4)(1073) = 4292$

$E_{b0-\lambda_1} = 0$ $\quad E_{b0-\lambda_2} = 0.53131$ $\quad E_b = (5.669 \times 10^{-8})(1073)^4$

$E_b = 7.515 \times 10^4\ W/m^2$

$q_{trans} = (0.53131)(0.9)(7.515 \times 10^4) = 35933\ W/m^2$

(b) $T = 550°C = 823\ K$ $\quad \lambda_2 T = (4)(823) = 3292$ $\quad E_b = (5.669 \times 10^{-8})(823)^4$

$E_b = 26007$ $\quad E_{b0-\lambda_2} = 0.33834$ $\quad E_{b0-\lambda_1} = 0$

$q_{trans} = (0.33834)(0.9)(26007) = 7919\ W/m^2$

(c) $T = 250°C = 523\ K$ $\quad \lambda_2 T = (4)(523) = 2092$ $\quad E_b = (5.669 \times 10^{-8})(523)^4$

$E_b = 4241$ $\quad E_{b0-\lambda_2} = 0.08182$ $\quad q_{trans} = (0.08182)(0.9)(4241) = 312.3\ \frac{W}{m^2}$

(d) $T = 70°C = 343\ K$ $\quad \lambda_2 T = (4)(343) = 1372$ $\quad E_b = (5.669 \times 10^{-8})(343)^4$

$E_b = 784.7$ $\quad E_{b0-\lambda_2} = 0.0069$ $\quad q_{trans} = (0.0069)(0.9)(784.7) = 4.87\ \frac{W}{m^2}$

8-2

(a) $\lambda_2 T = (5.5)(1073) = 5901.5$ $\quad E_{b0-\lambda_2} = 0.72921$

$q_{trans} = (0.72921)(0.85)(75150) = 46580\ W/m^2$

(b) $\lambda_2 T = (5.5)(823) = 4527$ $\quad E_{b0-\lambda_2} = 0.5684$

$q_{trans} = (0.5684)(0.85)(26007) = 12565\ W/m^2$

(c) $\lambda_2 T = (5.5)(523) = 2877$ $\quad E_{b0-\lambda_2} = 0.24497$

$q_{trans} = (0.24497)(0.85)(4241) = 883\ W/m^2$

(d) $\lambda_2 T = (5.5)(343) = 1887$ $\quad E_{b0-\lambda_2} = 0.05059$

$q_{trans} = (0.05059)(0.85)(784.7) = 33.74\ W/m^2$

8-3

(a) $\lambda_2 T = (52)(1073)$ $\quad E_{b0-\lambda_2} = 1.0$

$q_{trans} = (1.0)(0.92)(75150) = 69138 \ W/m^2$

(b) $\lambda_2 T = (52)(823) = 42796$ $\quad E_{b0-\lambda_2} = 0.998$

$q_{trans} = (0.998)(0.92)(26007) = 23880 \ W/m^2$

(c) $\lambda_2 T = (52)(523) = 27196$ $\quad E_{b0-\lambda_2} = 0.993$

$q_{trans} = (0.993)(0.92)(4241) = 3874 \ W/m^2$

(d) $\lambda_2 T = (52)(343) = 17836$ $\quad E_{b0-\lambda_2} = 0.97912$

$q_{trans} = (0.97912)(0.92)(784.7) = 706.8 \ W/m^2$

8-4 $\lambda_1 T = (4)(560) = 2240 \ \mu\text{-}°R$ $\quad \lambda_2 T = (15)(560) = 8400 \ \mu\text{-}°R$

$$\frac{E_b(0-\lambda_1)}{E_b(0-\infty)} = 0.00306 \qquad \frac{E_b(0-\lambda_2)}{E_b(0-\infty)} = 0.5890$$

$$E_b(\lambda_1 - \lambda_2) = (1.714 \times 10^{-9})(560)^4(0.5890 - 0.00306) = 98.9 \ Btu/hr \cdot ft^2$$

$$E(\lambda_1 - \lambda_2) = (0.6)(98.9) = 59.3 \ Btu/hr\,ft^2 = 187 \ W/m^2$$

8-6

		T = 300 K		T = 1000 K		T = 5000 K	
λ	α	λT	frac	λT	frac	λT	frac
0	0.05	0	0	0	0	0	0
1.2	0.5	360	0	1200	0.0021	6000	0.7377
3	0.4	900	0.74×10^{-4}	3000	0.273	15000	0.969
6	0.2	1800	0.039	6000	0.7377	30000	0.995
20	0	6000	0.7377	20000	0.9855	100 000	1.0
∞	0	–	–	–	–	–	–

8-6 contd

λ	α	T = 300 K Frac	α × frac	T = 1000 K	T = 5000 K
0 - 1.2	0.05		0	0.0021	0.7377
1.2 - 3	0.5		0.74×10^{-4}	0.271	0.231
3 - 6	0.4		0.039	0.465	0.026
6 - 20	0.2		0.699	0.248	0.005
20 - ∞	0		0	0	0

$\Sigma\ \alpha$ Frac = 0.155   0.371   0.164

8-7 T = 5795 K

λ	λT	Frac	λ	Frac contained
0	0		0 - 0.2	0.0015
0.2	1159	0.0015	0.2 - 0.4	0.1235
0.4	2318	0.125	0.4 - 1.0	0.596
1.0	5795	0.721	1.0 - 2.0	0.219
2.0	11590	0.94	over 2.0	0.06
∞		1.0		1.0

8-9 T = 1073 K

λ	λT (μ-K)	fraction 0-λT	fraction $\lambda_1 T - \lambda_2 T$
1	1073	7.74×10^{-4}	
2	2146	9.13×10^{-2}	8.35×10^{-3}
3	3219	0.32233	0.23103
4	4292	0.53131	0.20898
5	5365	0.67651	0.1452
6	6348	0.76538	0.08887

8-10 T = 1373 K

fraction	λT	λ (μ)
0.25	2897	2.11
0.5	4100	2.99
0.75	6149	4.48
0.98	18222	13.27

8-11
(a)

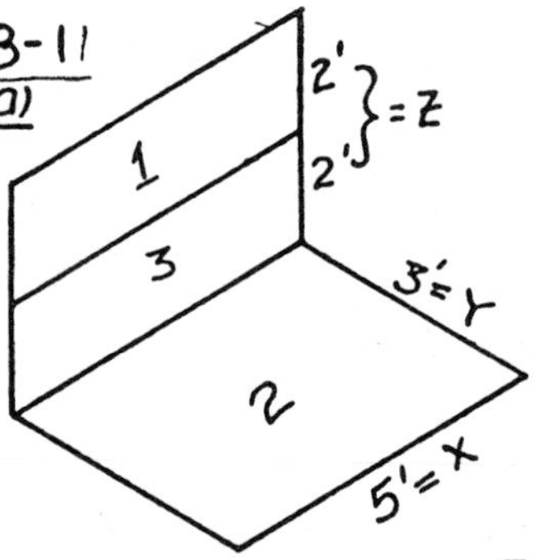

Find F_{1-2}

$\frac{Y}{X} = \frac{3}{5} = 0.6$

$\frac{Z}{X} = \frac{4}{5} = 0.8$

$F_{2-3,1} = F_{2-3} + F_{2-1}$

$F_{2-3,1} = 0.25$

$F_{2-3} = 0.17$

$F_{2-1} = F_{2-3,1} - F_{2-3} = 0.08$

$A_1 F_{1-2} = A_2 F_{2-1}$, $\therefore F_{1-2} = 0.105$

(b)

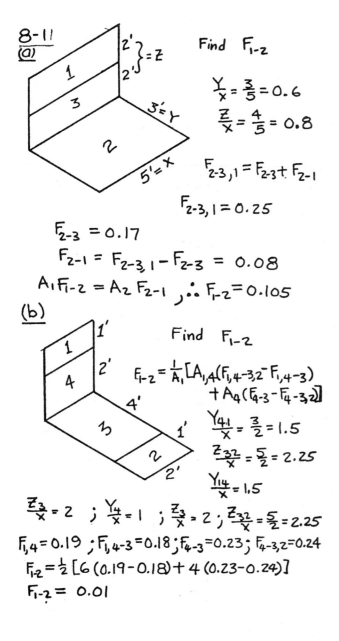

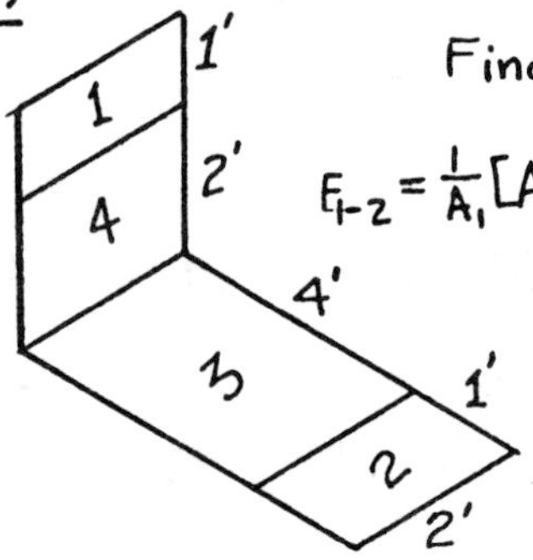

Find F_{1-2}

$F_{1-2} = \frac{1}{A_1}[A_{1,4}(F_{1,4-3,2} - F_{1,4-3}) + A_4(F_{4-3} - F_{4-3,2})]$

$\frac{Y_{41}}{X} = \frac{3}{2} = 1.5$

$\frac{Z_{32}}{X} = \frac{5}{2} = 2.25$

$\frac{Y_{14}}{X} = 1.5$

$\frac{Z_3}{X} = 2$; $\frac{Y_4}{X} = 1$; $\frac{Z_3}{X} = 2$; $\frac{Z_{32}}{X} = \frac{5}{2} = 2.25$

$F_{1,4} = 0.19$; $F_{1,4-3} = 0.18$; $F_{4-3} = 0.23$; $F_{4-3,2} = 0.24$

$F_{1-2} = \frac{1}{2}[6(0.19 - 0.18) + 4(0.23 - 0.24)]$

$F_{1-2} = 0.01$

8-11 contd

(c)

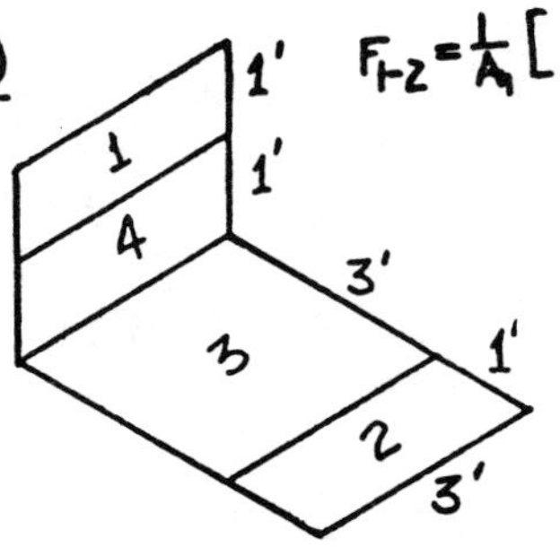

$$F_{1-2} = \frac{1}{A_1}\left[A_{1,4}(F_{1,4-3,2} - F_{1,4-3}) + A_4(F_{4-3} - F_{4-3,2})\right]$$

$\frac{Y_{3,2}}{X} = 1.33$; $\frac{Z_{1,4}}{X} = .666$

$\frac{Z_{1,4}}{X} = .666$; $\frac{Y_3}{X} = 1$

$\frac{Z_4}{X} = \frac{1}{3}$

$F_{1,4-3} = 0.26$

$F_{4-3} = 0.35$

$F_{4-3,2} = 0.36$

$F_{1,4-3,2} = 0.27$

$$F_{1-2} = \frac{1}{3}[6(0.27 - 0.26) + 3(0.36 - 0.35)] = 0.03$$

(d)

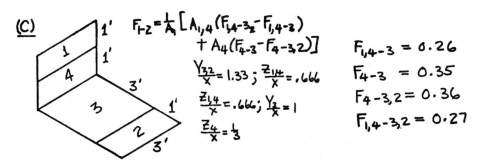

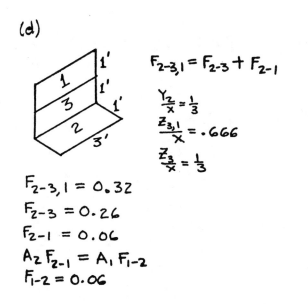

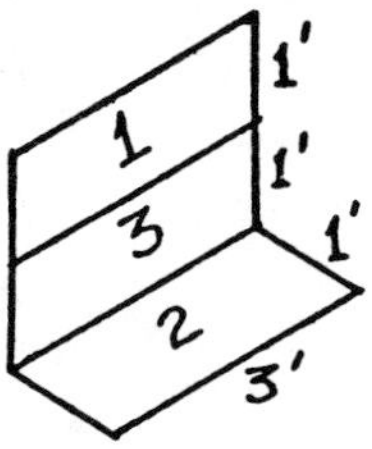

$$F_{2-3,1} = F_{2-3} + F_{2-1}$$

$\frac{Y_2}{X} = \frac{1}{3}$

$\frac{Z_{3,1}}{X} = .666$

$\frac{Z_3}{X} = \frac{1}{3}$

$F_{2-3,1} = 0.32$

$F_{2-3} = 0.26$

$F_{2-1} = 0.06$

$A_2 F_{2-1} = A_1 F_{1-2}$

$F_{1-2} = 0.06$

8-12

$r_1 = 5 cm$ $\quad r_2 = 2.5 cm$ $\quad L = 10 cm$

$r_2/L = 0.25$ $\quad L/r_1 = 2.0$ $\quad r_1/r_2 = 2.0$

$F_{12} = 0.06$ $\quad F_{21} = F_{12}\frac{r_1}{r_2} = 0.12$

8-13 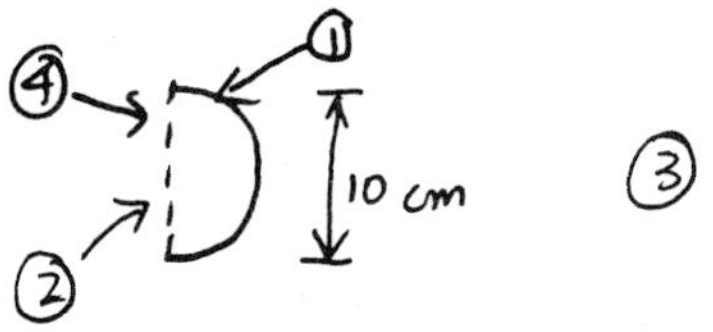

$F_{21} = 1.0 \qquad A_1 = 2\pi r^2 = (2)\pi(12.5)^2 = 981.7$

$A_2 = \pi r^2 = \pi(5)^2 = 78.5$

$F_{12} = (1.0)\left(\frac{78.5}{981.7}\right) = 0.08 \qquad A_4 = (\pi)(12.5^2 - 5^2) = 412.3$

$F_{41} = 1.0$

$F_{14} = F_{13} = 1.0\left(\frac{412.3}{981.7}\right) = 0.42$

$F_{12} + F_{11} + F_{13} = 1.0 \qquad F_{11} = 1.0 - 0.08 - 0.42 = 0.50$

Summary

$F_{11} = 0.50 \quad F_{12} = 0.08 \quad F_{13} = 0.42 \quad F_{21} = 1.0 \quad F_{23} = 0$

$F_{31} = F_{32} = 0 \quad F_{33} = 1.0$

8-14 $r_1 = 7.5$ cm $\quad r_2 = 12.5$ cm $\quad L = 7$ cm $\quad$ Open Ends 3 & 4

$L/r_2 = 0.56 \quad r_1/r_2 = 0.5 \quad F_{22} = 0.44 \quad F_{21} = 0.25$

$A_3 = A_4 = \pi(12.5^2 - 7.5^2) = 395.8 \text{ cm}^2$

$A_2 = \pi(25)(7) = 549.8 \text{ cm}^2 \qquad A_1 = \pi(12.5)(7) = 274.9 \text{ cm}^2$

$F_{21} + F_{22} + F_{23} + F_{24} = 1.0 \qquad F_{23} = \frac{1 - 0.44 - 0.25}{2} = 0.155$

$F_{12} = 0.25\left(\frac{549.8}{274.9}\right) = 0.5 \qquad F_{12} + F_{13} + F_{14} = 1.0 \qquad F_{13} = \frac{1 - 0.5}{2} = 0.25$

$F_{31} = 0.25\left(\frac{A_1}{A_3}\right) = (0.25)\left(\frac{274.9}{395.8}\right) = 0.174$

$F_{32} = 0.155\left(\frac{A_3}{A_2}\right) = (0.155)\left(\frac{395.8}{549.8}\right) = 0.112$

$F_{31} + F_{32} + F_{34} = 1.0$

$F_{34} = 1 - 0.174 - 0.112 = 0.714$

8-16

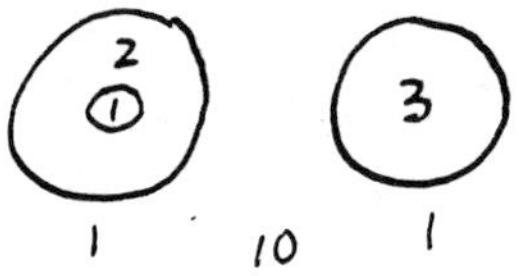

For 2 large disks $d/x = \frac{50}{10} = 5$

$F_{3-21} = 0.65 = F_{21-3}$

For hole to large disk $r_2/L = 10/10 = 1.0$

$L/r_1 = 10/25 = 0.4 \quad F_{31} = 0.15$

$F_{31} + F_{32} = F_{3-21}$

$F_{32} = 0.65 - 0.15 = 0.5$

8-17 $T_1 = 260°C = 533\text{ K} \quad E_{b1} = 4575\text{ W/m}^2 \quad T_2 = 90°C = 363\text{ K}$

$E_{b2} = 984 \quad A_1 = A_2 = 9\text{ m}^2 \quad F_{12} = 0.2 \quad \epsilon_1 = \epsilon_2 = 1.0$

$$q = \frac{(9)(4575 - 984)}{\dfrac{(9+9) - (2)(9)(0.2)}{9 - (9)(0.2)^2}} = 19391\text{ W}$$

8-18 $A_1 = A_2 = (1.2)^2 = 1.44\text{ m}^2 \quad T_1 = 800\text{K} \quad T_2 = 500\text{K}$

$E_{b1} = 23220\text{ W/m}^2 \quad E_{b2} = 3543\text{ W/m}^2 \quad F_{12} = 0.2$

$q_{12} = (1.44)(0.2)(23220 - 3543) = 5667\text{ W}$

8-19 $T_1 = 540°C = 813\text{ K} \quad T_2 = 300°C = 573\text{ K} \quad \epsilon_1 = 0.7 \quad \epsilon_2 = 0.5$

$T_3 = 30°C = 303\text{ K} \quad A_1 = A_2 = 0.2827\text{ m}^2 \quad F_{12} = 0.6 \quad F_{13} = F_{23} = 0.4$

$E_{b1} = 24767\text{ W/m}^2 \quad E_{b2} = 6111 \quad E_{b3} = 478 = J_3$

$$\frac{1-\epsilon_1}{\epsilon_1 A_1} = 1.516 \quad \frac{1-\epsilon_2}{\epsilon_2 A_2} = 3.537 \quad \frac{1}{A_1 F_{12}} = 5.896 \quad \frac{1}{A_1 F_{13}} = \frac{1}{A_2 F_{23}} = 8.843$$

$$\frac{24767 - J_1}{1.516} + \frac{J_2 - J_1}{5.896} + \frac{478 - J_1}{8.843} = 0 \quad \frac{J_1 - J_2}{5.896} + \frac{478 - J_2}{8.843} + \frac{6111 - J_2}{3.537} = 0$$

$J_1 = 18986\text{ W/m}^2 \quad J_2 = 8846\text{ W/m}^2 \quad q_1 = \dfrac{24767 - 18986}{1.516} = 3813\text{ W}$

$$q_2 = \frac{6111 - 8846}{3.537} = -773\text{ W}$$

8-20 $J_{2D} = \epsilon_2 E_{b2} = 3056 \quad \rho_{23} = 0.5 \quad \dfrac{1}{A_1 F_{12}(1-\rho_{23})} = 11.79$

$$\frac{1}{A_2 F_{23}(1-\rho_{23})} = 17.69 \qquad \frac{24767 - J_1}{1.516} + \frac{\dfrac{3.056}{0.5} - J_1}{11.79} + \frac{478 - J_1}{8.843} = 0$$

$$J_1 = 19719 \text{ W/m}^2 \qquad q_1 = \frac{24767 - 19719}{1.516} = 3330 \text{ W}$$

$$q_2 = \frac{6111 - 19719}{11.79} + \frac{6111 - 478}{17.69} = -836 \text{ W}$$

8-21 $E_{b1} = 459 \text{ W/m}^2 \quad E_{b2} = 401 \text{ W/m}^2 \quad F_{12} = 0.25 \quad A_1 = A_2 = 9$

$\epsilon_1 = 0.8 \quad \epsilon_2 = 0.8$

$$q = \frac{(9)(459 - 401)}{\dfrac{9 + 9 - (2)(9)(0.25)}{9 - (9)(0.25)^2} + (2)\left(\dfrac{1}{0.8} - 1\right)} = 248.6 \text{ W}$$

From symmetry $J_3 = \frac{1}{2}(E_{b1} + E_{b2}) = E_{b3} = 430$

$$T_3 = \left(\frac{E_{b3}}{\sigma}\right)^{1/4} = 295 \text{ K} = 22.1°\text{C}$$

8-22 $A_1 = A_2 = (1.5)^2 = 2.25 \text{ m}^2 \quad T_1 = 800°\text{C} = 1073 \text{K} \quad T_2 = 553 \text{K}$

$\epsilon_1 = 0.5 \quad \epsilon_2 = 0.8 \quad T_3 = 0 \text{ K} \quad F_{12} = 0.7 \quad F_{13} = F_{23} = 0.3 \quad E_{b1} = 75146 \text{ W/m}^2$

$E_{b2} = 5302 \text{ W/m}^2$

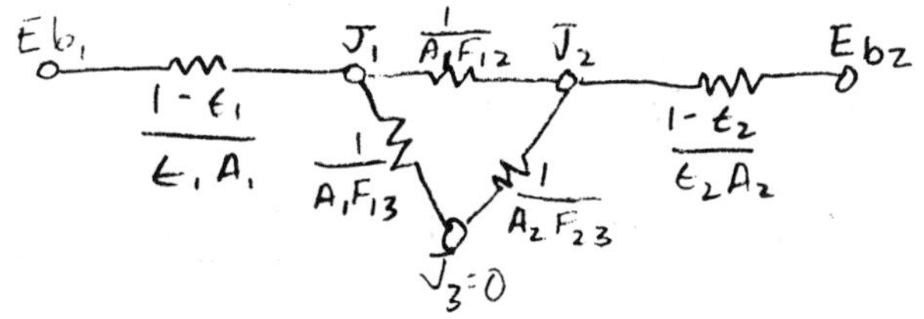

Node J_1

$$\frac{75146 - J_1}{0.4444} + \frac{J_2 - J_1}{0.6349} + \frac{0 - J_1}{1.481} = 0$$

Node J_2

$$\frac{J_1 - J_2}{0.6349} + \frac{0 - J_2}{1.481} + \frac{5302 - J_2}{0.1111} = 0$$

$J_1 = 41070 \text{ W/m}^2 \quad J_2 = 9992 \text{ W/m}^2$

$$q_1 = \frac{75146 - 41070}{0.4444} = 76671 \text{ W} \qquad q_2 = \frac{5302 - 9992}{0.1111} = -42210 \text{ W}$$

8-23 $A_1 = A_2 = (0.9)(0.6) = 0.54\ m^2$ $\quad \epsilon_1 = 0.6$ $\quad E_{b1} = 23220\ W/m^2$

$E_{b2} = 401\ W/m^2$ $\quad A_3 \to \infty$ $\quad F_{12} = 0.25$ $\quad F_{13} = 0.75 = F_{23}$

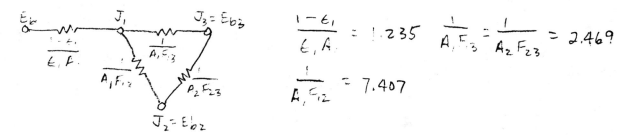

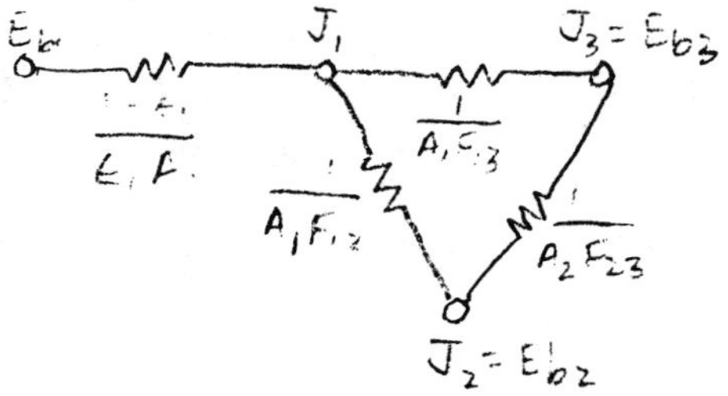

$$\frac{1-\epsilon_1}{\epsilon_1 A_1} = 1.235 \qquad \frac{1}{A_1 F_{13}} = \frac{1}{A_2 F_{23}} = 2.469$$

$$\frac{1}{A_1 F_{12}} = 7.407$$

$$q = \frac{23220 - 401}{1.235 + \dfrac{1}{\dfrac{1}{2.469} + \dfrac{1}{2.469 + 7.407}}} = 7108\ W = \frac{23220 - J_1}{1.235}$$

$J_1 = 14441\ W/m^2$ $\qquad \dfrac{14441 - J_2}{7.407} = \dfrac{J_2 - 401}{2.469}$; $\quad J_2 = 3911 = E_{b2} = \sigma T_2^4$

$T_2 = 512.5\ K = 239.5\ °C$

8-24 $q/L = (5.669 \times 10^{-8})\,\pi\,(0.05)(0.6)\left[366^4 - 293^4\right] = 56.5\ W/m = 17.22\ \frac{W}{ft}$

8-25 $q = \sigma A \epsilon\,(T_1^4 - T_2^4) = (5.669 \times 10^{-8})(0.6)(0.3)(0.8)\left[368^4 - 293^4\right]$

$= 89.55\ W$

8-27 A_1 = heater $\quad A_2$ = side walls $\quad A_4$ = room $\to \infty$ $\quad A_3$ = belt

$J_2 = E_{b2}$ $\quad T_1 = 698\ K$ $\quad T_3 = 393\ K$ $\quad T_4 = 298\ K$ $\quad E_{b1} = 13456\ W/m^2$

$E_{b3} = 1352$ $\quad E_{b4} = 447$ $\quad \epsilon_1 = 0.7$ $\quad \epsilon_3 = 0.8$ $\quad A_1 = A_3 = 3\ m^2$ $\quad A_2 = 1.8\ m^2$

$F_{12} = F_{32} = 0.28$ $\quad F_{31} = F_{13} = 0.67$ $\quad F_{14} = F_{34} = 1 - 0.67 - 0.28 = 0.05$

$F_{24} = (0.04)(2) = 0.08$ $\quad \dfrac{1}{A_3 F_{32}} = 1.19$ $\quad \dfrac{1}{A_1 F_{13}} = 0.4975$ $\quad \dfrac{1}{A_1 F_{12}} = 1.19$

$\dfrac{1}{A_2 F_{24}} = 6.944$ $\quad \dfrac{1}{A_1 F_{14}} = 6.667$ $\quad \dfrac{1}{A_3 F_{34}} = 6.667$ $\quad \dfrac{1-\epsilon_1}{\epsilon_1 A_1} = 0.1429$

8-27 contd

$$\frac{1-\epsilon_3}{A_3\epsilon_3} = 0.0833 \qquad \frac{13456-J_1}{0.1429} + \frac{J_2-J_1}{1.19} + \frac{J_3-J_1}{0.4975} + \frac{447-J_1}{6.667} = 0$$

$$\frac{J_1-J_2}{1.19} + \frac{J_3-J_2}{1.19} + \frac{447-J_2}{6.944} = 0 \qquad \frac{J_1-J_3}{0.4975} + \frac{J_2-J_3}{1.19} + \frac{1352-J_3}{0.0833} + \frac{447-J_3}{6.667} = 0$$

Solving: $J_1 = 10515$ $\quad J_2 = 6185$ $\quad J_3 = 2841$

$$q_1 = \frac{13456-10515}{0.1429} = 20580 \text{ W}$$

8-28 $T_2 = 1200\text{K}$ $\quad \epsilon_2 = 0.75$ $\quad \epsilon_3 = 0.3$ $\quad q_3 = 0$ $\quad T_1 = 293\text{K}$

$E_{b2} = \sigma T_2^4 = 1.176\times10^5$ $\quad E_{b1} = \sigma T_1^4 = 417.8$

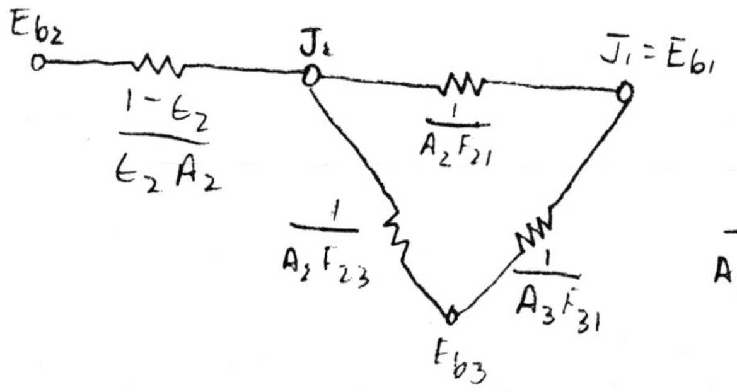

$$\frac{1-\epsilon_2}{\epsilon_2 A_2} = 42.44 \qquad \frac{1}{A_2F_{23}} = 167.53$$

$$\frac{1}{A_2F_{21}} = 530.52 \qquad \frac{1}{A_3F_{31}} = 361.84$$

$$R_{eq} = 42.44 + \frac{1}{\frac{1}{530.52} + \frac{1}{167.53 + 361.84}} = 307.44$$

$$q = \frac{1.176\times10^5 - 417.8}{307.4} = 381.2 \text{ W} = \frac{1.176\times10^5 - J_2}{42.44}$$

$$J_2 = 1.014\times10^5 \qquad \frac{1.014\times10^5 - E_{b3}}{167.53} = \frac{1.014\times10^5 - 417.8}{167.53 + 361.84}$$

$E_{b3} = 6.944\times10^4$ $\quad T_3 = 1052$ K

8-29 $L = 30$ cm $\quad r_1 = 4$ cm $\quad F_{12} = 0.8 \quad r_2 F_{21} = r_1 F_{12}$

Iteration Solution

r_2	r_1/r_2	L/r_2	F_{21}	$F_{12} = F_{21}\frac{r_2}{r_1}$	$0.8 - F_{12}$
8	0.5	3.75	0.45	0.9	−0.1
10	0.4	3.0	0.35	0.87	−0.07
12	0.33	2.5	0.27	0.81	−0.01
16	0.25	1.875	0.18	0.72	+0.08
12.44	0.3215	2.41	Interpolated		0

8-30 $d = 75$ cm $\quad x = 50$ cm $\quad d/x = 1.5 \quad q/A|_1 = 7000 \quad \epsilon_1 = 0.8 \quad \epsilon_2 = 0.6$

$T_2 = 400$ K $\quad F_{12} = F_{21} = 0.3 \quad F_{13} = F_{23} = 0.7 \quad E_{b2} = 1451$

$A_1 = A_2 = \pi\left(\frac{0.75}{2}\right)^2 = 0.4418 \qquad A_3 = \pi(0.5)(0.75)\ 1.178$

$F_{31} = F_{32} = 0.263 \qquad F_{33} = 1-(0.2)(0.263) = 0.474$

$\frac{1-\epsilon_1}{\epsilon_1 A_1} = 0.5659 \qquad \frac{1-\epsilon_2}{\epsilon_2 A_2} = 1.509 \qquad \frac{1}{A_1 F_{12}} = 7.545 \qquad \frac{1}{A_1 F_{13}} = \frac{1}{A_2 F_{23}} = 3.234$

$J_1 - [F_{12} J_2 + F_{13} J_3] = 7000 \qquad J_2 - 0.4[F_{21} J_1 + F_{23} J_3] = 871$

$J_3(1 - 0.474) - [F_{31} J_1 + F_{32} J_2] = 0$

Solutions: $\quad J_1 = 16943 \quad J_2 = 6131 \quad J_3 = 11537 = E_{b3}$

$T_3 = 672$ K $\qquad \dfrac{E_{b1} - 16943}{\frac{1-0.8}{0.8}} = 7000 \qquad E_{b1} = 44943$

$T_1 = 944$ K

8-31

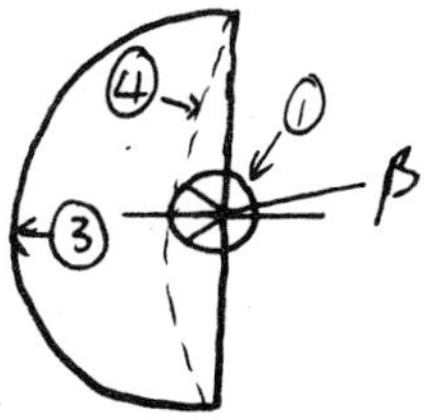

(2)

$T_1 = 1253\ K \quad T_2 = 288\ K \quad F_{13} = 0.5 \quad F_{12} = 0.5$

$\beta = 2\alpha \quad \sin\alpha = \frac{3.75}{25} \quad \alpha = 8.627° \quad \beta = 17.25°$

$A_4 = 2(0.25 \cos\alpha) + \pi(0.075)\left(\frac{2\alpha}{360}\right) = 0.5056$

$F_{43} = 1.0 \quad A_3 = \pi(0.25) = 0.7854$

$F_{34} = F_{31} + F_{32} = \frac{(0.5056)(1.0)}{0.7854} = 0.6437$

$A_1 = \pi(0.075) = 0.2356 \qquad F_{31} = \frac{(0.2356)(0.5)}{0.7854} = 0.15 \qquad F_{32} = 0.6437 - 0.15 = 0.4937$

$\frac{1-\epsilon_1}{\epsilon_1 A_1} = 1.061$

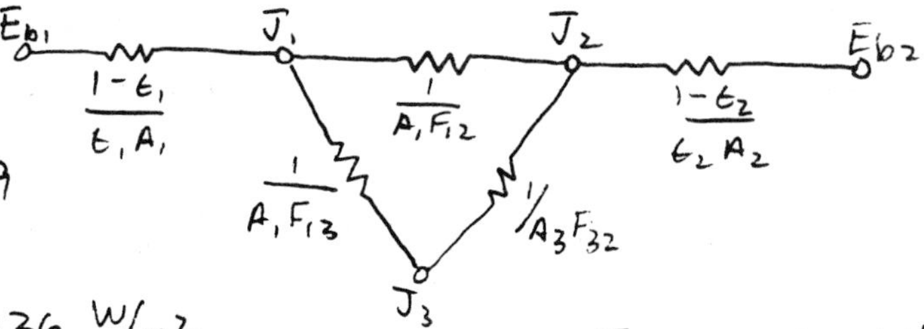

$\frac{1}{A_1F_{12}} = 8.489 \qquad \frac{1}{A_3F_{32}} = 2.579$

$\frac{1}{A_1F_{13}} = 8.489 \qquad E_{b1} = 139736\ W/m^2 \qquad E_{b2} = 390\ W/m^2$

$$q = \frac{139736 - 390}{1.061 + \dfrac{1}{\dfrac{1}{8.489} + \dfrac{1}{8.489 + 2.579}}} = 23758\ W/m$$

q w/o reflector $= \epsilon_1 A_1 (E_{b1} - E_{b2}) = 26264\ W/m$

8-32

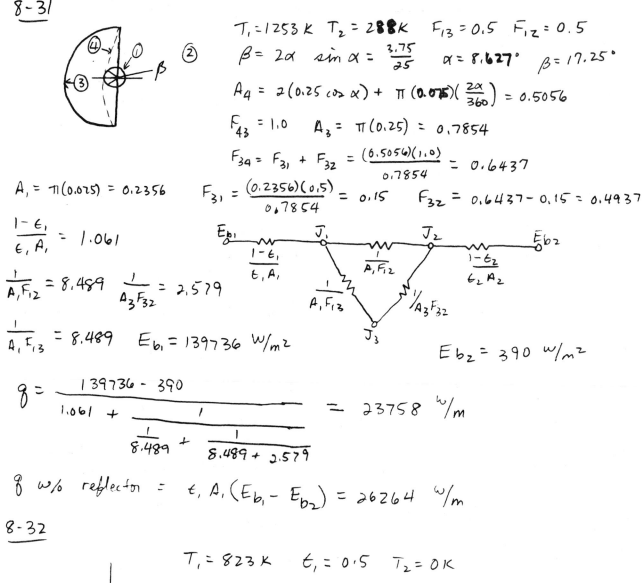

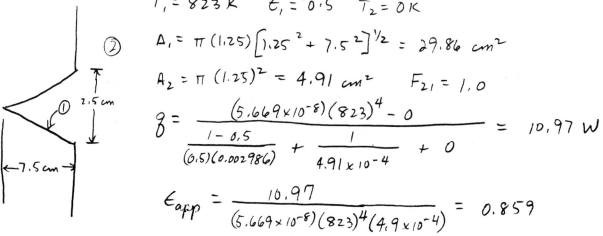

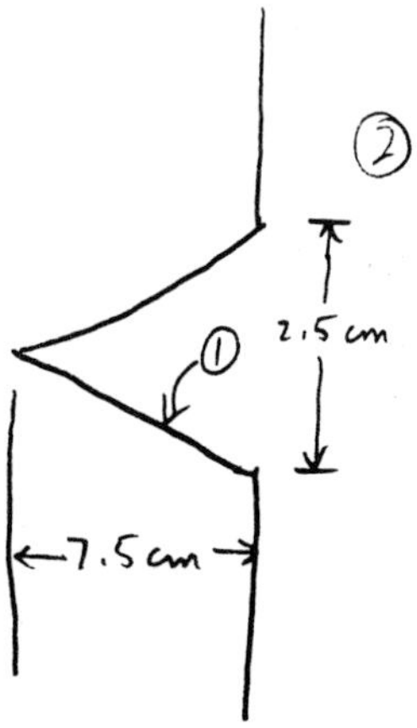

(2)

$T_1 = 823\ K \quad \epsilon_1 = 0.5 \quad T_2 = 0\ K$

$A_1 = \pi(1.25)\left[1.25^2 + 7.5^2\right]^{1/2} = 29.86\ cm^2$

$A_2 = \pi(1.25)^2 = 4.91\ cm^2 \qquad F_{21} = 1.0$

$$q = \frac{(5.669\times10^{-8})(823)^4 - 0}{\dfrac{1-0.5}{(0.5)(0.002986)} + \dfrac{1}{4.91\times10^{-4}} + 0} = 10.97\ W$$

$$\epsilon_{app} = \frac{10.97}{(5.669\times10^{-8})(823)^4(4.9\times10^{-4})} = 0.859$$

8-33

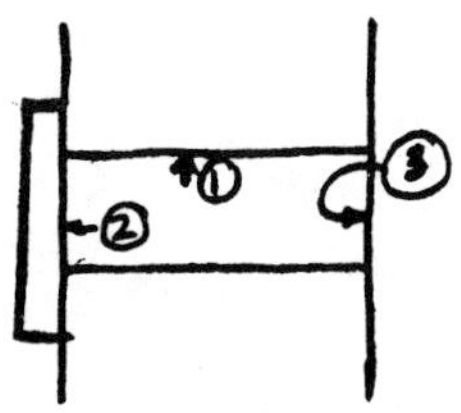

$T_1 = 260°C = 533°K$ $\quad \epsilon_3 = 1.0$

$T_2 = 425°C = 698°K$ $\quad \epsilon_2 = 0.5$

$\epsilon_1 = 0.07$ $\quad T_3 = 0°K$

$F_{23} = 0.04$ $\quad F_{21} = 0.96$

$A_1 = \pi(2.5)(7.5) = 58.9\ cm^2$

$A_2 = A_3 = \pi(1.25)^2 = 4.91\ cm^2$

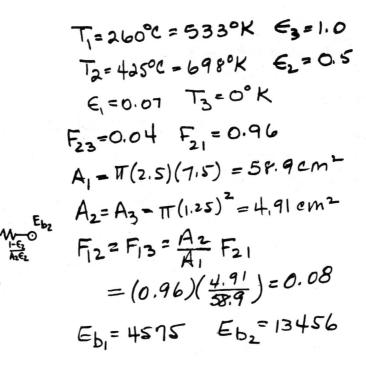

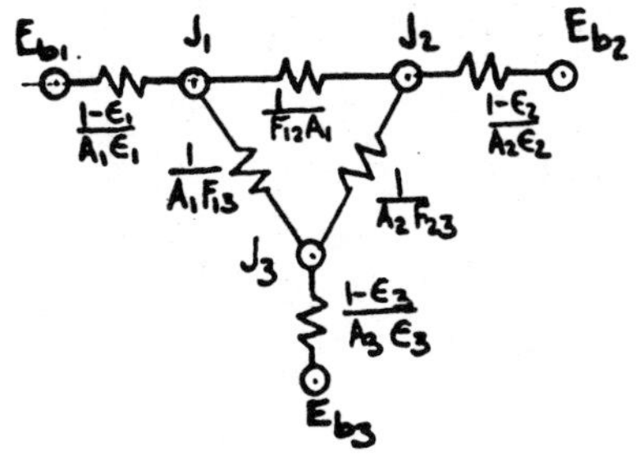

$F_{12} = F_{13} = \frac{A_2}{A_1} F_{21}$

$= (0.96)\left(\frac{4.91}{58.9}\right) = 0.08$

$E_{b1} = 4575$ $\quad E_{b2} = 13456$

Node J_1

$$\frac{4575 - J_1}{2256} + \frac{J_2 - J_1}{2121} + \frac{0 - J_1}{2122} = 0$$

Node J_2

$$\frac{13456 - J_2}{2037} + \frac{J_1 - J_2}{2121} + \frac{0 - J_2}{50916} = 0$$

$J_1 = 4484$ $\quad J_2 = 8879$

$q_1 = \frac{4575 - 4484}{2256} = 0.0403$

$q_2 = \frac{13456 - 8879}{2037} = 2.247$ $\quad q_3 = 2.287\ W$

7-34

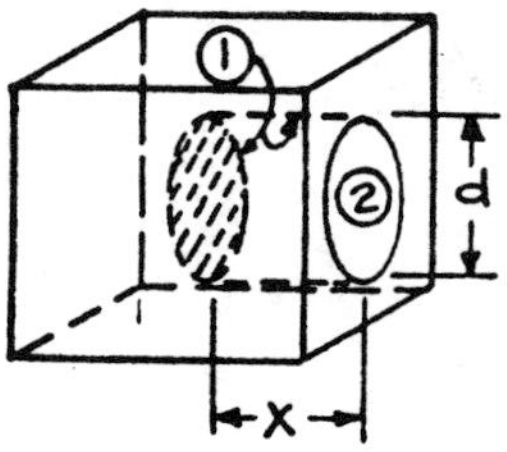

Plot apparent emissivity ϵ_{ap} Vs $\frac{x}{d}$ and ϵ

$$q_{1-2} = \frac{E_{b_1} - 0}{\frac{1-\epsilon_1}{A_1\epsilon_1} + \frac{1}{A_2}}$$

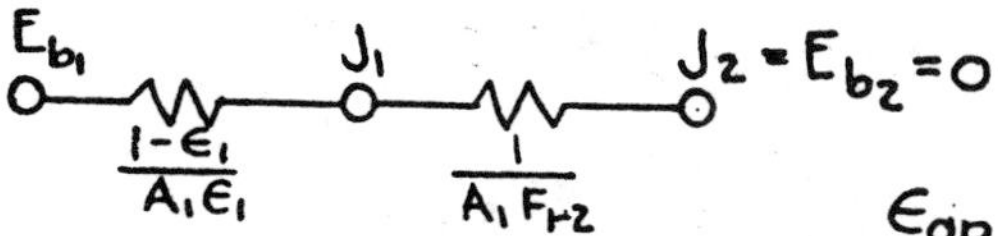

$$\epsilon_{ap} = \frac{q_{1-2}}{(q_2)_b} = \frac{E_{b_1}}{\left[\frac{(1-\epsilon_1)A_2 + A_1\epsilon_1}{A_1A_2\epsilon_1}\right]E_{b_1}A_2}$$

$$A_1F_{1-2} = A_2F_{2-1} = A_2$$

$$F_{2-1} = 1$$

$$\epsilon_{ap} = \frac{4}{\left(\frac{1}{\frac{x}{d}+\frac{1}{4}}\right)\left(\frac{1-\epsilon_1}{\epsilon_1}\right)+4}$$

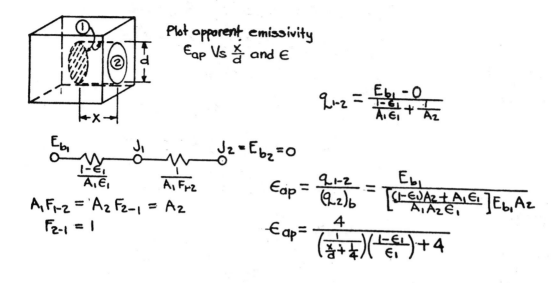

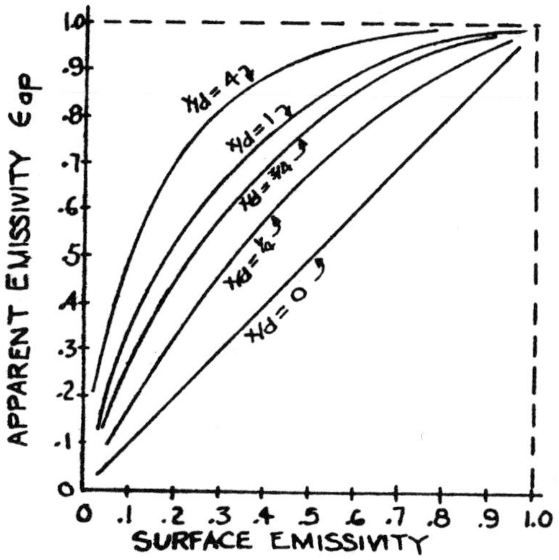

8-35

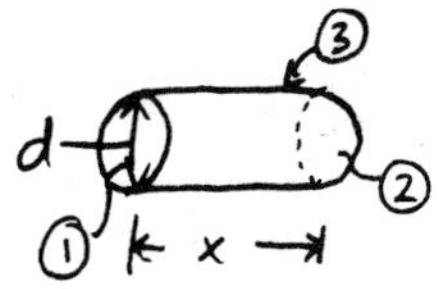

$\epsilon_3 = \epsilon \qquad E_{b1} = E_{b2} = 0 \qquad \epsilon_2 = \epsilon_1 = 1.0$

$F_{13} = F_{23} = 1 - F_{12} \qquad A_2 = A_1 = \frac{\pi d^2}{4} \qquad A_3 = \pi x d$

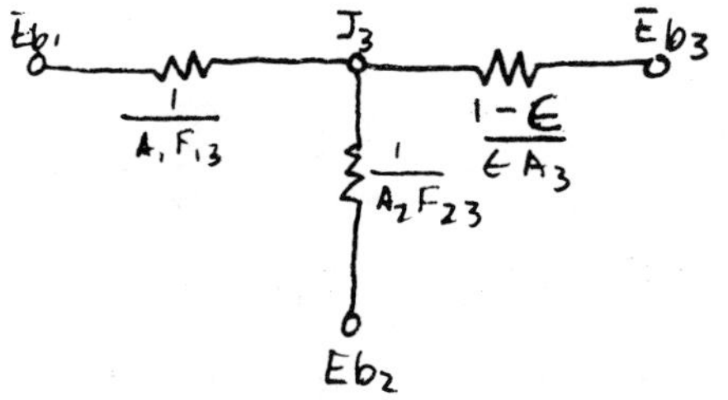

$$q_1 = \frac{\frac{1}{2}(E_{b3} - 0)}{\frac{1-\epsilon_3}{\epsilon_3 A_3} + \frac{1}{A_1(1-F_{12}) + A_2(1-F_{12})}}$$

$$\epsilon_{App1} = q_1 / A_1 E_{b3}$$

$$\epsilon_{app1} = \epsilon_{app2} = \frac{\frac{1}{2}}{\left(\frac{1-\epsilon}{\epsilon}\right)\left(\frac{d}{4x}\right) + \frac{1}{2(1-F_{12})}}$$

F_{12} is obtained from Fig. 8-13. Note that E_{app} approaches 1.0 as $\frac{x}{d} \to \infty$

As $\frac{x}{d} \to 0 \quad E_{app} \to 0$

8-36 $r_1 = 0.5 \quad L = 1 \quad T_1 = 800\ K \quad \epsilon_1 = 0.65 \quad r_2 = 1.0 \quad q_2 = 0 \quad T_3 = 300\ K$

$r_1/r_2 = 0.5 \quad L/r_2 = 1.0 \quad F_{21} = 0.25 \quad F_{22} = 0.23 \quad F_{23} = 0.52 \quad F_{12} = F_{13} = 0.5$

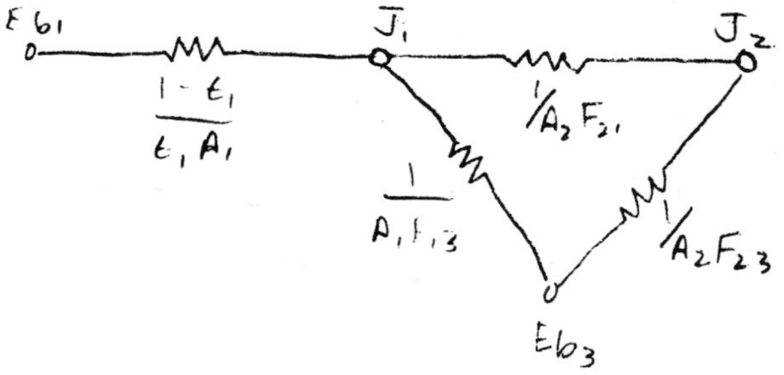

$E_{b1} = (5.669 \times 10^{-8})(800)^4$

$= 23220\ W/m^2$

$E_{b3} = (5.669 \times 10^{-8})(300)^4$

$= 459\ W/m^2$

$A_1 = \pi(1)(1) = 3.142\ m^2 \qquad A_2 = \pi(2)(1) = 6.283\ m^2$

$\frac{1-\epsilon_1}{\epsilon_1 A_1} = 0.1712 \qquad \frac{1}{A_2 F_{21}} = 0.6366 \qquad \frac{1}{A_2 F_{23}} = 0.3061 \qquad \frac{1}{A_1 F_{13}} = 0.6366$

$$q = \frac{23220 - 459}{0.1712 + \frac{1}{\frac{1}{0.6366} + \frac{1}{0.6366 + 0.3061}}} = 34980\ W$$

8-37 $T_1 = 973\,K$ $E_{b1} = 50811\ W/m^2$ $\epsilon_1 = 0.8$ $A_1 = (2)(3) = 6\ m^2$

② Surface = side walls $A_2 = (10)(2) = 20\ m^2$ ③ = room $T_3 = 303\ K$

$E_{b3} = 478\ W/m^2$

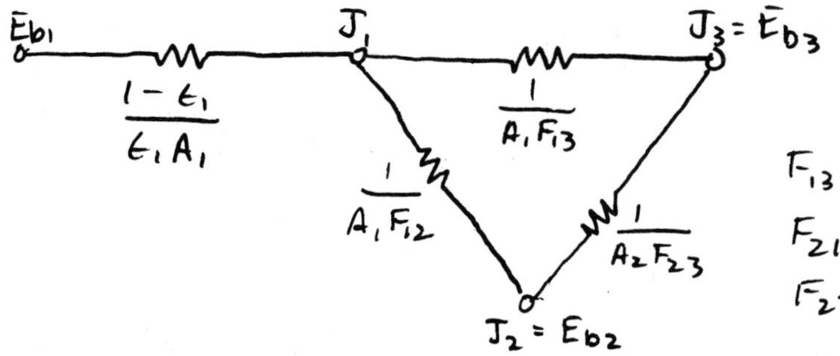

$\frac{X}{D} = 1.0$ $\frac{Y}{D} = 1.5$

$F_{13} = 0.25$ $F_{12} = 0.75$

$F_{21} = \frac{6}{20}(0.75) = 0.225$

$F_{23} = 0.225$

$\frac{1-\epsilon_1}{\epsilon_1 A_1} = 4.1666 \times 10^{-2}$ $\frac{1}{A_1F_{13}} = 0.6667$ $\frac{1}{A_1F_{12}} = 0.2222$ $\frac{1}{A_2F_{23}} = 0.2222$

$$q = \frac{50811 - 478}{0.0417 + \dfrac{1}{\dfrac{1}{0.6667} + \dfrac{1}{0.2222 + 0.2222}}}$$

$= 1.633 \times 10^5\ W$

8-38 $T_1 = 1173\,K$ $E_{b1} = 1.073 \times 10^5\ W/m^2$ $\epsilon_1 = 0.6$ $T_3 = 303\ K$

Surface ② is sphere in radiant balance. $E_{b3} = 478\ W/m^2$

$d_1 = 0.03$ $d_2 = 0.09$ $\epsilon_2 = 0.3$

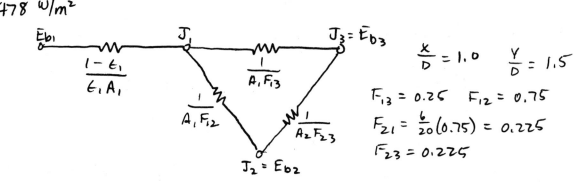

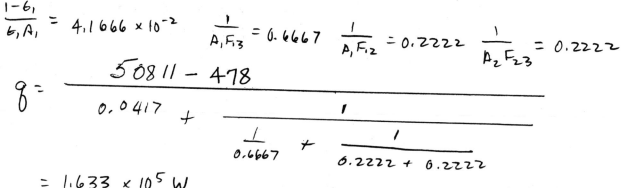

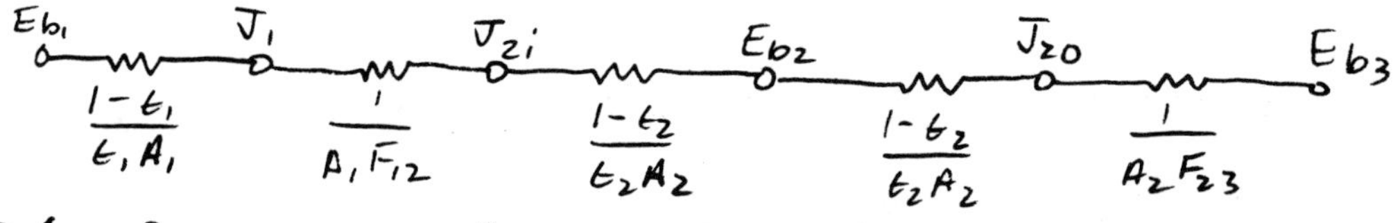

$A_1 = 4\pi r_1^2 = 2.83 \times 10^{-3}\ m^2$ $A_2 = 4\pi r_2^2 = 0.0254\ m^2$

$F_{12} = 1.0$ $F_{23} = 1.0$ $\frac{1-\epsilon_1}{\epsilon_1 A_1} = 235.6$ $\frac{1}{A_1F_{12}} = 353.4$ $\frac{1-\epsilon_2}{\epsilon_2 A_2} = 91.86$

$\frac{1}{A_2F_{23}} = 39.37$ $q = \frac{\Delta E_b}{\Sigma R} = \frac{107320 - 478}{235.6 + 353.4 + (2)(91.86) + 39.37}$

$q = 131.6\ W$

8-39 $T_1 = 873$ $\epsilon_1 = 0.65$ $q_2 = 0$ $T_3 = 303\,K$ $F_{12} = 0.2$ $F_{13} = F_{23} = 0.8$

$A_1 = A_2 = (0.6)^2 = 0.36$ $\frac{1-\epsilon_1}{\epsilon_1 A_1} = 1.496$ $\frac{1}{A_1 F_{13}} = 3.472 = \frac{1}{A_2 F_{23}}$

$\frac{1}{A_1 F_{12}} = 13.889$ $E_{b1} = 32928$ $E_{b3} = 478$

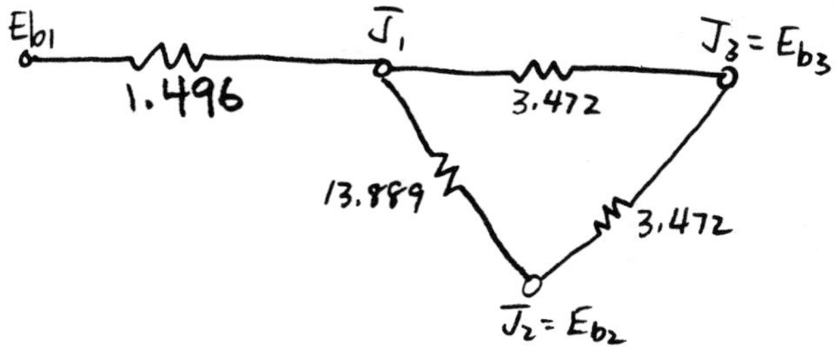

$$q = \frac{32928 - 478}{1.496 + \dfrac{1}{\dfrac{1}{3.472} + \dfrac{1}{13.889 + 3.472}}} = 7393\text{ W}$$

$$= \frac{32928 - J_1}{1.496} \qquad J_1 = 21868\ \frac{\text{W}}{\text{m}^2} \qquad \frac{21868 - J_2}{13.889} = \frac{21868 - 478}{13.889 + 3.472}$$

$J_2 = 4756 = E_{b2}$ $T_2 = 538\,K$

8-40 $r_1 = 5\text{cm}$ $r_2 = 10\text{cm}$ $L = 10\text{cm}$ $T_1 = 973\,K$ $T_3 = 303\,K$ $\epsilon_1 = 0.6$

$\epsilon_2 = 0.7$ $q_2 = 0$ $r_1/r_2 = 0.5$ $L/r_2 = 1.0$ $F_{21} = 0.25$ $F_{12} = 0.50$ $F_{22} = 0.22$

$F_{13} = 0.5$ $F_{23} = 0.53$ $A_1 = \pi(0.1)^2 = 0.03146$ $A_2 = \pi(0.2)(0.1) = 0.06283$

$\frac{1-\epsilon_1}{\epsilon_1 A_1} = 21.19$ $\frac{1}{A_1 F_{12}} = 63.57$ $\frac{1}{A_1 F_{13}} = 63.57$ $\frac{1}{A_2 F_{23}} = 30.03$

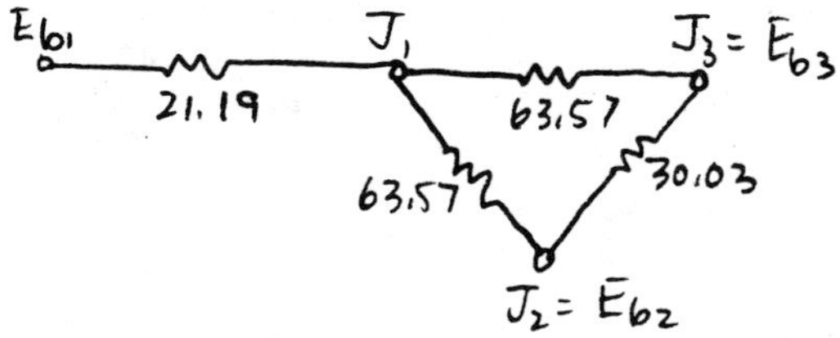

$$q = \frac{50811 - 478}{21.19 + \dfrac{1}{\dfrac{1}{63.57} + \dfrac{1}{63.57 + 30.03}}} = 852\text{ W}$$

8-41 $A_1 = A_3 = 25\ m^2$ $\quad A_2 = (4)(5)(4) = 80\ m^2$ $\quad T_1 = 301\ K$ $\quad T_3 = 293\ K$

$E_{b1} = 465.3$ $\quad E_{b3} = 417.8\ W/m^2$ $\quad \epsilon_1 = 0.62$ $\quad \epsilon_3 = 0.75$ $\quad \frac{X}{D} = \frac{Y}{D} = 1.25$

$F_{13} = 0.35$ $\quad F_{12} = 0.65$ $\quad \frac{1-\epsilon_1}{\epsilon_1 A_1} = 0.02452$ $\quad \frac{1-\epsilon_3}{\epsilon_3 A_3} = 0.01333$

$\frac{1}{A_1F_{12}} = 0.06154 = \frac{1}{A_3F_{32}}$ $\quad \frac{1}{A_1F_{13}} = 0.01428$

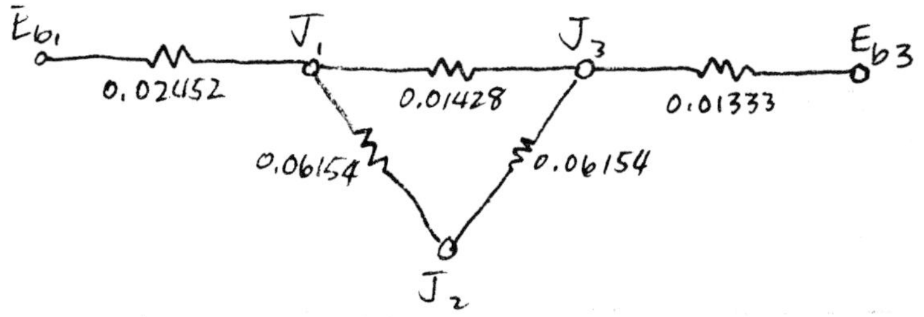

$$q = \frac{465.3 - 417.8}{0.02452 + \dfrac{1}{\dfrac{1}{0.01428} + \dfrac{1}{2(0.06154)}} + 0.01333} = 489.1\ W$$

$$= \frac{E_{b1} - J_1}{0.02452} = \frac{J_3 - E_{b3}}{0.01333} \qquad J_1 = 453.3\ W/m^2 \quad J_3 = 424.3\ W/m^2$$

from Symmetry: $J_2 = \frac{J_1 + J_3}{2} = 438.8\ W/m^2 = \sigma T_2^{\,4}$ $\quad T_2 = 296.6\ K$
$= 23.6°C$

8-42

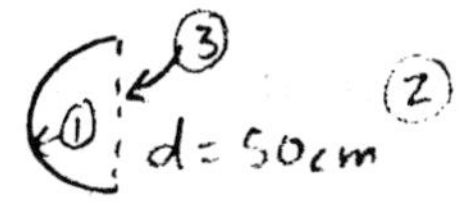

$T_1 = 1000\ K$ $\quad T_2 = 300\ K$ $\quad \epsilon_1 = 0.7$

$F_{31} = 0.5$ $\quad F_{13} = F_{12} = 0.5\ \frac{A_3}{A_1} = 0.5\left(\frac{d}{\pi d/2}\right) = 0.3183$

$E_{b1} = 56690\ W/m^2$ $\quad E_{b2} = 459\ W/m^2$ $\quad$ For unit length $A_1 = \pi d = 1.571$

$\frac{1-\epsilon_1}{\epsilon_1 A_1} = 0.2728$ $\quad \frac{1}{A_1F_{12}} = 2.0$

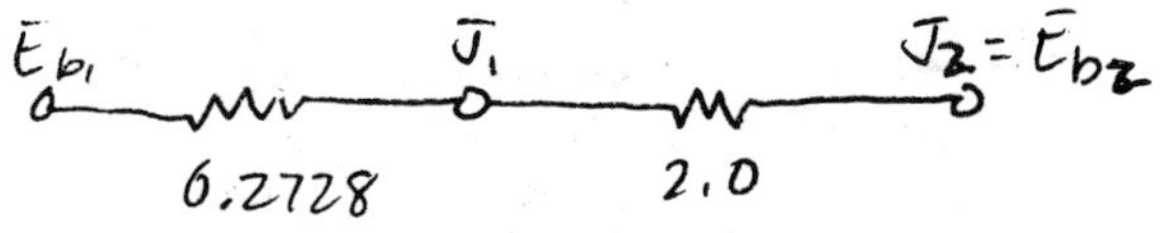

$$q/L = \frac{56690 - 459}{0.2728 + 2.0} = 2.47 \times 10^4\ W$$

7-43 $d/x = \frac{50}{12.5} = 4.0 \quad F_{12} = 0.59 \quad F_{13} = F_{23} = 0.41$

$A_1 = A_2 = \pi(0.25)^2 = 0.1963\ m^2 \quad \epsilon_1 = \epsilon_2 = 0.8 \quad T_1 = 1000K \quad T_2 = 500K$

$T_3 = 300K \quad E_{b1} = 56690 \quad E_{b2} = 3543\ W/m^2 \quad E_{b3} = 459$

$$\frac{1-\epsilon_1}{\epsilon_1 A_1} = \frac{1-\epsilon_2}{\epsilon_2 A_2} = 1.274 \quad \frac{1}{A_1 F_{12}} = 8.634 \quad \frac{1}{A_1 F_{13}} = \frac{1}{A_2 F_{23}} = 12.425$$

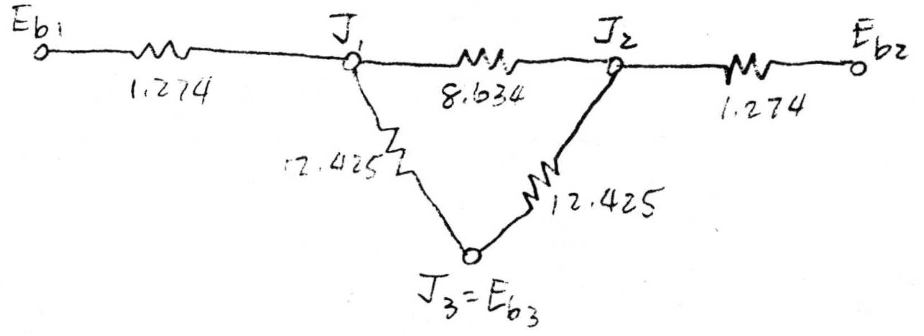

$$\frac{56690 - J_1}{1.274} + \frac{J_2 - J_1}{8.634} + \frac{459 - J_1}{12.425} = 0$$

$$\frac{J_1 - J_2}{8.634} + \frac{459 - J_2}{12.425} + \frac{3543 - J_2}{1.274} = 0$$

$J_1 = 46373\ W/m^2 \qquad J_2 = 8346\ W/m^2$

$$q_1 = \frac{56690 - J_1}{1.274} = 8098\ W$$

$$q_2 = \frac{3543 - J_2}{1.274} = -3770\ W$$

8-44 Disk = ① shield = ② room = ③ $d/x = 50/25 = 2.0$

$F_{13} = 0.28$ $F_{12} = 0.72$ $T_1 = 1273\text{K}$ $\epsilon_1 = 0.55$ $\epsilon_2 = 0.1$ $T_3 = 30°\text{C}$

$E_{b1} = 148870 \text{ W/m}^2$ $E_{b3} = 459 \text{ W/m}^2$ $A_1 = \pi(0.25)^2 = 0.1963 \text{ m}^2$

$A_2 = \pi(0.5)(0.25) = 0.3927 \text{ m}^2$ $F_{21} = F_{12}\dfrac{A_1}{A_2} = 0.36 = F_{23}$ (inside)

F_{23} (outside) $= 1.0$ $\dfrac{1-\epsilon_1}{\epsilon_1 A_1} = 4.168$ $\dfrac{1-\epsilon_2}{\epsilon_2 A_2} = 22.918$

$\dfrac{1}{A_1 F_{12}} = 7.075$ $\dfrac{1}{A_1 F_{13}} = 18.194$ $\dfrac{1}{A_2 F_{23}} = 7.074$ (inside) $\dfrac{1}{A_2 F_{23}} = 2.546$ (outside)

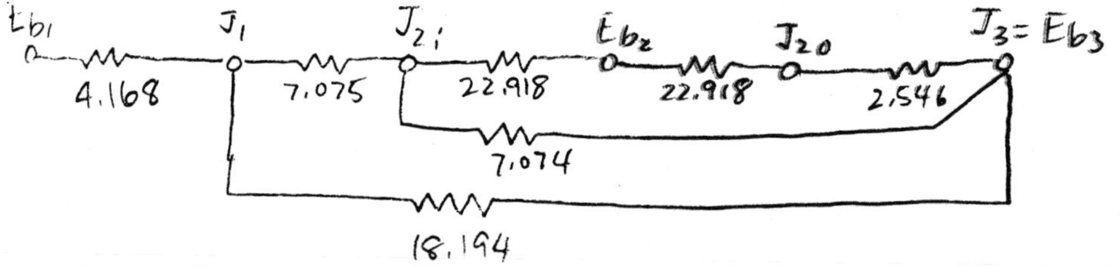

$q = \dfrac{E_{b1} - E_{b3}}{\Sigma R}$ $\dfrac{1}{R_1} = \dfrac{1}{(22.918)(2) + 2.546} + \dfrac{1}{7.074}$

$\dfrac{1}{R_2} = \dfrac{1}{R_1 + 7.075} + \dfrac{1}{18.194}$ $\Sigma R = 4.168 + R_2$

$R_1 = 6.1716$ $R_2 = 7.648$ $\Sigma R = 11.817$

$q = \dfrac{148870 - 478}{11.817} = 1.256 \times 10^4 \text{ W} = \dfrac{148870 - J_1}{4.168}$; $J_1 = 96529 \text{ W/m}^2$

$\dfrac{96529 - J_{2i}}{7.075} = \dfrac{J_{2i} - 459}{6.1716}$ $J_{2i} = 45218 \text{ W/m}^2$

$\dfrac{J_{2i} - J_{2o}}{(2)(22.918)} = \dfrac{J_{2i} - 459}{48.382}$ $J_{2o} = 2380$

$E_{b2} = \dfrac{J_{2i} + J_{2o}}{2} = 23800 \text{ W/m}^2 = \sigma T_2^4$

$T_2 = 805 \text{ K} = 532°\text{C}$

8-46 (1500) $\alpha = \epsilon\sigma T^4$ $\quad \epsilon = \alpha \quad T = 403.3\text{ K} = 130.3°C$

8-47 (a) Dark side

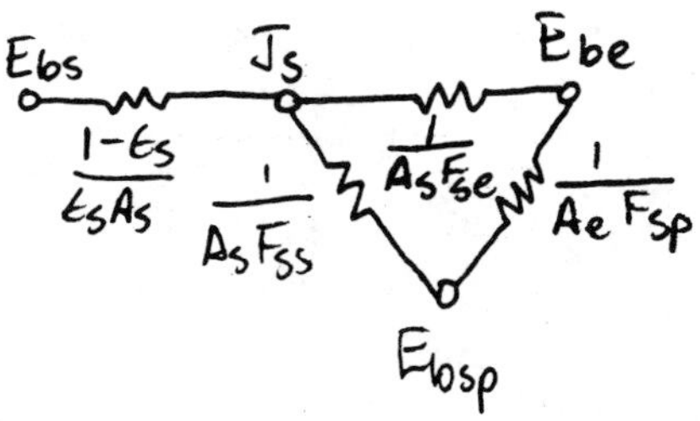

at equilibrium $J_s = E_{bs}$

$\sin\phi = \dfrac{6.45}{6.95}$

$\phi = 70.3°$

$\cos\phi = 0.3367$

$$\text{solid angle} = \frac{1-\cos\phi}{2} = 0.3316 = F_{se} \qquad F_{ss} = 0.6684 \qquad F_{esp} \approx 1.0$$

$A_s = \pi(1)^2 = \pi\ m^2 \qquad E_{be} = (5.669\times10^{-8})(288)^4 = 390\ W/m^2$

$$E_{bsp} = 0 \qquad \frac{390 - J_s}{1/0.3316} + \frac{0 - J_s}{1/0.6684} = 0 \qquad J_s = 129.3 = E_s = \sigma T_s^4$$

$T_s = 218.5\text{ K} = -5.4°C$

(b) BRIGHT SIDE $\quad A_{view} = \pi(1/2)^2 = \pi/4$ for sun

For polished Aluminum $\epsilon = 0.048 \qquad q\ \text{absorbed} = (0.048)(\pi/4)(1400) = 52.779\ W$

$$\frac{E_{bs} - J_s}{\dfrac{1-\epsilon_s}{\epsilon_s A_s}} = 52.779 \qquad 52.779 + \frac{390 - J_s}{1/\pi(0.3316)} + \frac{0 - J_s}{1/\pi(0.6684)} = 0$$

$$J_s = 146.1 \qquad E_{bs} = 146.1 + \frac{(52.779)(1-0.048)}{(0.048)(\pi)} = 479.3\ W/m^2$$

$$T_s = \left(\frac{479.3}{5.669\times10^{-8}}\right)^{1/4} = 303.2\text{ K} = 30.2°C$$

8-48 $A_1 = A_2 = 0.09\ m^2$ $T_1 = 423\,K$ $\epsilon_1 = \epsilon_2 = 0.8$ $T_3 = 293\,K$

$h = 1.42\left(\frac{\Delta T}{L}\right)^{1/4}$ $F_{12} = 0.55$ $F_{13} = F_{23L} = 0.45$ $F_{23R} = 1.0$

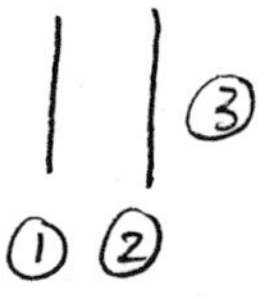

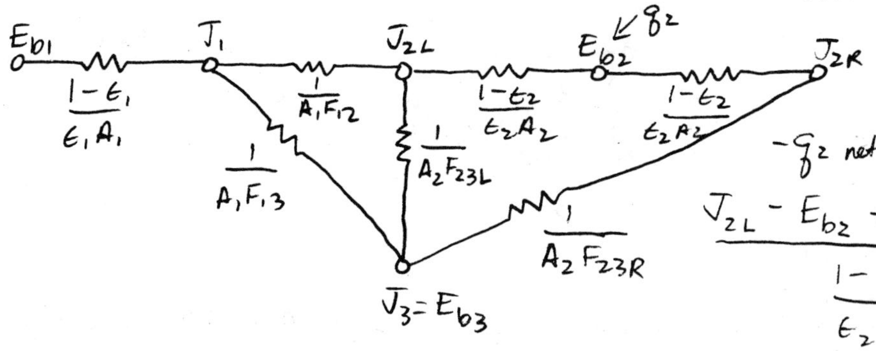

$-q_2$ net rad $= q_2$ conv

$$\frac{J_{2L} - E_{b2} + J_{2R} - E_{b2}}{\frac{1-\epsilon_2}{\epsilon_2 A_2}} = 2hA_2(T_2 - 293)$$

$\frac{1-\epsilon_1}{\epsilon_1 A_1} = 2.778$ $\frac{1}{A_1F_{12}} = 20.20$ $\frac{1-\epsilon_2}{\epsilon_2 A_2} = 2.778$ $\frac{1}{A_1F_{13}} = 24.69 = \frac{1}{A_2F_{23L}}$

$\frac{1}{A_2F_{23R}} = 11.11$ $E_{b1} = 1815\ W/m^2$ $E_{b3} = 418\ W/m^2$ $\frac{J_{2R} - E_{b2}}{\frac{1-\epsilon_2}{\epsilon_2 A_2}} = \epsilon_2 A_2 (418 - E_{b2})$

(J_1) $\frac{1815 - J_1}{2.778} + \frac{J_{2L} - J_1}{20.20} + \frac{418 - J_1}{24.69} = 0$ (J_2) $\frac{J_1 - J_{2L}}{20.20} + \frac{418 - J_{2L}}{24.69} + \frac{E_{b2} - J_{2L}}{2.778} = 0$

$J_1 = 1489.5 + 0.11 J_{2L}$ $J_{2L} = 0.11 J_1 + 334.4 + \frac{E_{b2}}{2.778} = 504.34 + 0.3644\,E_{b2}$

$$\frac{504.34 - 0.6356\,E_{b2}}{2.778} + (0.8)(0.09)(418 - E_{b2}) = (2)(1.42)\left(\frac{1}{0.3}\right)^{1/4}(T - 293)^{5/4}$$

Solution by iteration! $E_{b2} = \sigma T_2^4$ $T_2 = 304.8\,K = 31.8°C$

8-49

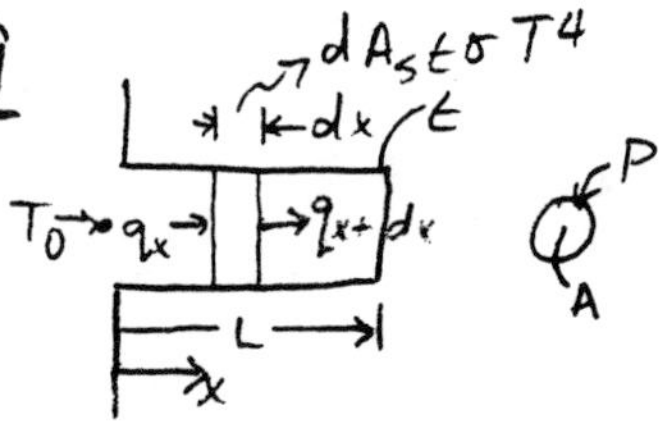

$q_x = q_{x+dx} + dA_s\epsilon\sigma T^4$

$$-kA\frac{dT}{dx} = -kA\left[\frac{dT}{dx} + \frac{d^2T}{dx^2}dx\right] + P\,dx\,\epsilon\sigma T^4$$

$kA\frac{d^2T}{dx^2} - P\epsilon\sigma T^4 = 0$ $\frac{d^2T}{dx^2} - \frac{P\epsilon\sigma}{kA}T^4 = 0$

Boundary conditions

① at $x = 0$ $T = T_0$

② at $x = L$ $-kA\frac{dT}{dx} = A\epsilon\sigma T^4$

8-50 $T_1 = 1200\ K$ $T_3 = 300\ K$ $E_{b1} = 1.1755 \times 10^5$ $E_{b3} = 459$

$\epsilon_1 = 0.2$ $\epsilon_2 = 0.5$ $\epsilon_3 = 0.8$ $F_{12} = F_{23} = 1.0$ $\frac{1-\epsilon_1}{\epsilon_1} = 4.0$

$\frac{1-\epsilon_2}{\epsilon_2} = 1.0$ $\frac{1-\epsilon_3}{\epsilon_3} = 0.25$

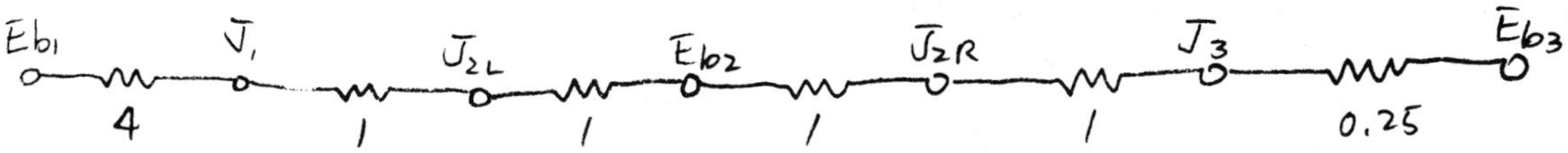

$$\frac{117550 - E_{b2}}{4+1+1} = \frac{117550 - 459}{4+1+1+1+1+0.25}$$

$E_{b2} = 32393\ W/m^2 = \sigma T_2^4$ $T_2 = 869\ K$

8-51 $T_1 = 800\ K$ $\epsilon_1 = 0.3$ $T_3 = 400\ K$ $\epsilon_3 = 0.5$ $\epsilon_2 = 0.05$

$E_{b1} = 23220$ $E_{b3} = 1451\ W/m^2$

(a) q/A (w/o shield) $= \dfrac{23220 - 1451}{\frac{1}{0.3} + \frac{1}{0.5} - 1} = 5024\ W/m^2$

$\frac{1-\epsilon_1}{\epsilon_1} = 2.333$ $\frac{1-\epsilon_2}{\epsilon_2} = 19$ $\frac{1-\epsilon_3}{\epsilon_3} = 1.0$

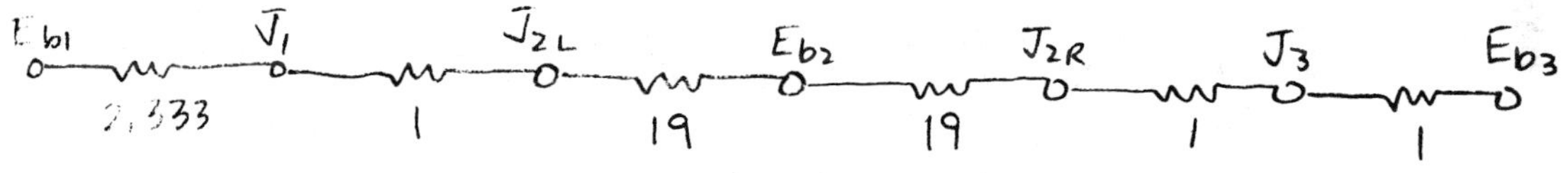

(b) $q = \dfrac{23220 - 1451}{42.333} = 514.2\ W/m^2$

a reduction of 90%

(c) $\dfrac{23220 - E_{b2}}{2.333 + 1 + 19} = 514.2$

$E_{b2} = 11736\ W/m^2 = \sigma T_2^4$ $T_2 = 675\ K$

8-52

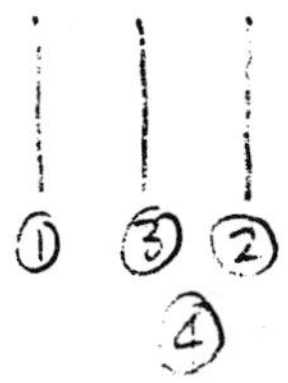

$A_1 = A_2 = A_3 = 1.44\ m^2$ $\quad \epsilon_1 = 0.4$ $\quad \epsilon_2 = 0.6$ $\quad A_4 \to \infty$

$T_1 = 1033\ K$ $\quad T_2 = 573\ K$ $\quad \epsilon_3 = 0.05$ $\quad T_4 = 313\ K$

(a)

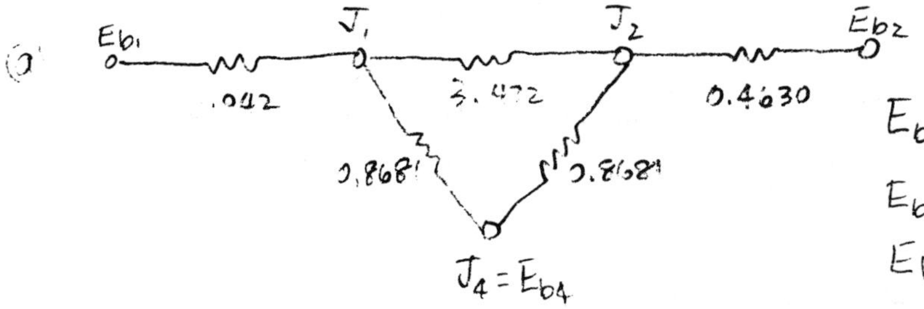

$E_{b1} = 64552$ $\quad F_{12} = 0.2$

$E_{b2} = 6111$ $\quad F_{14} = F_{24} = 0.8$

$E_{b4} = 544$

Node J_1

$$\frac{64552 - J_1}{1.042} + \frac{J_2 - J_1}{3.472} + \frac{544 - J_1}{0.8681} = 0 \qquad J_1 = 26791\ W/m^2$$

Node J_2

$$\frac{J_1 - J_2}{3.472} + \frac{544 - J_2}{0.8681} + \frac{6111 - J_2}{0.4630} = 0 \qquad J_2 = 5984\ W/m^2$$

$$q_1 = \frac{64552 - 26791}{1.042} = 36239\ W \qquad q_2 = \frac{6111 - 5984}{0.463} = 274\ W$$

(b)

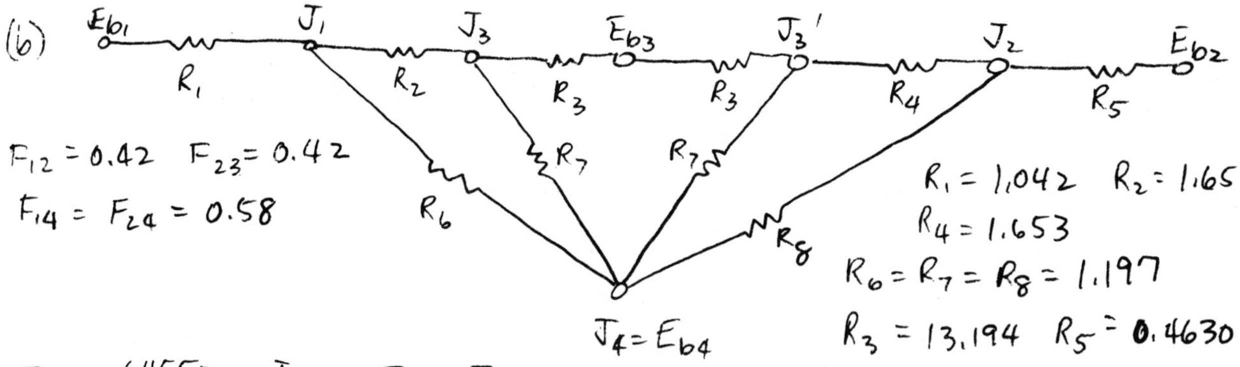

$F_{12} = 0.42$ $\quad F_{23} = 0.42$

$F_{14} = F_{24} = 0.58$

$R_1 = 1.042$ $\quad R_2 = 1.653$

$R_4 = 1.653$

$R_6 = R_7 = R_8 = 1.197$

$R_3 = 13.194$ $\quad R_5 = 0.4630$

$$J_1: \quad \frac{64552 - J_1}{1.042} + \frac{J_3 - J_1}{1.653} + \frac{544 - J_1}{1.197} = 0 \qquad J_1 = 28895$$

$$J_3: \quad \frac{J_1 - J_3}{1.653} + \frac{J_3' - J_3}{(2)(13.194)} + \frac{544 - J_3}{1.197} = 0 \qquad J_3 = 11602$$

$$J_3': \quad \frac{J_3 - J_3'}{(2)(13.194)} + \frac{J_2 - J_3'}{1.653} + \frac{544 - J_3'}{1.197} = 0 \qquad J_3' = 1855$$

8-52 contd

J_2 : $\frac{J_3' - J_2}{1.653} + \frac{6111 - J_2}{0.4630} + \frac{544 - J_2}{1.197} = 0 \qquad J_2 = 4195$

$q_1 = \frac{64552 - 28895}{1.042} = 34220 \text{ W} \qquad q_2 = \frac{6111 - 4195}{0.463} = 4138 \text{ W}$

$E_{b3} = \frac{J_3 + J_3'}{2} : 6729 = \sigma T_3^4 \qquad T_3 = 587 \text{ K} = 314 \,°\text{C}$

8-53

$T_1 = 923 \text{ K} \quad T_3 = 298 \text{ K} \quad \epsilon_1 = 0.8 \quad \epsilon_2 = 0.2 \quad A_1/L = 0.07854$

$A_2/L = 0.9425 \quad E_{b1} = 41145 \text{ W/m}^2 \quad E_{b3} = 447$

$\frac{1-\epsilon_1}{\epsilon_1 A_1} = 3.183 \quad \frac{1-\epsilon_2}{\epsilon_2 A_2} = 4.244 \quad \frac{1}{A_1 F_{12}} = 12.732 \quad \frac{1}{A_2 F_{23}} = 1.061$

$\frac{1-\epsilon_3}{\epsilon_3 A_3} \to 0$ w/o shield $q/L = \frac{41145 - 447}{3.183 + 12.732} = 2557 \text{ W/m}$

w/ shield : $q/L = \frac{41145 - 447}{3.183 + 12.732 + (2)(4.244) + 1.061} = 1598 \text{ W/m}$

$\frac{41145 - E_{b2}}{41145 - 447} = \frac{3.183 + 12.732 + 4.244}{3.183 + 12.732 + (2)(4.244) + 1.061} \qquad E_{b2} = 8926$

$8926 = \sigma T_2^4 \qquad T_2 = 630 \text{ K} = 357 \,°\text{C}$

8-54

$T_1 = 1073 \text{ K} \quad T_3 = 373 \text{ K} \quad \epsilon_1 = 0.8 \quad \epsilon_3 = 0.4 \quad d_1 = 4 \text{ cm}$

$d_3 = 8 \text{ cm} \quad d_2 = 6 \text{ cm} \quad \epsilon_2 = 0.3 \quad E_{b1} = 75146 \text{ W/m}^2 \quad E_{b3} = 1097 \text{ W/m}^2$

w/o shield $A_1 = \pi d_1 = 0.1257$ (per unit length)

$q = \frac{0.1257(75146 - 1097)}{\frac{1}{0.8} + 0.5\left(\frac{1}{0.4} - 1\right)} = 4654 \text{ W/m}$

with shield $A_2 = 0.1886 \quad A_3 = 0.2514 \text{ m}^2\text{/m} \quad F_{12} = F_{23} = 1.0$

$\frac{1-\epsilon_1}{\epsilon_1 A_1} = 1.989 \quad \frac{1}{A_1 F_{12}} = 7.955 \quad \frac{1}{A_2 F_{23}} = 5.302 \quad \frac{1-\epsilon_2}{\epsilon_2 A_2} = 12.371$

$\frac{1-\epsilon_3}{\epsilon_3 A_3} = 5.967$

8-54 contd

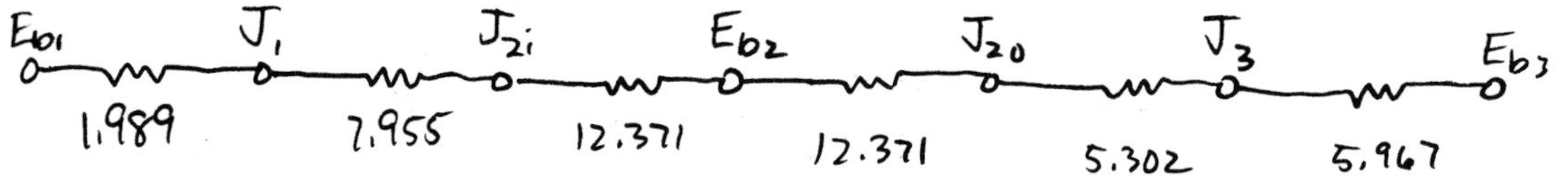

$$q/L = \frac{E_{b1} - E_{b3}}{\Sigma R} = \frac{75146 - 1097}{45.953} = 1611 \text{ W/m}$$

a reduction of 65%

8-55 $T_1 = 673K$ $L = 10$ cm $d_1 = 5$ cm (cast iron) $\epsilon_1 = 0.44$

② monel $d_2 = 10$cm $L = 10$cm $\epsilon_2 = 0.41$ ③ Room $T_3 = 293K$

$E_{b1} = 11630 \text{ W/m}^2$ $E_{b3} = 418 \text{ W/m}^2$ $L/r_2 = 2$ $r_1/r_2 = 0.5$ $F_{21} = 0.42$

$F_{12} = 0.84$ $F_{13} = 0.16$ $F_{22} = 0.32$ $F_{23} = 0.26$ $A_1 = \pi(0.05)(0.1) = 0.01571$

$A_2 = \pi(0.1)^2 = 0.03142$ F_{23} (outside) $= 1.0$ $\frac{1-\epsilon_1}{\epsilon_1 A_1} = 81.014$ $\frac{1-\epsilon_2}{\epsilon_2 A_2} = 45.8$

$\frac{1}{A_1 F_{12}} = 75.778$ $\frac{1}{A_1 F_{13}} = 397.84$ $\frac{1}{A_2 F_{23}} = 12.241$ $\frac{1}{A_2 F_{23\,(out)}} = 31.827$

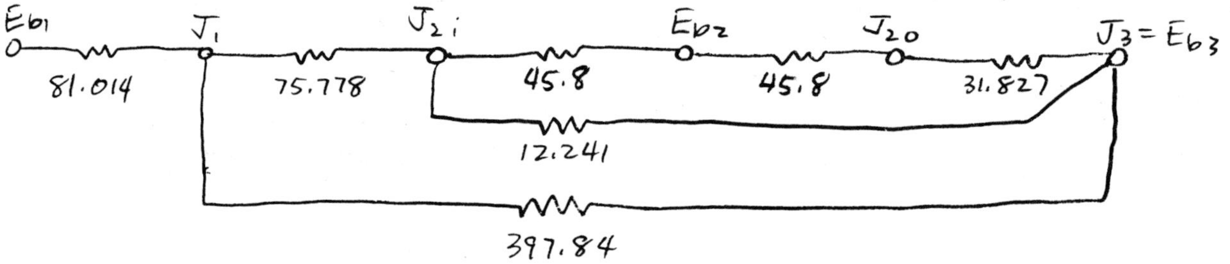

$$q = \frac{E_{b1} - E_{b3}}{\Sigma R} \qquad \frac{1}{R_1} = \frac{1}{12.241} + \frac{1}{2(45.8) + 31.827} \qquad R_1 = 11.136$$

$$1/R_2 = \frac{1}{R_1 + 75.778} + \frac{1}{397.84} \qquad R_2 = 7.331$$

$$\Sigma R = 81.014 + R_2 = 152.345$$

$$q = \frac{11630 - 418}{152.345} = 73.6 \text{ W}$$

8-56 $d_1 = 0.3 \quad d_2 = 0.6 \quad \epsilon_1 = 0.5 \quad \epsilon_2 = 0.3 \quad T_1 = 1033\ K \quad T_2 = 643\ K$

$A_1 = \pi(0.3) = 0.9425 \quad \tau_g = 0.7 \quad A_2 = \pi(0.6) = 1.885 \quad \alpha_g = \epsilon_g = 0.3$

$F_{12} = 1.0 \quad F_{1g} = F_{2g} = 1.0$ Resistance shown on figure $E_{b1} = 64552$

$E_{b2} = 9691$

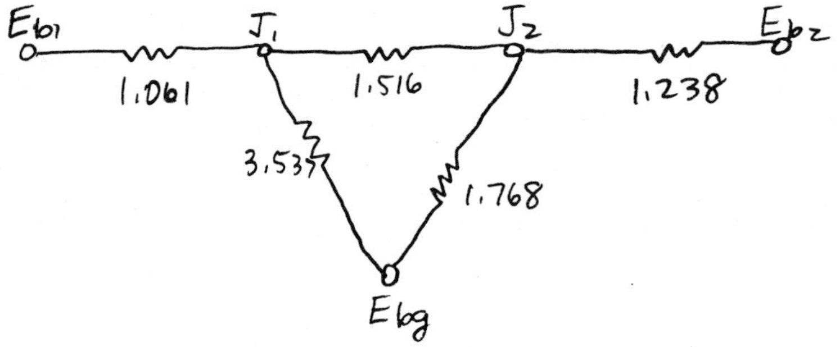

$$q = \frac{64552 - 9691}{1.061 + \dfrac{1}{\dfrac{1}{5.16} + \dfrac{1}{3.537 + 1.768}} + 1.238} = 15773\ \text{W/m}$$

$$15773 = \frac{64552 - J_1}{1.061} = \frac{J_2 - 9691}{1.238} \qquad J_1 = 47816 \quad J_2 = 29218$$

$$\frac{47816 - E_{bg}}{47816 - 29218} = \frac{3.537}{3.537 + 1.768} \qquad E_{bg} = 35416 = \sigma T_g^4 \quad T_g = 889\ K = 616\ °C$$

8-58 Same Network as 8-56 $\quad A_1 = A_2 = 1.0 \quad F_{12} = 1.0$

$$\frac{1-\epsilon_1}{\epsilon_1 A_1} = 1.0 \quad \frac{1-\epsilon_2}{\epsilon_2 A_2} = 2.333 \quad \frac{1}{A_1 F_{12} \tau_g} = 1.429 \quad \frac{1}{A_1 F_{1g} \alpha_g} = \frac{1}{A_2 F_{2g} \alpha_g} = 3.333$$

$$q = \frac{64552 - 9691}{1 + \dfrac{1}{\dfrac{1}{1.429} + \dfrac{1}{(2)(3.333)}} + 2.333} = 12164\ \text{W/m}^2$$

$$= \frac{64552 - J_1}{1} = \frac{J_2 - 9691}{2.333} \qquad J_1 = 52388 \quad J_2 = 38070$$

$$E_{bg} = \frac{J_1 + J_2}{2} = \sigma T_g^4 \qquad T_g = 945\ K = 672\ °C$$

8-59 Same network as 8-56 $Re = \frac{(1.6)(6)(0.6-0.3)}{5.4\times10^{-5}} = 53330$

$Pr = \frac{(1670)(5.4\times10^{-5})}{0.11} = 0.82$ $h_1 = \frac{0.11}{0.3}(0.023)(53330)^{0.8}(0.82)^{0.4} = 47.11$

$h_2 = \frac{1}{2}h_1 = 23.55$ $T_g = 1373\ K$ $E_{bg} = 2.015\times10^5$

(J₁) $\frac{64552 - J_1}{1.061} + \frac{J_2 - J_1}{1.516} + \frac{2.015\times10^5 - J_1}{3.537} = 0$ $J_1 = 95980$

(J₂) $\frac{J_1 - J_2}{1.516} + \frac{2.015\times10^5 - J_2}{1.768} + \frac{9691 - J_2}{1.238} = 0$ $J_2 = 91051$

Inner Surface:

$\text{Cooling} = \frac{95980 - 64552}{1.061} + (47.11)(\pi)(0.3)(1373 - 1033) = 44717\ \frac{W}{m}$

Outer Surface:

$\text{Cooling} = \frac{91051 - 9691}{1.238} + (23.55)\pi(0.6)(1373 - 643) = 98124\ W/m$

8-60 $E_{b1} = 37194$ $E_{b2} = 14513$ $A_1/L = 0.1571$ $A_2/L = 0.3142$

$\frac{1-\epsilon_1}{\epsilon_1 A_1} = 9.548$ $\frac{1-\epsilon_2}{\epsilon_2 A_2} = 2.122$ $\frac{1}{A_1 F_{2g}\epsilon_g} = 42.44$ $\frac{1}{A_2 F_{2g}\epsilon_g} = 21.22$

$\frac{1}{A_1 F_{12}(1-\epsilon_g)} = 7.489$ $R = 9.548 + \frac{1}{\frac{1}{7.489} + \frac{1}{42.44 + 21.22}} = 18.37$

$q = \frac{37194 - 14513}{18.37} = 1235\ W/m\cdot\text{length}$

$= \frac{37194 - J_1}{9.548} = \frac{J_2 - 14513}{21.22}$ $J_1 = 25406$ $J_2 = 17134$

$E_{bg} = 17134 + \frac{1}{3}(25406 - 17134) = 19891$ $T_g = 770\ K$

8-61 0.7 m cube 15% CO_2 85% N_2 $T_g = 1600\ K$ $T_w = 523\ K$

$\sigma T_g^4 = 3.715 \times 10^5\ W/m^2$ $\sigma T_w^4 = 4241$ $L_e = 0.42\ m$ $p_c L_e\ 6.38\ \frac{kN}{m}$

$\epsilon_c(CO_2) = 15198\ W/m^2$ $\epsilon_c = \epsilon_g = 0.10$ $p_c L_e \frac{T_w}{T_g} = 2.085\ \frac{kN}{m}$ $\epsilon_c = 0.062$

$C_c = 1.0$ $a_g(T_w) = \alpha_c = (0.062)\left(\frac{1600}{523}\right)^{0.65} = 0.128$

$q/A = (0.10)(3.715 \times 10^5) - (0.128)(4241) = 36607\ W/m^2$

$q = (36607)(6)(0.7)^2 = 107.6\ kW$

8-69

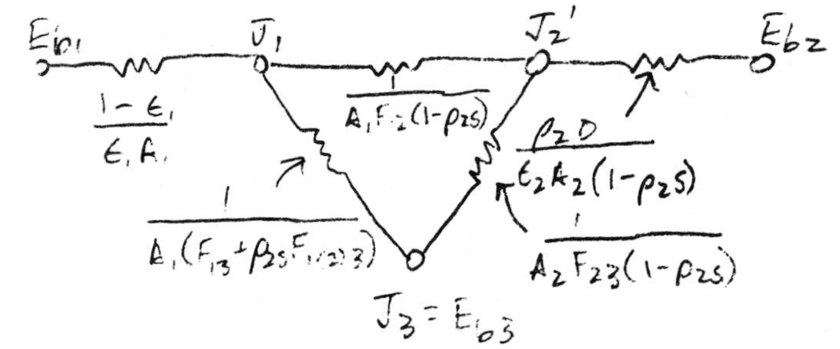

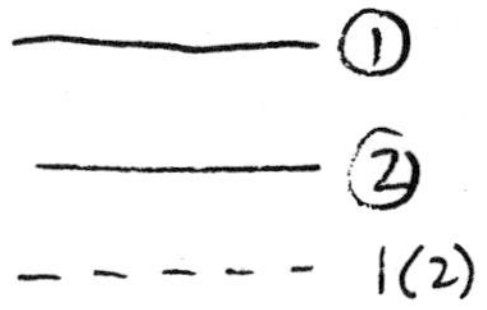

$T_1 = 813 K$ $\epsilon_1 = 0.3$ $T_2 = 533 K$ $T_3 = 293 K$ $\rho_{D2} = 0.2$ $F_{1(2)2} = 0.6$

$F_{1(2)1} = 0.37$ $F_{1(2)3} = 0.23$ $\rho_{s2} = 0.4$ $\epsilon_2 = 1 - \rho_2 = 0.4$

$A_1 = A_2 = \pi(0.05)^2 = 7.854 \times 10^{-3}\ m^2$ $F_{23} = 0.4$ $F_{13} = 0.4$ $E_{b1} = 24767\ \frac{W}{m^2}$

$E_{b2} = 4575$ $E_{b3} = 418$ $\frac{1-\epsilon_1}{\epsilon_1 A_1} = 297.1$ $\frac{1}{A_1 F_{12}(1-\rho_{2s})} = 3537$

$\frac{1}{A_1(F_{13} + \rho_{2s}F_{1(2)3})} = 258.8$ $\frac{\rho_2 D}{\epsilon_2 A_2(1-\rho_{2s})} = 106.1$ $\frac{1}{A_2 F_{23}(1-\rho_{2s})} = 530.5$

$J_2' = J_{D2}/1-\rho_{2s}$

$$\frac{24767 - J_1}{297.1} + \frac{J_2' - J_1}{353.7} + \frac{418 - J_1}{258.8} = 0 \qquad J_1 = 9878\ W/m^2$$

$$\frac{J_1 - J_2'}{353.7} + \frac{418 - J_2'}{530.5} + \frac{4575 - J_2'}{106.1} = 0 \qquad J_2 = 5081\ W/m^2$$

$$q_1 = \frac{24767 - 9878}{297.1} = 50.11\ W \qquad q_2 = \frac{4575 - 5081}{106.1} = -4.76\ W$$

8-70

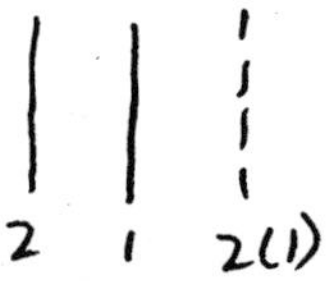

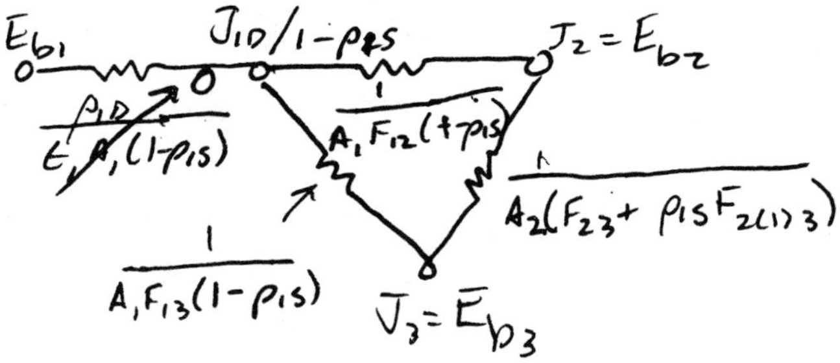

$T_1 = 823K \quad \epsilon_1 = 0.6 \quad T_3 = 283K \quad \rho_{1D} = 0 \quad \rho_{1S} = 0.4 \quad A_1 = A_2 = 0.54\,m^2$

$F_{12} = F_{21} = 0.25 \quad F_{13} = 0.75 \quad J_{1D} = \epsilon_1 E_{b1} = 15604$

$$\frac{J_{1D}}{1-\rho_{1S}} = 26007\ W/m^2 \quad \frac{1}{A_1 F_{12}(1-\rho_{1S})} = 12.346 \quad \frac{1}{A_1 F_{13}(1-\rho_{1S})} = 4.15$$

$$\frac{1}{A_2(F_{23}+\rho_{1S}F_{2(1)3})} = 2.281$$

$$q = \frac{26007 - 363.6}{\dfrac{1}{\dfrac{1}{4.15} + \dfrac{1}{12.346+2.281}}} = 7985\ W$$

$$\frac{26007 - J_2}{26007 - 363.6} = \frac{12.346}{12.346 + 2.281} \quad J_2 = 4362 = \sigma T_2^4$$

$T_2 = 526.7\ K = 253.7\ °C$

8-71

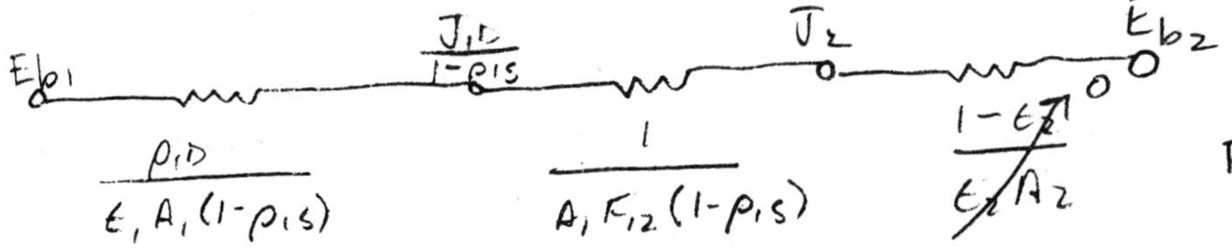

$F_{12} = 1$

$$q = \frac{E_{b1} - E_{b2}}{\Sigma R} = \epsilon_1 A_1 (E_{b1} - E_{b2}) = \frac{\sigma(T_1^4 - T_2^4)}{\dfrac{\rho_{1D}}{\epsilon_1 A_1(1-\rho_{1S})} + \dfrac{1}{A_1(1-\rho_{1S})}}$$

$$= \frac{A_1 \sigma (1-\rho_{1S})(T_1^4 - T_2^4)}{\dfrac{\rho_{1D}}{\epsilon_1} + 1}$$

Same answer with Diffuse

8-72

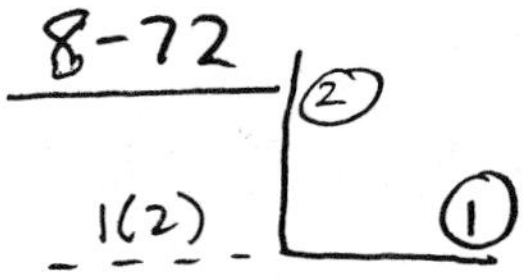

$T_1 = 643\text{ K}$ $\epsilon_1 = 0.6$ $A_1 = (0.3)(0.6) = 0.18\text{ m}^2$ $T_3 = 363\text{ K}$

$\rho_{2S} = 0.7$ $\rho_{2D} = 0.1$ $\epsilon_2 = 0.2$ $F_{12} = F_{21} = 0.25$

$F_{13} = 0.75$ $F_{23L} = 1.0$ $F_{23R} = 0.75$ $F_{1(2)3} = F_{12} = 0.25$ $\dfrac{1-\epsilon_1}{\epsilon_1 A_1} = 3.704$

$$\frac{1}{A_1 F_{12}(1-\rho_{2S})} = 74.07 \qquad \frac{\rho_{2D}}{\epsilon_2 A_2 (1-\rho_{2S})} = 9.259 \qquad \frac{1}{A_1(F_{13}+\rho_{2S}F_{1(2)3})} = 6.006$$

$$\frac{1}{A_2 F_{23R}(1-\rho_{2S})} = 24.69 \qquad \frac{1}{A_2 F_{23L}(1-\rho_{2S})} = 18.52 \qquad E_{b1} = 9691\text{ W/m}^2$$

$E_{b3} = 984\text{ W/m}^2$

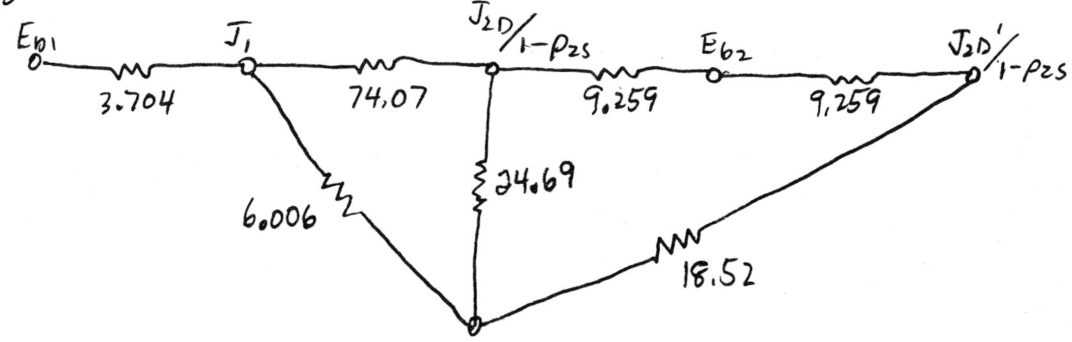

$$q = \frac{E_{b1} - E_{b3}}{\Sigma R} = \frac{9691 - 984}{9.33} = 933\text{ W}$$

w/o Reflector ; $q = (0.6)(0.18)(9691 - 984) = 940.4\text{ W}$

$$\frac{9691 - J_1}{9691 - 984} = \frac{3.704}{9.33} \qquad J_1 = 6234\text{ W/m}^2$$

$$\frac{6234 - \dfrac{J_{2D}}{1-\rho_{2S}}}{6234 - 984} = \frac{74.07}{74.07 + 14.84} \qquad \frac{J_{2D}}{1-\rho_{2S}} = 1859$$

$$\frac{1859 - E_{b2}}{1859 - 984} = \frac{9.259}{37.038} \qquad E_{b2} = 1640 = \sigma T_2^{\,4}$$

$T_2 = 412\text{ K} = 139\text{ °C}$

8-73

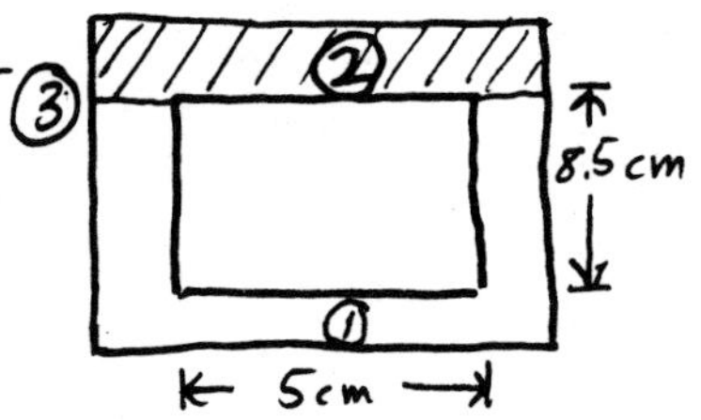

$T_1 = 533\ K \quad \epsilon_1 = 0.6 \quad \rho_{2S} = 0.1$

$T_3 = 293\ K \quad \rho_{2D} = 0.1 \quad \tau_{2D} = 0.3 \quad \tau_{2S} = 0.3$

$\epsilon_2 = 0.2 \quad E_{b1} = 4575\ W/m^2 \quad E_{b2} = 418\ \frac{W}{m^2}$

$A_1 = 75\ cm^2 = 7.5 \times 10^{-3}\ m^2 \quad A_2 = 25\ cm^2 = 2.5 \times 10^{-3}\ m^2 \quad A_3 \to \infty$

$F_{21} = 1.0 \quad F_{12} = \frac{25}{75} = 0.3333 = F_{13} \quad \frac{1-\epsilon_1}{\epsilon_1 A_1} = 88.89$

$\frac{1}{A_2 F_{23T}(1-\tau_2-\rho_{2S})} = 1333.3 \quad \frac{1}{A_1 F_{13} \tau_{2S}} = 1333.33 = \frac{1}{A_1 F_{12} F_{23T} \tau_{2D}}$

$\frac{1}{A_1 F_{12}(1-\tau_2-\rho_{2S})} = 1333.3 \quad \frac{\rho_{2D}}{\epsilon_2 A_2 (1-\tau_2-\rho_{2S})} = 666.6$

$q = \frac{4575 - 418}{660.32} = 6.295\ W \quad$ w/o window: $q = \frac{4575-418}{88.89 + 400} = 8.503\ W$

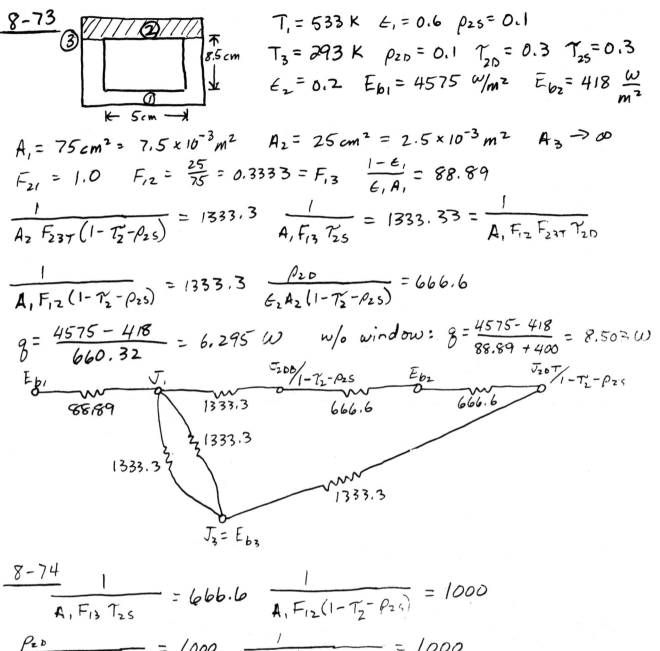

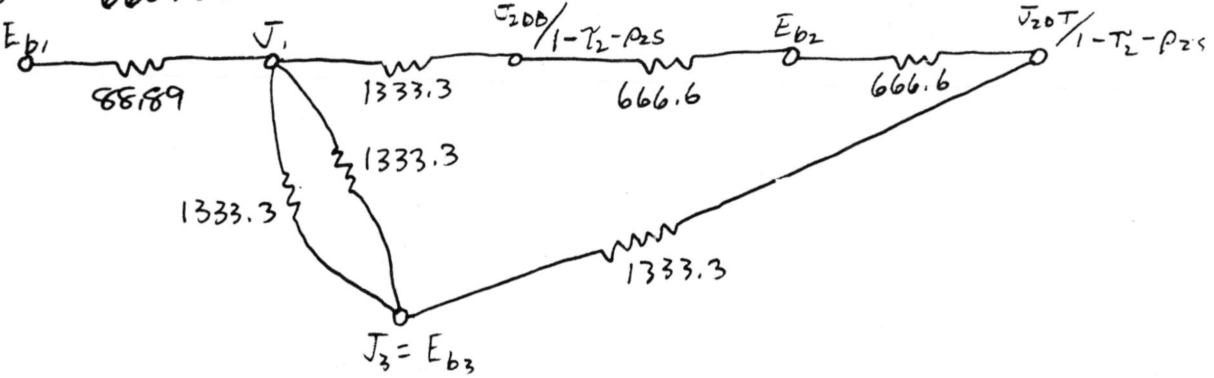

8-74

$\frac{1}{A_1 F_{13} \tau_{2S}} = 666.6 \quad \frac{1}{A_1 F_{12}(1-\tau_2-\rho_{2S})} = 1000$

$\frac{\rho_{2D}}{\epsilon_2 A_2 (1-\tau_2-\rho_{2S})} = 1000 \quad \frac{1}{A_2 F_{23T}(1-\tau_2-\rho_{2S})} = 1000$

$q = \frac{4575 - 418}{660.32} = 6.295\ W$

8-77

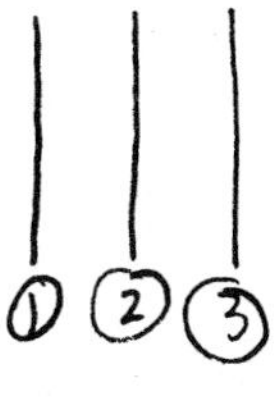

$T_1 = 1073\ K \quad T_3 = 308\ K \quad \epsilon_1 = 0.5 \quad \epsilon_3 = 0.8 \quad \rho_{2D} = 0.4$

$\epsilon_2 = 0.2 \quad F_{12} = F_{23} = 1.0$ etc. $\quad \dfrac{1-\epsilon_1}{\epsilon_1} = 1.0 \quad \dfrac{1}{F_{12}(1-\rho_{2s})} = 1.67$

$\dfrac{\rho_{2D}}{\epsilon_2(1-\rho_{2s})} = 3.333 \qquad \dfrac{1-\epsilon_3}{\epsilon_3} = 0.25 \qquad E_{b1} = 75146\ W/m^2$

$E_{b3} = 510\ W/m^2$

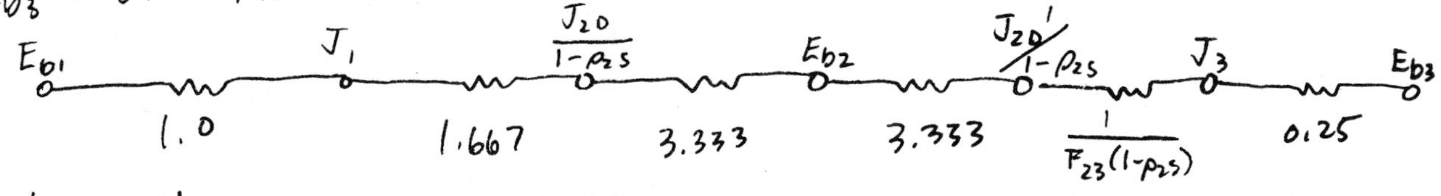

with shield: $q/A = \dfrac{75146 - 510}{1 + 1.667 + (2)(3.3333) + 1.667 + 0.25} = 6634\ W/m^2$

w/o shield: $q/A = \dfrac{75146 - 510}{1 + 1 + 0.25} = 33172\ W/m^2$

same for diffuse shield.

8-86 $d_1 = d_2 = 30\ cm \quad x = 5\ cm \quad d/x = 6 \quad F_{12} = F_{21} = 0.7 \quad F_{13} = F_{23} = 0.3$

$T_3 = 293\ K \quad q_1/A = 1 \times 10^5\ W/m^2 \quad \epsilon_1 = 0.9 \quad \epsilon_2 = 0.5$

$J_1 - F_{12}F_{2T} = 10^5 \qquad J_{2T} - (1-0.5)(F_{23}E_{b3} + F_{21}J_1) = 0.5E_{b2}$

$J_{2B} - (1-0.5)(E_{b3}) = 0.5E_{b2} \qquad E_{b2} = \dfrac{J_{2T} + J_{2B}}{2}$

Solution: $E_{b2} = 4678 \quad T_2 = 953\ K \quad J_{2T} = 69961$

$J_1 = 1.49 \times 10^5$

$10^5 = \dfrac{E_{b1} - 1.49 \times 10^5}{\frac{1-0.9}{0.9}} \qquad E_{b1} = 10.49 \times 10^5$

$T_1 = 2074\ K$

8-88 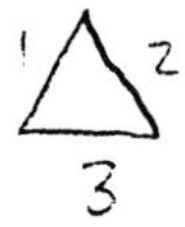

$T_1 = 1100\,K$ $\quad \epsilon_1 = 0.6$ $\quad T_2 = 2100\,K$ $\quad \epsilon_2 = 0.8$

$q/A)_3 = 1000\ W/m^2$ $\quad \epsilon_3 = 0.7$

$F_{12} = F_{13} = F_{21} = F_{23} = F_{31} = F_{32} = 0.5$ $\quad E_{b1} = 83000$ $\quad E_{b2} = 1.103 \times 10^6$

$J_1 - 0.4[F_{12}J_2 + F_{13}J_3] = 0.6\,E_{b1}$ $\qquad J_1 = 384950$

$J_2 - 0.2[F_{21}J_1 + F_{23}J_3] = 0.8\,E_{b2}$ $\qquad J_2 = 989500$

$J_3 - [F_{31}J_1 + F_{32}J_2] = 1000$ $\qquad J_3 = 686270$

$$q/A)_1 = \frac{83000 - 384950}{\frac{1-0.6}{0.6}} = -4.53 \times 10^5$$

$$q/A)_2 = \frac{1.103 \times 10^6 - 9.985 \times 10^5}{\frac{1-0.8}{0.8}} = +4.54 \times 10^5$$

$$1000 = \frac{E_{b3} - 6.863 \times 10^5}{\frac{1-0.7}{0.7}} \qquad E_{b3} = 6.886 \times 10^5$$

$T_3 = 1867\,K$

8-89 $A_1 = A_2 = 1\,m^2$ $\quad T_1 = 573\,K$ $\quad \epsilon_1 = 0.5$ $\quad E_{b1} = 6111$ $\quad q_2 = 0$

$\epsilon_2 = 0.7$ $\quad T_3 = 303\,K$ $\quad E_{b3} = 478$ $\quad F_{12} = F_{21} = 0.2$

$F_{13} = F_{23} = 0.8$ $\quad F_{11} = F_{22} = 0$ $\quad F_{31} = F_{32} \approx 0$ $\quad F_{33} \cong 1.0$ $\quad E_{b3} = J_3$

$J_1 - (1-0.5)[0.2J_2 + 0.8E_{b3}] = 0.5\,E_{b1}$

$J_2 - [0.2J_1 + 0.8E_{b3}] = 0$

Solution:

$J_1 = 3352\ W/m^2$ $\qquad J_2 = 1053 = \sigma T_2^4$ $\qquad T_2 = 369\,K$

$$q_1 = \frac{E_{b1} - J_1}{\frac{1-\epsilon_1}{\epsilon_1 A_1}} = \frac{(0.5)(6111 - 478)(1)}{1-0.5} = 5633\ W$$

8-90 $d_1 = d_2 = 60$ cm $x = 15$ cm $T_1 = 813$ K $T_2 = 573$ K $\epsilon_1 = 0.7$ $\epsilon_2 = 0.5$

$T_3 = 303$ K $A_1 = A_2 = 0.2827$ m^2 $F_{12} = 0.6 = F_{21}$ $F_{13} = F_{31} = 0.4$

$E_{b1} = 24767$ W/m^2 $E_{b2} = 6111$ $F_{11} = F_{22} = 0$ $F_{31} = F_{32} = 0$ $F_{34} \rightarrow 1.0$

$E_{b3} = J_3 = 478$

$$J_1 = (1-0.7)\left[0.6 J_2 + 0.4 E_{b3}\right] + 0.7 E_{b1} \qquad J_1 = 18980 \text{ W/m}^2$$

$$J_2 = (1-0.5)\left[0.6 J_1 + 0.4 E_{b3}\right] + 0.5 E_{b2} \qquad J_2 = 8840 \text{ W/m}^2$$

$$q_1 = \frac{\epsilon_1 A_1 (E_{b1} - J_1)}{1-\epsilon_1} = \frac{(0.7)(0.2827)(24767 - 18980)}{0.3} = 3817 \text{ W}$$

$$q_2 = \frac{\epsilon_2 A_2 (E_{b2} - J_2)}{1-\epsilon_2} = \frac{(0.5)(0.2827)(6111 - 8840)}{0.5} = -771 \text{ W}$$

8-91 $d_1 = d_2 = 1$ m $x = 0.25$ $d/x = 4$ $F_{12} = F_{21} = 0.6$ $F_{13} = F_{23} = 0.4$

$A_1 = A_2 = \pi(0.5)^2 = 0.7854$ $T_1 = 573$ K $T_3 = 303$ K $\epsilon_1 = 0.5$ $E_{b1} = 6111$

$E_{b3} = 478$ $\frac{1-\epsilon_1}{\epsilon_1 A_1} = 1.273$ $\frac{1}{A_1 F_{12}} = 2.122$ $\frac{1}{A_1 F_{13}} = \frac{1}{A_2 F_{23}} = 3.183$

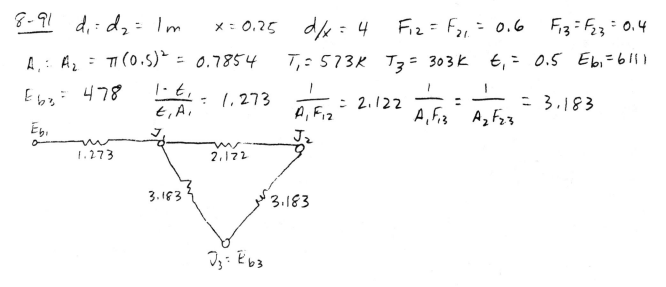

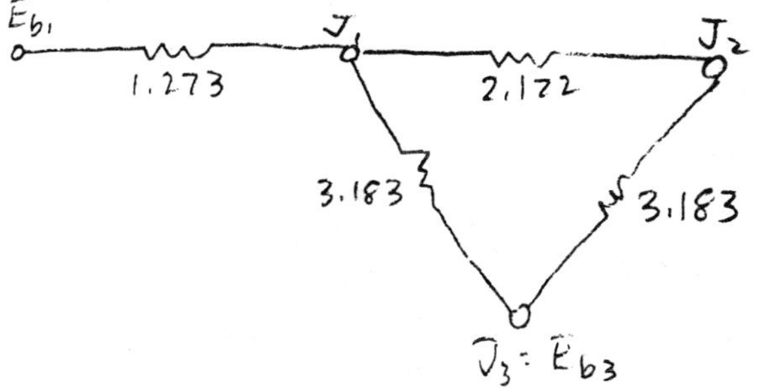

$$q = \frac{6111 - 478}{1.273 + \cfrac{1}{\cfrac{1}{3.183} + \cfrac{1}{2.122 + 3.183}}} = 1727 \text{ W}$$

8-92 $d_1 = d_2 = 50\text{ cm}$ $x = 10\text{ cm}$ $F_{12} = F_{21} = 0.65$ $F_{13} = F_{23} = 0.35$

$T_1 = 350\text{ K}$ $\epsilon_1 = \epsilon_2 = 0.6$ $E_{b1} = 850.7$ $q/A|_2 = 10000\text{ W/m}^2$

$q/A|_3 = 0$ $F_{11} = F_{22} = 0$ $F_{33} \to 1.0$ $A_1 = A_2 = \pi(0.25)^2 = 0.1963\text{ m}^2$

$A_3 = \pi(0.5)(0.1) = 0.1571\text{ m}^2$ $F_{31} = F_{32} = 0.35\left(\frac{A_1}{A_3}\right) = 0.4373$

$F_{33} = 1 - (2)(0.4373) = 0.125$

$J_1 = (1-0.6)\left[0.65 J_2 + 0.35 J_3\right] + 0.6 E_{b1}$ $J_1 = 7514.31\text{ W/m}^2$

$J_2 = \left[0.65 J_1 + 0.35 J_3\right] + 10000$ $J_2 = 19632.83\text{ W/m}^2$

$J_3 = \frac{1}{1-0.125}\left[0.4373 J_1 + 0.4373 J_2\right]$ $J_3 = 13566.78\text{ W/m}^2$

$J_3 = E_{b3}$ $T_3 = \left(\frac{J_3}{(5.669\times10^{-8})}\right)^{1/4} = 699.4\text{ K}$

$E_{b2} = J_2 + \left(\frac{1-0.6}{0.6}\right)(10000) = 26299.5\text{ W/m}^2$

$T_2 = \left(\frac{E_{b2}}{5.669\times10^{-8}}\right)^{1/4} = 825.3\text{ K}$

8-93 $T_1 = 1000\text{ K}$ $\epsilon_1 = 0.5$ $E_{b1} = 56690$ $A_1 = \pi(0.5)^2 = 0.7854$

$A_2 = 2\pi(2)^2 = 25.13\text{ m}^2$ $T_2 = 300\text{ K}$ $\epsilon_2 = 0.8$ $E_{b2} = 459$ $q_3 = 0$

$A_3 = \pi(2^2 - 0.5^2) = 11.781\text{ m}^2$ $F_{12} = F_{32} = 1.0$ $F_{21} = 1.0\left(\frac{A_1}{A_2}\right) = 0.0313$

$F_{23} = 1.0\left(\frac{A_3}{A_2}\right) = 0.4688$ $\frac{1-\epsilon_1}{\epsilon_1 A_1} = 1.273$ $\frac{1-\epsilon_2}{\epsilon_2 A_2} = 0.00995$

$\frac{1}{A_1 F_{12}} = 1.273$ $\frac{1}{A_3 F_{32}} = 0.0849$ $\frac{1}{A_1 F_{13}} = \infty$

8-93 contd

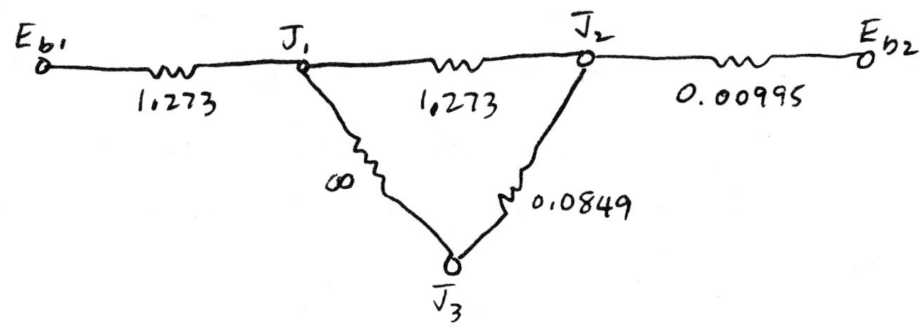

$$q = \frac{56690 - 459}{1.273 + 1.273 + 0.00995} = 22000 \text{ W}$$

$$q = \frac{J_2 - E_{b2}}{0.00995} \qquad J_2 = 677.9 \text{ W/m}^2$$

$$q = A_2 (G_2 - J_2) \qquad G_2 = 1553 \text{ W/m}^2$$

8-94 $d/x = 1.0$ $F_{12} = 0.17$ ① disk ② room ③ shield

$F_{13} = 0.83$ $\epsilon_1 = 1.0$ $\epsilon_3 = 0.3$ $T_1 = 773\text{K}$ $E_{b1} = 20241 \text{ W/m}^2$

$T_2 = 313 \text{ K}$ $E_{b2} = 544$ $A_1 = \pi (0.05)^2 = 0.007854 \text{ m}^2$

$A_3 = \pi(0.1)(0.1) = 0.03142$ $F_{31} = 0.83\left(\frac{A_1}{A_3}\right) = 0.2075 = F_{32}$

$\frac{1-\epsilon_1}{\epsilon_1 A_1} = 0$ $\frac{1-\epsilon_3}{\epsilon_3 A_3} = 74.26$ $\frac{1}{A_1 F_{12}} = 749$ $\frac{1}{A_1 F_{13}} = 153.4$

$\frac{1}{A_3 F_{32}} = 153.4$ $F_{32}\text{ (out)} = 1.0$ $\frac{1}{A_3 F_{32}}$ outside $= 31.83$

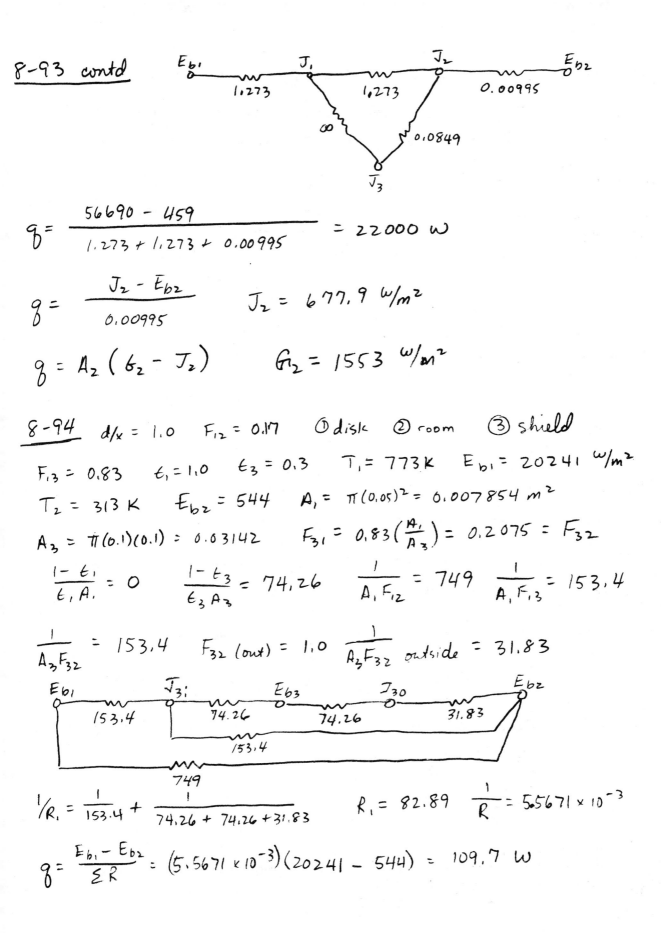

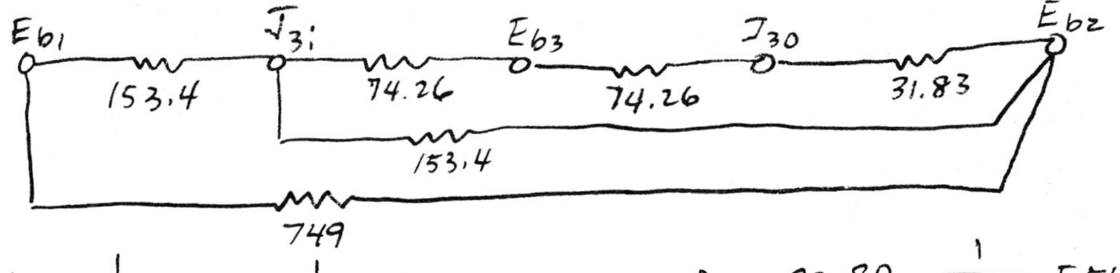

$$1/R_1 = \frac{1}{153.4} + \frac{1}{74.26 + 74.26 + 31.83} \qquad R_1 = 82.89 \qquad \frac{1}{R} = 5.5671 \times 10^{-3}$$

$$q = \frac{E_{b1} - E_{b2}}{\Sigma R} = (5.5671 \times 10^{-3})(20241 - 544) = 109.7 \text{ W}$$

8-95 $d/x = 1.0 \quad F_{12} = 0.17 \quad$ ① disk ② room $\quad T_1 = 1273\ K$

$T_2 = 293\ K \quad \epsilon_1 = 0.55 \quad E_{b1} = 1.489 \times 10^5 \quad E_{b2} = 418\ W/m^2$

$F_{13} = 0.83 \quad A_1 = \pi(0.25)^2 = 0.1963 \quad A_3 = \pi(0.5)^2 = 0.7854$

$F_{31} = F_{32} = 0.83/4 = 0.2075 \quad \frac{1-\epsilon_1}{\epsilon_1 A_1} = 4.168 \quad \frac{1}{A_1 F_{12}} = 29.966$

$\frac{1}{A_1 F_{13}} = \frac{1}{A_3 F_{32}} = 6.137$

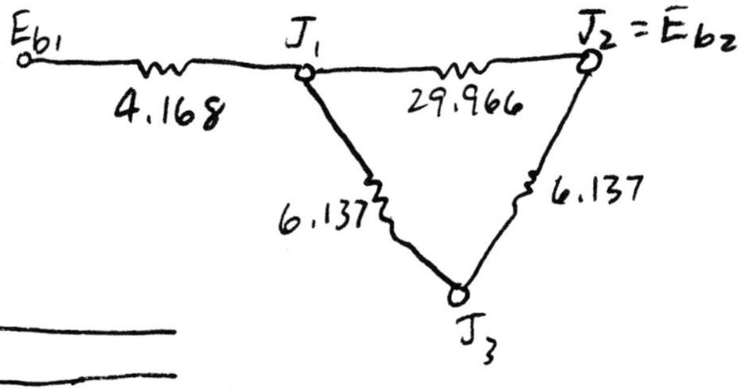

$$q = \frac{148900 - 418}{4.168 + \cfrac{1}{\cfrac{1}{29.966} + \cfrac{1}{2(6.137)}}}$$

$= 11532\ W$

8-96 Bottom = ① $\quad$ Sides = ② $\quad$ Room = ③ $\quad A_1 = 7.854 \times 10^{-5}$

$A_2 = \pi(0.005 + 0.01)\left[(0.01^2 - 0.005^2) + 0.03^2\right]^{1/2} = 0.001433$

$T_1 = 773\ K \quad E_{b1} = 20241 \quad E_{b3} = 478 = J_3 \quad T_3 = 303\ K \quad \epsilon_1 = 0.6$

For figure: $r_2/L = 0.167 \quad L/r_1 = 3$ open hole $A_3 = 3.142 \times 10^{-4}$

$F_{13} = (0.05)(2)^2 = 0.8 \quad \frac{1-\epsilon_1}{\epsilon_1 A_1} = 8488 \quad F_{23} = 1 - 0.05 = 0.95$

$F_{12} = 1 - 0.2 = 0.8 \quad 1/A_1 F_{13} = 63662 \quad 1/A_1 F_{12} = 15915 \quad \frac{1}{A_2 F_{23}} = 735$

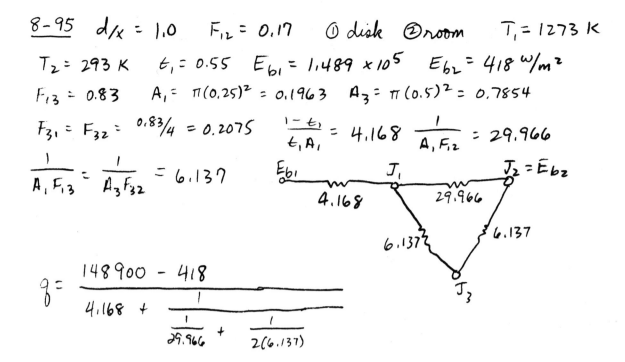

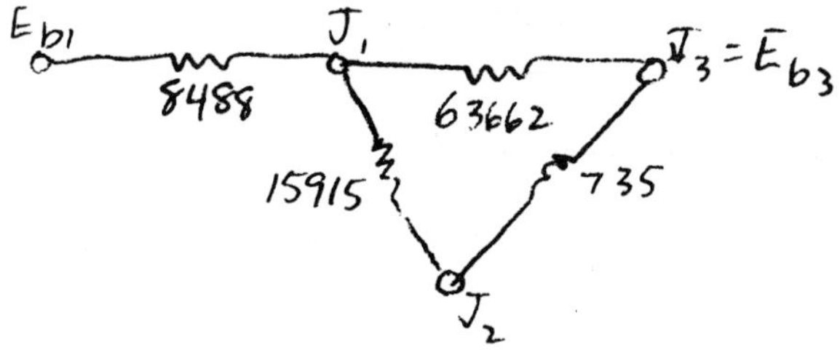

$$q = \frac{20241 - 478}{8488 + \cfrac{1}{\cfrac{1}{63662} + \cfrac{1}{15915 + 735}}} = 0.911\ W$$

8-97

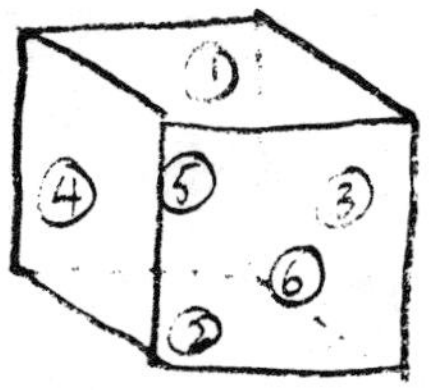

$T_1 = 373K \quad T_2 = 313 K \quad \epsilon_1 = 0.7 \quad \epsilon_2 = 0.5$

$\epsilon_3 = \epsilon_4 = \epsilon_5 = \epsilon_6 = 0.6 \quad T_3 = T_4 = 343 K$

$T_5 = T_6 = 60°C = 333K \quad E_{b1} = 1097 \ W/m^2$

$E_{b2} = 544 \quad E_{b3} = E_{b4} = 785 \quad E_{b5} = E_{b6} = 735$

$F_{12} = 0.2 \quad F_{13} = F_{14} = F_{15} = F_{16} = 0.2$ all shape factors are 0.2

$A_{ij} = 0$ From symmetry $J_3 = J_4 \quad J_5 = J_6$

$$J_1 - (1-0.7)(0.2)(J_2 + 2J_3 + 2J_5) = (0.7)(1097)$$

$$J_2 - (1-0.5)(0.2)(J_1 + 2J_3 + 2J_5) = (0.5)(544)$$

$$J_3 - (1-0.6)(0.2)(J_1 + J_2 + J_3 + 2J_5) = (0.6)(785)$$

$$J_5 - (1-0.6)(0.2)(J_1 + J_2 + J_5 + 2J_3) = (0.6)(735)$$

Solving:

$J_1 = 995.215 \ W/m^2 \quad J_2 = 682.195 \ W/m^2 \quad J_3 = J_4 = 790.51$

$J_5 = J_6 = 762.68 \ W/m^2$

$$\frac{1-\epsilon_1}{\epsilon_1 A_1} = 0.42816 \quad \frac{1-\epsilon_2}{\epsilon_2 A_2} = 1.0 \quad \frac{1-\epsilon_3}{\epsilon_3 A_3}, \text{ etc} = 0.6667$$

$$q_1 = \frac{E_{b1} - J_1}{0.4286} = 238 \ W \qquad q_2 = \frac{E_{b2} - J_2}{1.0} = -138 \ W$$

$$q_3 = q_4 = \frac{E_{b3} - J_3}{0.6667} = -8.3 \ W$$

$$q_5 = q_6 = \frac{E_{b5} - J_5}{0.6667} = -41.5 \ W$$

8-98 $r_1 = 10\,cm$ $r_2 = 5\,cm$ $L = 10\,cm$ $r_2/L = 0.5$ $L/r_1 = 1.0$ $F_{12} = 0.11$

③ room $F_{13} = 1 - 0.11 = 0.89$ $F_{13\,bottom} = 1.0$ $T_2 = 700K$ $\epsilon_2 = 0.8$

$\epsilon_1 = 0.4$ $T_3 = 298\,K$ $F_{21} = (0.11)(2)^2 = 0.44$ $E_{b2} = 13611$ $E_{b3} = 447$

$F_{23} = 1 - 0.44 = 0.56$ $A_1 = 0.007854$ $A_2 = 0.001963\,m^2$ $\frac{1-\epsilon_1}{\epsilon_1 A_1} = 191$

$\frac{1-\epsilon_2}{\epsilon_2 A_2} = 127$ $1/A_2F_{23} = 910$ $1/A_1F_{12} = 1157$ $1/A_1F_{13} = 143$ $1/A_1F_{13\,bot} = 127$

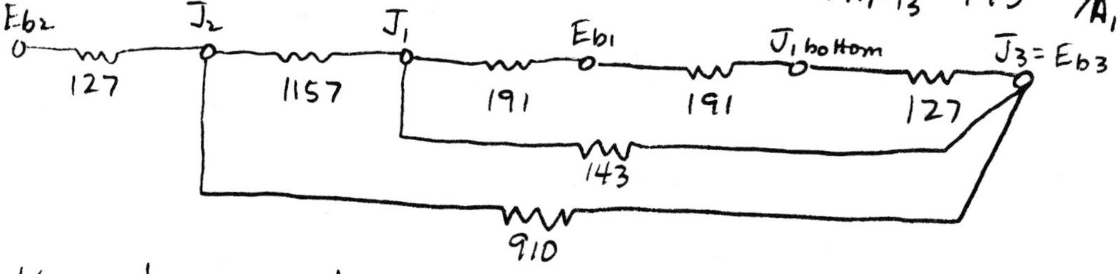

$$1/R_1 = \frac{1}{143} + \frac{1}{(2)(191) + 127} = 0.008958 \qquad R_1 = 111.6$$

$$1/R_2 = \frac{1}{910} + \frac{1}{1157 + R_1} \qquad R_2 = 529.9 \qquad R = 127 + R_2 = 656.9$$

$$q_2 = \frac{E_{b2} - E_{b3}}{R} = \frac{13611 - 447}{656.9} = 20.04\ W$$

8-99 $T_1 = 1073K$ $\epsilon_1 = 0.63$ $r_1 = r_2 = 25\,cm$ $q_2 = 80\,kW/m^2$

$\epsilon_2 = 0.75$ $d = 12.5\,cm$ $T_3 = 303\,K$ $E_{b1} = 75146\ W/m^2$ $E_{b3} = 478 = J_3$

$F_{12} = 0.6$ $F_{13} = 0.4$ $A_1 = A_2 = \pi(0.25)^2 = 0.1963\,m^2$ $F_{21} = 0.6$

$F_{23} = 0.4$

$$J_1 - (1 - 0.63)[0.6 J_2 + (0.4)(478)] = (0.63)(75146) \qquad J_1 = 75237\ W/m^2$$

$$J_2 - [0.6 J_1 + (0.4)(478)] = 80000 \qquad J_2 = 1.253 \times 10^5$$

$$q_1 = \frac{\epsilon_1 A_1}{1-\epsilon_1}(E_{b1} - J_1) = -30.4\ W \qquad q_2 = \frac{\epsilon_2 A_2}{1-\epsilon_2}(E_{b2} - J_2) = 15704\ W$$

$$E_{b2} = 1.5197 \times 10^5 = \sigma T_2^4 \qquad T_2 = 1280\ K = 1007\,°C$$

$$q(\text{room}) = q_1 + q_2 = 15674\ W$$

8-100 $J_1 - [0.6 J_2 + (0.4)(478)] = 100{,}000$

$J_2 - [0.6 J_1 + (0.4)(478)] = 80{,}000$

Solution $J_1 = 2.317 \times 10^5$ W/m²

$J_2 = 2.192 \times 10^5$ W/m²

$q_1 = \frac{\epsilon_1 A_1}{1-\epsilon_1}(E_{b1} - J_1) = 19630$ W $\quad E_{b1} = 2.904 \times 10^5$ W/m² $= \sigma T_1^4$

$T_1 = 1504$ K $= 1231°C$

$q_2 = \frac{\epsilon_2 A_2}{1-\epsilon_2}(E_{b2} - J_2) = 15704$ W $\quad E_{b2} = 2.459 \times 10^5$ W/m² $= \sigma T_2^4$

$T_2 = 1443$ K $= 1170°C$

$q_{room} = q_1 + q_2 = 35064$ W

8-101 Heater = surface ③ Room is ④ $d_1 = 10$ cm $\epsilon_1 = 0.4$

$d_2 = 20$ cm $\epsilon_2 = 0.6$ $L = 10$ cm $\epsilon_3 = 0.8$ $q/A|_3 = 90$ kW/m²

$T_4 = 303$ K $A_1 = \pi(0.1)^2 = 0.03142$ m² $A_2 = 0.06283$ m²

$A_3 = \pi(0.1^2 - 0.05^2) = 0.02356$ m² $F_{21} = 0.32$ $F_{22} = 0.23$

$F_{24} + F_{23} = 1 - 0.32 - 0.23 = 0.45$ $F_{24i} = F_{23} = 0.225$

$F_{12} = (2)(0.32) = 0.64$ $F_{13} = F_{14i} = \frac{1}{2}(1 - 0.64) = 0.18$ $F_{32} = 0.6$

$F_{34} = 0.4$ $F_{31} = 0.24$

Both inside and outside of cylinder exchange heat with room and are in radiant balance

Open ends of ① $F = 0.175$ $A_{end} = 0.007854$ m² $F_{end-①} = 0.825$

$F_{①-end} = 0.2065$ $F_{14i} = (2)(0.206) = 0.413$ $E_{b4} = 478$ W/m²

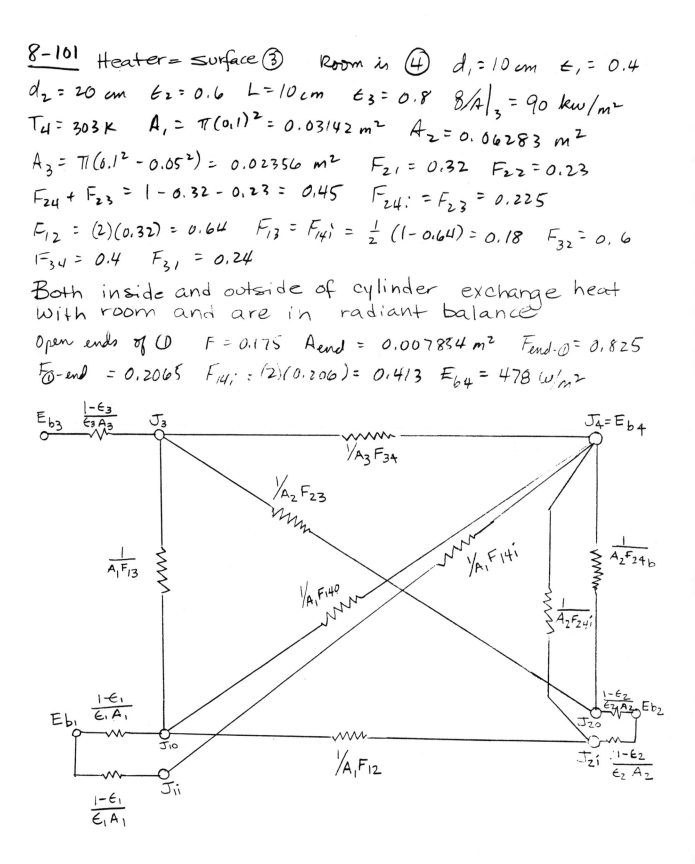

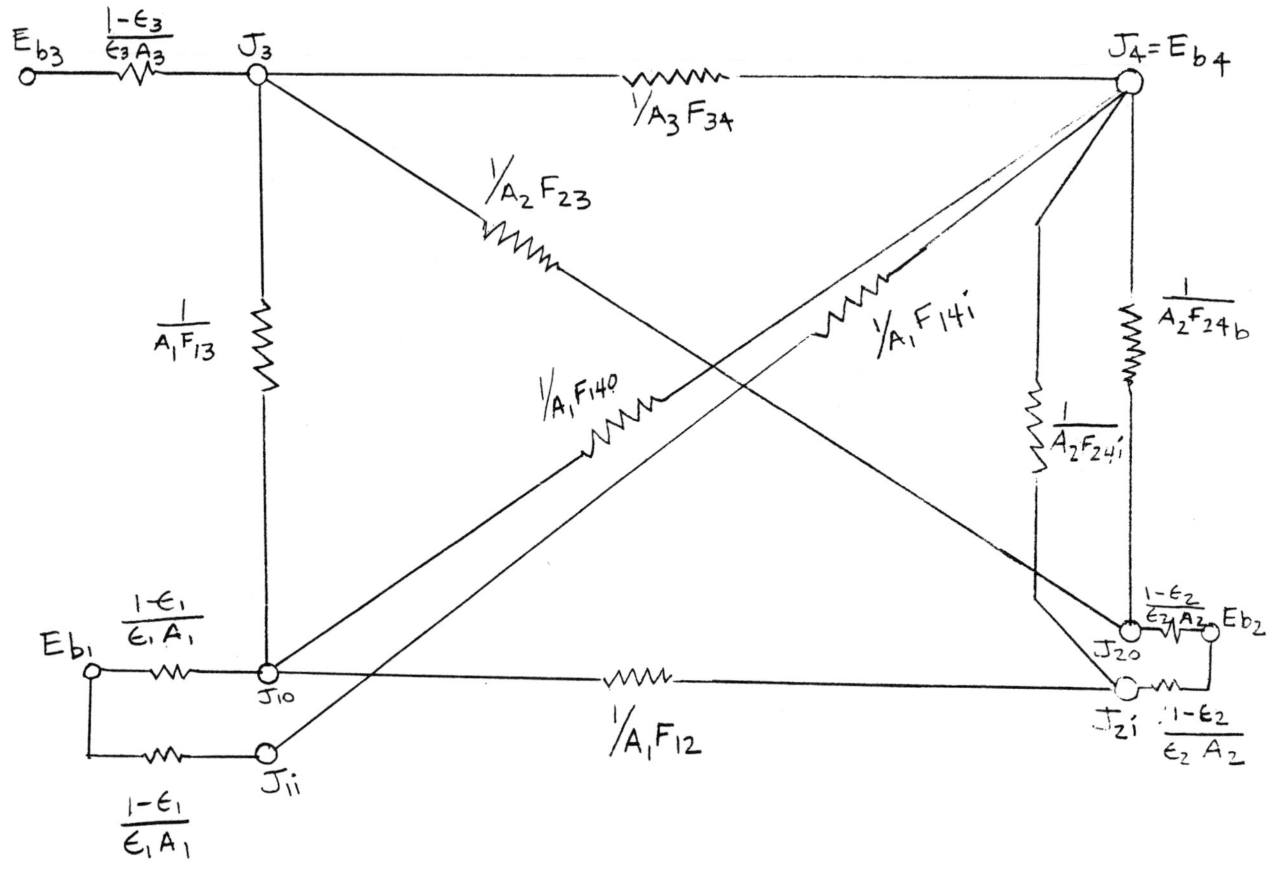

8-101 contd

$$\frac{1-\epsilon_1}{\epsilon_1 A_1} = 47.74 \quad \frac{1-\epsilon_2}{\epsilon_2 A_2} = 10.61 \quad \frac{1}{A_1 F_{13}} = 176.8 \quad \frac{1}{A_2 F_{23}} = 70.74$$

$$\frac{1}{A_1 F_{14o}} = 176.8 \quad \frac{1}{A_1 F_{14i}} = 77.06 \quad \frac{1}{A_1 F_{12}} = 49.73 \quad \frac{1}{A_3 F_{34}} = 106.1$$

$$\frac{1}{A_2 F_{24i}} = 70.74 \quad \frac{1}{A_2 F_{24o}} = 15.92 \quad q_3 = \left(\frac{q}{A}\right) A_3 = (90000)(0.02356) = 2120\ \text{W}$$

NODE J_3

$$2120 + \frac{J_{1o} - J_3}{176.8} + \frac{J_{2i} - J_3}{70.74} + \frac{478 - J_3}{106.1} = 0$$

NODE J_{1o}

$$\frac{J_{1i} - J_{1o}}{(2)(47.74)} + \frac{J_3 - J_{1o}}{176.8} + \frac{478 - J_{1o}}{176.8} + \frac{J_{2i} - J_{1o}}{49.73} = 0$$

NODE J_{1i}

$$\frac{J_{1o} - J_{1i}}{(2)(47.74)} + \frac{478 - J_{1i}}{77.06} = 0$$

NODE J_{2o}

$$\frac{J_{2i} - J_{2o}}{(2)(10.61)} + \frac{478 - J_{2o}}{15.92} = 0$$

NODE J_{2i}

$$\frac{J_{2o} - J_{2i}}{(2)(10.61)} + \frac{J_{1o} - J_{2i}}{49.73} + \frac{J_3 - J_{2i}}{70.74} + \frac{478 - J_{2i}}{70.74} = 0$$

8-101 contd

Solution

$J_{1o} = 81792 \text{ W/m}^2$ $\quad J_{1i} = 36584 \text{ W/m}^2$ $\quad J_{2o} = 47927 \text{ W/m}^2$

$J_{2i} = 111173 \text{ W/m}^2$ $\quad J_3 = 142337 \text{ W/m}^2$

Then $E_{b1} = \dfrac{J_{1o} + J_{1i}}{2} = \sigma T_1^4 = 59188 \text{ W/m}^2 \quad T_1 = 1011 \text{ K}$

$E_{b2} = \dfrac{J_{2o} + J_{2i}}{2} = \sigma T_2^4 = 79550 \text{ W/m}^2 \quad T_2 = 1088 \text{ K}$

$$q_3 = 2120 = \frac{E_{b3} - 142337}{\dfrac{1 - 0.8}{(0.8)(0.02356)}} \qquad E_{b3} = 1.648 \times 10^5 = \sigma T_3^4$$

$T_3 = 1306 \text{ K}$

8-102 $T_1 = 600\text{ K}$ $F_{12} = 1.0$ $E_{b1} = 7347\ \text{W/m}^2$ $\epsilon_1 = 0.75$

③ room $E_{b3} = 447\ \text{W/m}^2$ $\epsilon_2 = 0.8$ $T_3 = 298\text{ K}$

Per unit length: $A_1 = \pi(0.02) = 0.06283$ $A_2 = 0.1571$ $T_\infty = 308\text{ K}$

$$\frac{1-\epsilon_1}{\epsilon_1 A_1} = 5.305 \quad 1/A_1F_{12} = 15.916 \quad \frac{1-\epsilon_2}{\epsilon_2 A_2} = 1.591 \quad 1/A_2F_{23} = 6.365$$

$$q = \frac{(5.669\times10^{-8})(600^4 - T_2^4)}{5.305 + 15.916 + 1.591} = \frac{(5.669\times10^{-8})(T_2^4 - 298^4)}{1.591 + 6.365} + \pi(0.05)(180)(T_2 - 308)$$

By iteration: $T_2 = 318\text{ K}$ $q = 297\ \text{W/m}$

8-103 $T_1 = 500\text{ K}$ $T_3 = 300\text{ K}$ $\epsilon_1 = 0.8$ $\epsilon_2 = 0.4$ $E_{b1} = 3543\ \text{W/m}^2$

$A_1 = 0.03$ $A_2 = 0.04$ $E_{b3} = 459\ \text{W/m}^2$ $F_{21} = 0.18$ $F_{23} = 0.82$ $F_{13} = 0.76$

$$\frac{1-\epsilon_1}{\epsilon_1 A_1} = 8.333 \quad 1/A_1F_{12} = 138.89 \quad 1/A_1F_{13} = 43.86 \quad 1/A_2F_{23} = 30.49$$

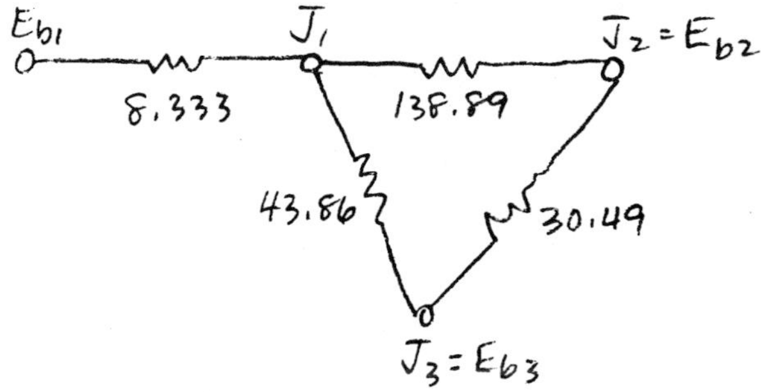

$$q = \frac{3543 - 459}{8.333 + \dfrac{1}{\dfrac{1}{43.86} + \dfrac{1}{138.89 + 30.49}}} = 71.44\ \text{W}$$

$$= \frac{E_{b1} - J_1}{8.333} \qquad J_1 = 2948\ \text{W/m}^2$$

$$\frac{2948 - J_2}{138.89} = \frac{2948 - 459}{138.89 + 30.49} \qquad J_2 = 907\ \text{W/m}^2 = E_{b2} = \sigma T_2^4$$

$T_2 = 356\text{ K} = 83°\text{C}$

8-106 $\alpha A \alpha_{solar} = \epsilon_{lowtemp} A \sigma (T^4 - T_\infty^4)$ $(1500)(0.15) = (0.04)(5.669\times10^{-8}) T^4$

$T = 561\ K = 288\ °C$

8-107 $T_\infty = -70\ °C = 203\ K$ $\alpha_{solar} = 0.46$ $\alpha_{low\ temp} = 0.95$

$(1070)(0.46) = (0.95)(5.669\times10^{-8})(T^4 - 203^4)$ $T = 322.6\ K = 49.6\ °C$

8-108 $Re = 34562$ $T_\infty = 27°C = 300\ K$ $\bar{h} = \frac{(0.664)(0.02749)}{0.3}(34562)^{1/2}(0.7)^{1/3}$

$\bar{h} = 10.04\ W/m^2\cdot°C$ $(1100)(0.6) = (10.04)(T - 300) + (0.09)(5.669\times10^{-8})(T^4 - 300^4)$

$T = 316.5\ K = 43.5°C$

8-110 $\alpha_s (950) = \sigma \epsilon (T^4 - 300^4) + 12(T - 300)$

$f(T) = 0 = -902.5 + (0.6)(5.669\times10^{-8})(T^4 - 300^4)$

T	$f(T)$
350	−67.6
360	+113.3
T = 353.7	0

8-111 $\alpha_{solar} = 0.94$ $\alpha_{lowtemp} = 0.21$

$(800)(0.94) = (0.21)(5.669\times10^{-8})(T^4 - 298^4)$

$T = 516\ K = 243\ °C$

8-112 $A = 0.37$ $\alpha = 50°$ $I_o = 1395 \sin 50 = 1069\ W/m^2$ $n = 3.5$

$m = \csc 50 = 1.305$ $a_{ms} = 0.128 - 0.054 \log(1.305) = 0.1217$

$I_c = 1069 \exp[-(0.1217)(3.5)(1.305)] = 613\ W/m^2$ absorption $= (1 - 0.37)(613)$

absorption $= 386\ W/m^2$

8-113 $\alpha = 30°$ $n = 4.0$ $I_0 = 1395 \sin 30 = 697.5 \text{ W/m}^2$ $m = \csc 30° = 2$

$a_{ms} = 0.128 - 0.054 \log(2) = 0.1117$ $I_c = (697.5)\exp[-(0.117)(2)(4.0)]$

$I_c = 285.3 \text{ W/m}^2$ For clear air at $\alpha = 90°$ $m = 1$ $n = 2$

$I_c = (1395)\exp[-(0.128)(2)] = 1080 \text{ W/m}^2$

8-124 $T_t = 443 \text{ K}$ $T_w = 698 \text{ K}$ $\epsilon_t = 0.43$ $h = 150 \text{ W/m}^2 \cdot °\text{C}$

$(5.669 \times 10^{-8})(0.43)[698^4 - 443^4] = (150)(443 - T_g)$

$T_g = 410.7 \text{ K} = 137.7 °\text{C}$

8-125 $T_s = -40°\text{C} = 233 \text{ K}$ $\epsilon = 0.8$

$hA(T_w - T_{aw})_{lam} + hA(T_w - T_{aw})_{turb} = -\sigma A \epsilon (T_w^4 - T_s^4)$

laminar: $h = 56.26$ $x_c = 0.222 \text{ m}$ $T_{aw} = 584 \text{ K}$

Turbulent: $h = 87.45$ 0.222 to 0.7 m $T_{aw} = 605 \text{ K}$

$(56.26)(0.222)(T_w - 584) + (87.45)(0.7 - 0.222)(T_w - 605) =$

$-(5.669 \times 10^{-8})(0.7)(0.8)[T_w^4 - (233)^4]$ $T_w = 549 \text{ K} = 276 °\text{C}$

8-126 $(28)A(T_a - 273) = (5.669 \times 10^{-8})A(1.0)[273^4 - 203^4]$

$T_a = 280.8 \text{ K} = 7.8 °\text{C}$

8-127 $h = 1.32\left(\frac{\Delta T}{d}\right)^{1/4}$ $T_f = 650°\text{C} = 923 \text{ K}$ $T_a = 560°\text{C} = 833 \text{K}$

$q/A)_{conv} = 1.32(\Delta T)^{5/4}\left(\frac{1}{d}\right)^{1/4}$

$(5.669 \times 10^{-8})(0.6)[923^4 - T_t^4] = 1.32\left(\frac{1}{0.0032}\right)^{1/4}(T_t - 833)^{5/4}$

$3.4014 \times 10^{-8} T_t^4 + 5.55(T_t - 833)^{5/4} - 24687 = 0$ $T_t = 911 \text{ K} = 638 °\text{C}$

8-129 $T_w = 373\,K$ $T_{sun} = 673\,K$ $\epsilon_w = 0.6$ assume $T_f = 350\,K$

$\rho = 0.998$ $\mu = 2.075 \times 10^{-5}$ $k = 0.03003$ $Pr = 0.7$

$$Re = \frac{(0.998)(7)(0.003)}{2.075 \times 10^{-5}} = 1010 \qquad C = 0.683 \qquad n = 0.466$$

$$h = \frac{0.03003}{0.003}(0.683)(1010)^{0.466}(0.7)^{1/3} = 152.5\ \mathrm{W/m^2 \cdot {}^\circ C}$$

$$(0.6)(5.669 \times 10^{-8})\left[673^4 - 373^4\right] = (152.5)(373 - T_g) \qquad T_g = 331.5\,K = 58.5^\circ C$$

$T_f = 352\,K$

8-130 $A_1 F_{12} = A_2 F_{21}$ $F_{21} = (0.86)(1/2) = 0.43$ $F_{22} = 1 - 0.43 - 0.24 = 0.33$

$$J_1 - (1-\epsilon_1)\left[F_{12} J_{2i} + F_{13} E_{b3}\right] = \epsilon_1 E_{b1}$$

$$J_{2i}\left[1 - F_{22}(1-\epsilon_2)\right] - (1-\epsilon_2)\left[F_{21} J_1 + F_{23i} E_{b3}\right] = \epsilon_2 E_{b2}$$

$$J_{2o} - (1-\epsilon_2)\left[F_{23o} E_{b3}\right] = \epsilon_2 E_{b2}$$

$$h A_2 (T_3 - T_2) = \frac{E_{b2} - J_{2o}}{\frac{1-\epsilon_2}{\epsilon_2 A_2}} + \frac{E_{b2} - J_{2i}}{\frac{1-\epsilon_2}{\epsilon_2 A_2}}$$

$$h(T_3 - T_2) = \frac{2\epsilon_2}{1-\epsilon_2} E_{b2} - \frac{\epsilon_2}{1-\epsilon_2}(J_{2o} + J_{2i})$$

Solution yields:

$T_2 = 406\,K$ $E_{b2} = 1540$ $J_{2i} = 23644$ $J_{2o} = 675$ $J_1 = 4943$

$$q_1 = \frac{56690 - 4943}{\frac{1 - 0.8}{0.8(0.06283)}} = 1824\ \mathrm{W}$$

8-131 $T_t = 1023$ K $\quad h = 20$ W/m²·°C $\quad T_\infty = 923$ K

$$(5.669 \times 10^{-8})(0.7)(T_f^4 - 1023^4) = (20)(1023 - 923)$$

$$T_f = 1035 \text{ K} = 762°\text{C}$$

8-132 $T_t = 328$ K $\quad h = 30$ W/m²·°C $\quad T_w = 373$ K $\quad \epsilon = 0.94$

$$(5.669 \times 10^{-8})(0.94)(373^4 - 328^4) = (30)(328 - T_a)$$

$$T_a = 314 \text{ K} = 41°\text{C}$$

8-133 $T_f = \dfrac{20 + 150}{2} = 85°\text{C} = 358\text{K}$ $\quad \nu = 21.58 \times 10^{-6}$ $\quad k = 0.0306$

$Pr = 0.7 \quad Re = \dfrac{(25)(0.5)}{21.58 \times 10^{-6}} = 5.792 \times 10^5$

$$Nu_d = 0.3 + \frac{(0.62)(5.79 \times 10^5)^{1/2}(0.7)^{1/3}}{\left[1 + \left(\frac{0.4}{0.7}\right)^{2/3}\right]^{1/4}} \left[1 + \left(\frac{5.79 \times 10^5}{282000}\right)^{5/8}\right]^{4/5}$$

$$= 601.7 \quad h = \frac{(601.7)(0.0306)}{0.5} = 36.82 \text{ W/m}^2\cdot°\text{C}$$

$$q_c = (36.82)\pi(0.5)(150 - 20) = 7519 \text{ W/m}$$

$$q_r = (5.669 \times 10^{-8})(0.7)\pi(0.5)(423^4 - 293^4) = 1536 \text{ W/m}$$

$$q_{total} = 7519 + 1536 = 9055 \text{ W/m}$$

8-134

(a) $F_{21} = 1.0$

$F_{12} = \frac{A_2}{A_1} = \frac{d}{\pi d/2} = \frac{2}{\pi} = 0.637$

$F_{11} = 1 - A_2/A_1 = 1 - 2/\pi = 0.363$

(b) $F_{12} = A_1/A_{opening} = 0.637$

$F_{11} = 1 - 0.637 = 0.363$

$F_{21} \approx 0$ because $A_2 \to \infty$

(c) Parallel disk $d/x = 2/1 = 2$, bottom = ②

$F_{23} = 0.37$

$F_{21} = 1 - 0.37 = 0.63$

$A_2 F_{21} = A_1 F_{12}$

$F_{12} = \frac{\pi(1)^2 (0.63)}{\pi(2)(1)} = 0.315 = F_{13}$

$F_{11} + F_{12} + F_{13} = 1.0$

$F_{11} = 1 - 0.63 = 0.37$

(d) $F_{12} = 1.0$, $F_{13} = 0$

$A_1 = \pi (0.5)^2 = \pi/4$

$A_2 = 2\pi (1.5)^2 = 4.5\pi$

$A_2 F_{21} = A_1 F_{12}$

$F_{21} = \frac{(\pi/4)(1.0)}{4.5\pi} = 0.0555$

$A_{open} = \pi (1.5)^2 = 2.25\pi$

$F_{open-2} = 1.0$

$A_{open} F_{open-2} = A_2 F_{2-open}$

8-134 (contd)

$$F_2 \text{ open} = \frac{(2.25)(1)}{4.5} = 0.5 = F_{21} + F_{23}$$

$$F_{23} = 0.5 - 0.0555 = 0.4444$$

$$F_{21} + F_{22} + F_{23} = 1.0$$

$$F_{22} = 0.5$$

(e) Parallel Disks

$r_1 = 0.5$, $r_3 = 1.5$, $L = 2.0$, $r_3/L = 0.75$ $L/r_1 = 4.0$

$$F_{13} = 0.35$$

$$F_{12} = 1 - 0.35 = 0.65$$

$$A_1 = \pi(0.5)^2 = \pi/4$$

$$A_2 = \pi(3)(2) = 6\pi$$

$$A_{3'} = \pi(1.5)^2 = 2.25\pi$$

$F_{33'} = 0.28$ Top = surface 3' $d/x = 3/2$

$$F_{3'2} = 1 - 0.28 = 0.72$$

$$A_{3'} - F_{3'2} = A_2 F_{23'}$$

$$F_{23'} = \left(\frac{2.25}{6}\right)(0.72) = 0.27 = F_{2\text{-bottom}}$$

$$A_1 F_{12} = A_2 F_{21}$$

$$F_{21} = \frac{\pi/4}{6\pi}(0.65) = 0.0271$$

$$F_{21} + F_{22} + F_{23} = 1.0$$

$$F_{23} = 0.27 + 0.27 - 0.0271 = 0.513$$

$$F_{22} = 1 - 0.0271 - 0.513 = 0.46$$

8-134 (contd)

(f) $F_{12} = 0.5$

$F_{21} \to 0$ because $A_2 \to \infty$

(g) $A_1 = \sqrt{2}(1.5) = 2.121$

A_3 for opening $= 2.121$

$A_2 = 3.0$

By symmetry $F_{21} = F_{23} = 0.5$

$A_1 F_{12} = A_2 F_{21}$

$F_{12} = \frac{3}{2.121}(0.5) = 0.707$

(h) Perpendicular Rectangles

$Y/X = Z/X$

Y/X	F_{12}
0.6	0.23
0.4	0.25
0.2	0.27
0.1	0.285

$F_{12} = 0.293 = F_{21}$

$F_{13} = F_{23} = 1 - 0.293 = 0.707$

8-134 (contd)

(i) $A_2 = 4\pi(0.5)^2 = \pi$

$A_1 = 2\pi(1.5)^2 = 4.5\pi$

$F_{21} = 0.5$

$A_1 F_{12} = A_2 F_{21}$

$F_{12} = \frac{0.5}{4.5} = 0.1111$

$A_{3'}$ = tangential frustum of cone plus portion of sphere top

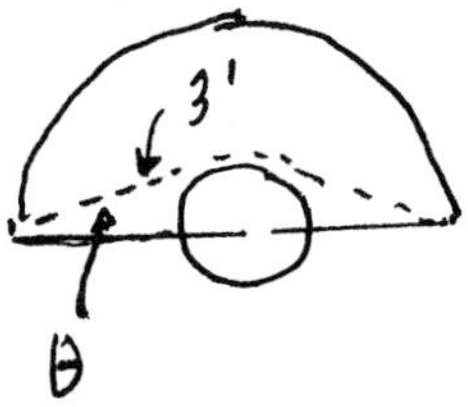

$\sin\theta = \frac{0.5}{1.5}$ $\theta = 19.47°$

tangent = $1.5\cos\theta = 1.414$

r at tangent $= 0.5\sin\theta = 0.1667$

Cone area $= 2\pi\left[\frac{1.5 + 0.1667}{2}\right](1.414) = 7.404$

Portion of sphere area $= 4\pi(0.5)^2(2)(19.47/360)$
$= 0.3398$

$A_{3'} = 7.404 + 0.3398 = 7.744$

$F_{3'-1} = 1.0$ $A_{3'} F_{3'-1} = A_1 F_{1-3'}$

$F_{1-3'} = \frac{(1.0)(7.744)}{4.5\pi} = 0.548 = F_{12} + F_{13}$

$F_{13} = 0.548 - 0.1111 = 0.4367$

$F_{11} + F_{12} + F_{13} = 1.0$; $F_{11} = 1 - 0.1111 - 0.4367 =$
0.452

8-135 $\lambda_1 T = (0.4)(3400) = 1360\,\mu\cdot K = 2448\,\mu\cdot {}^\circ R$

$\lambda_2 T = (0.7)(3400) = 2380\,\mu\cdot K = 4284\,\mu\cdot {}^\circ R$

$E_b = (0 - \lambda_1 T) = 0.00644$

$E_b = (0 - \lambda_2 T) = 0.13626$

Fraction between 0.4 and 0.7μ =

0.12982

8-136 $F_{ii} = 0$,

Perpendicular sides $F_{ij} = 0.2$

Parallel sides $F_{ij} = 0.2$

8-137 $F_{ii} = 0 \quad F_{ij} = 1/3$

8-138 $E_b = 5.669 \times 10^{-8}(3400)^4 = 7.576 \times 10^6\ W/m^2$

Between 0.4 and $0.7\ \mu = (7.576 \times 10^6)(0.12982)$

$= 9.835 \times 10^5$

$400\ W = A(7.576 \times 10^6$

$A = 5.286 \times 10^{-5}\ m^2 = 0.5286\ cm^2$

8-139

$$E_b = 5.669 \times 10^{-8}(3000)^4 = 4.592 \times 10^6 \text{ W/m}^2$$

λ	ϵ	λT	Fraction
0	0.6	0	
2	↓ 0.2	6000	0.73777
8	↓	24000	0.99075

$$E = (4.592 \times 10^6)[(0.6)(0.73777) + (0.2)(0.99075 - 0.73777)]$$

$$= 2.265 \times 10^6 \text{ W/m}^2$$

8-141

$T = 5800$ K

λ	λT	Fraction
0.25	1450	0.0099
0.5	2900	0.25055
1.5	8700	0.88066
2.5	14500	0.96604

Fraction transmitted through plain glass $= (0.9)(0.96604 - 0.0099) = 0.86053$

Fraction through tinted glass

$$= (0.9)(0.88066 - 0.25055) = 0.56710$$

8-142 $h = 12 \ W/m^2-°C$

$T_\infty = 400 K$

For $T_w = 510 K$ $E = 1100 \ W/m^2$

$E_{bw} = \sigma T_w^4 = 3835$

$$\epsilon = \frac{E}{E_b} = 0.287$$

$A(G-J) = hA(T_w - 400)$

$G = 2200 \ W/m^2$

$J = 450 + \epsilon \sigma T_w^4$

$$2200 - 450 - (0.289)\sigma T_w^4 = 12(T_w - 400)$$

$T_w = 475 K$

$J = 450 + (0.287)(5.669 \times 10^{-8})(475)$

$= 1278 \ W/m^2$

8-143

$$E_b = \sigma(1600)^4 = 3.715 \times 10^5 \ \text{W/m}^2$$

λ	λT	Fraction
0.6	960	0.0001
5	8000	0.85624

$$E = (3.715 \times 10^5)\left[(0.08)(0.0001) + (0.4)(0.85624 - 0.0001) + (0.7)(1 - 0.85624)\right]$$

$$= 1.646 \times 10^5 \ \text{W/m}^2$$

8-145

$T_1 = 500\text{K}$ $\quad \epsilon_1 = 0.5$

$T_2 = 300\text{K}$ $\quad \epsilon_2 = 0.7$

$A_1 = A_2 = (3)^2 = 9$

$F_{12} = 0.2$

Eq. (8-41)

$$q = \frac{(5.669 \times 10^{-8})(9)(500^4 - 300^4)}{\frac{18(1-0.2)}{9(1-0.2^2)} + \left(\frac{1}{0.5} - 1\right)\left(\frac{1}{0.7} - 1\right)}$$

$$= \frac{27750}{3.095} = 8968 \ \text{W}$$

8-146

$T_1 = 800K, \quad \epsilon_1 = 0.6 \qquad A_1 = \pi (0.05)^2 = 7.85 \times 10^{-3}\ m^2$

$T_2 = 300K, \quad \epsilon_2 = 0.3 \qquad A_2 = 2\pi (0.05)^2 = 0.0157\ m^2$

$E_{b1} = \sigma T_1^4 = 23220 \qquad E_{b2} = \sigma T_2^4 = 459\ W/m^2$

$d/x = 10/15 = 2/3 \qquad F_{12} = 0.12$

$F_{13} = 1 - 0.12 = 0.88$

$\frac{1-\epsilon_1}{\epsilon_1 A_1} = 84.93 \qquad \frac{1-\epsilon_2}{\epsilon_2 A_2} = 148.6 \qquad \frac{1}{A_1 F_{12}} = 1062$

$\frac{1}{A_1 F_{13}} = 144.8 \qquad \frac{1}{A_2 F_{23}} = 144.8$

$F_{2'-2} = 1.0 \qquad F_{2-2'} = 0.5$

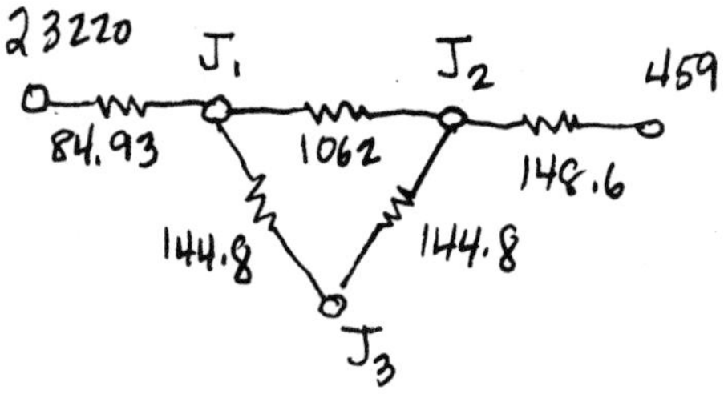

$F_{22} = 0.5$

$F_{21} = \frac{1}{2}(0.12) = 0.06$

$F_{23} = 1 - 0.5 - 0.06 = 0.44$

$$q = \frac{23220 - 459}{84.93 + \frac{1}{\frac{1}{1062} + \frac{1}{(2)(144.8)}} + 148.6} = 49.4\ W$$

8-147 $E_{b1} = 23220$ $E_{b2} = 459\ W/m^2$

Per Unit Length:

$$q/L = \pi(0.05)(0.7)(23220 - 459) = 2503\ W/m$$

With shield, as surface 3, $F_{13} = 1.0$, $\epsilon_3 = 0.2$

$$\frac{1-\epsilon_1}{\epsilon_1 \pi(0.05)} = 2.73 \qquad \frac{1}{\pi(0.05)(1.0)} = 6.37$$

$$\frac{1-\epsilon_3}{\epsilon_3 \pi(0.1)} = 12.73 \qquad \frac{1}{\pi(0.1)} = 3.18$$

q with shield per unit length;

$$q/L = \frac{23220 - 459}{2.73 + 6.37 + (2)(12.73) + 3.18}$$

$$= 603\ W$$

Reduced by 76%

8-148

$T_1 = 450\,K,\ \epsilon_1 = 0.5 \qquad A_1 = \pi(0.25)^2 = 0.1963\,m^2 = A_2$

$T_2 = 600\,K,\ \epsilon_2 = 0.6$

$T_3 = 1000\,K,\ \epsilon_3 = 0.7 \qquad A_3 = \pi(0.5)(0.5) = 0.7854\,m^2$

$d/x = 1.0 \quad F_{12} = 0.17 \quad F_{13} = 0.83 = F_{23}$

$= F_{21}$

$$F_{31} = F_{32} = (0.83)\left(\frac{0.1963}{0.7854}\right) = 0.2075$$

$$F_{33} = 1 - (2)(0.2075) = 0.585$$

$E_{b1} = 2325\ W/m^2 \quad E_{b2} = 7347\ W/m^2 \quad E_{b3} = 56690\ W/m^2$

$$\begin{cases} J_1 - 0.5[0.17 J_2 + 0.83 J_3] = 0.5(2325) \\ J_2 - 0.4[0.17 J_1 + 0.83 J_3] = 0.6(7347) \\ J_3[1 - 0.585(0.3)] - 0.3[0.2075 J_1 + 0.2075 J_2] = 0.7(56690) \end{cases}$$

$$\begin{cases} J_1 - 0.085 J_2 - 0.415 J_3 = 1162.5 \\ -0.068 J_1 + J_2 - 0.332 J_3 = 4408.2 \\ -0.06225 J_1 - 0.06225 J_2 + 0.8245 J_3 = 39683 \end{cases}$$

Solving,

$J_1 = 24614\ W/m^2$

$J_2 = 23261$

$J_3 = 51744$

8-148 (contd)

$$q_1 = \frac{0.5}{0.5}(0.1963)(2325 - 24614) = -4375\ W = 0.1963(J_1 - G_1)$$

$G_1 = 46901\ W/m^2$

$$q_2 = \frac{0.6}{0.4}(0.1963)(7347 - 23261) = -4686\ W = 0.1963(J_2 - G_2)$$

$G_2 = 47133\ W/m^2$

$$q_3 = \frac{0.7}{0.3}(0.7854)(56690 - 51744) = 9064\ W = 0.7854(J_3 - G_3)$$

$G_3 = 40203\ W/m^2$

8-149

Behaves as if $T_2 = 300K$, $\epsilon_2 = 1.0$

$E_{b2} = 459\ W/m^2 = J_2$

Do not need Eqn. for surface 2.

Inserting $J_2 = 459$ in Eqns for 1 and 3 gives

$$J_1 - 0.415\ J_3 = 1201.5$$

$$-0.06225\ J_1 + 0.8245\ J_3 = 39712$$

Solving, $J_1 = 21875$

$J_3 = 49817$

$$q_1 = \frac{0.5}{0.5}(0.1963)(2325 - 21875) = -3838\ W = 0.1963(J_1 - G_1)$$

$G_1 = 41427\ W/m^2$

$$q_3 = \frac{0.7}{0.3}(0.7854)(56690 - 49817) = 12595\ W = 0.7854(J_3 - G_3)$$

$G_3 = 33781\ W/m^2$

$$q_2 = -12595 + 3838 = -8757\ W$$

$G_2 \to 0$ because $A_2 \to \infty$

8-150

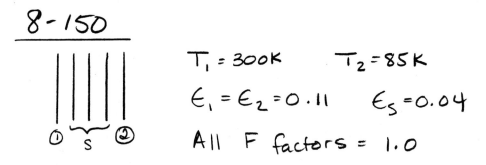

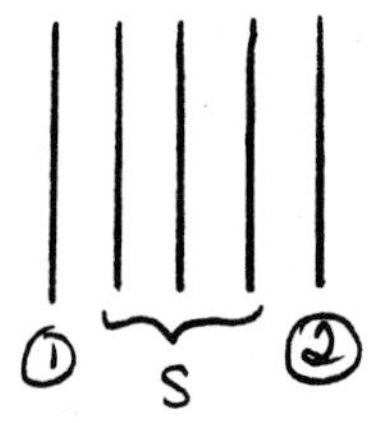

$T_1 = 300K \qquad T_2 = 85K$

$\epsilon_1 = \epsilon_2 = 0.11 \qquad \epsilon_s = 0.04$

All F factors = 1.0

Work problem per unit area

$$\frac{1-\epsilon_1}{\epsilon_1} = \frac{1-\epsilon_2}{\epsilon_2} = 8.091$$

$$\frac{1-\epsilon_s}{\epsilon_s} = \frac{0.96}{0.04} = 24$$

4 space resistances = 1.0

Total resistance = $(2)(8.091) + (4)(1.0) + (6)(24) = 164.2$

$$q = \frac{(5.669 \times 10^{-8})(300^4 - 85^4)}{164.2} = 2.779 \text{ W/m}^2$$

$$= k_{eff} \frac{(300-85)}{0.008} \qquad k_{eff} = 1.034 \times 10^{-4} \text{ W/m-°C}$$

$$R = 77.4 \quad \text{°C-m}^2\text{/W}$$

8-151 $T_1 = 750\text{K}$ $\epsilon_1 = 0.75$ $A_1 = \pi(0.2)^2 = 0.1256$

$T_2 = 600\text{ K}$ $\epsilon_2 = 0.4$ $A_2 = \pi[0.2^2 - 0.1^2] = 0.09425$

$q_3 = 0$ $A_3 = \pi(0.4)(0.3) = 0.377\text{ m}^2$

$A_{hole} = \pi(0.1)^2 = 0.03146$

$d_1/x = 40/30 = 1.33$ $F_{1-2'} = 0.25$

To hole $30/20 = 1.5$ $10/30 = 1/3$ $F_{1-hole} = 0.07 = F_{14}$

$F_{12} = 0.25 - 0.07 = 0.18$

$F_{13} = 1 - 0.25 = 0.75$

$T_4 = \text{room} = 300\text{K}$

$$F_{31} = \frac{(0.75)(0.1256)}{0.337} = 0.25$$

$F_{33} = 1 - (2)(0.25) = 0.5$

$$F_{21} = \frac{(0.1256)(0.18)}{0.09425} = 0.24$$

$$F_{hole-1} = \frac{(0.1256)(0.07)}{0.03146} = 0.28$$

$F_{hole-3} = 1 - 0.28 = 0.72$

$$F_{3-hole} = F_{34} = \frac{(0.72)(0.03146)}{0.377} = 0.06$$

$F_{32} = 0.25 - 0.06 = 0.19$

$F_{23} = (0.19)(0.377)/0.09425 = 0.76$

$E_{b1} = 17937\text{ W/m}^2$ $E_{b2} = 7347$ $E_{b4} = 459 = J_4$

$$\left\{\begin{array}{l} J_1 - 0.25[0.18J_2 + 0.07(459) + 0.75J_3] = 0.75(17937) \\ J_3(1-0.5) - [0.25J_1 + 0.19J_2 + 0.06(459)] = 0 \\ J_2 - 0.6[0.24J_1 + 0.76J_3] = 0.4(7347) \end{array}\right\}$$

8-151 (contd)

$$\begin{cases} J_1 - 0.045 J_2 - 0.1875 J_3 = 13461 \\ 0.25 J_1 - 0.19 J_2 + 0.5 J_3 = 27.54 \\ -0.144 J_1 + J_2 - 0.456 J_3 = 2939 \end{cases}$$

Solving,

$J_1 = 16264 \text{ W/m}^2$

$J_2 = 10904 \text{ W/m}^2$

$J_3 = 12330 \text{ W/m}^2 = G_3$

$J_3 = E_{b3} = \sigma T_3^4 \qquad T_3 = 683 K$

$$q_1 = \frac{(0.75)(0.1256)}{0.25}(17937 - 16264) = 630.4 \text{ W} = 0.1256 (J_1 - G_1)$$

$G_1 = 11245 \text{ W/m}^2$

$$q_2 = \frac{(0.4)(0.09425)}{0.6}(7347 - 10904) = -223.5 \text{ W} = 0.09425 \times (J_2 - G_2)$$

$G_2 = 13275 \text{ W/m}^2$

$q_4 = 630.4 - 223.5 = 406.9 \text{ W}$

8-152

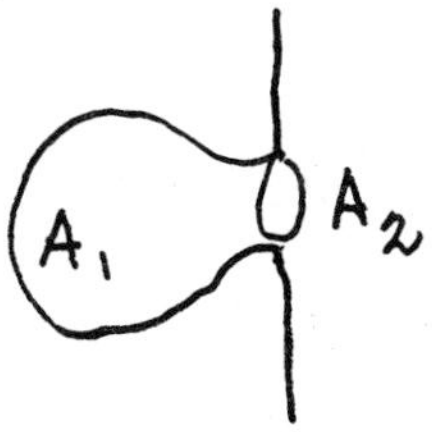

$$q = \epsilon_{app} A_2 E_{b_1} = \frac{E_{b_1} - 0}{\frac{1-\epsilon_1}{\epsilon_1 A_1} + \frac{1}{A_2}}$$

$$\epsilon_{app} = \frac{1}{\frac{A_2}{A_1}\left(\frac{1}{\epsilon_1} - 1\right) + 1}$$

$$= \frac{\epsilon_1 A_1}{A_2 + \epsilon_1 (A_1 - A_2)}$$

As $A_2/A_1 \to 0$, $\epsilon_{app} \to 1.0$

and hole approaches a hohl raum

9-1

$$\bar{h}\left(\frac{\mu_f^2}{k_f^3\rho_f^2 g}\right)^{1/3} = 0.0077\,Re_f^{0.4} \qquad Re_f = \frac{4\bar{h}L(T_g - T_w)}{h_{fg}\,\mu_f}$$

$$\bar{h}\left(\frac{\mu_f^2}{k_f^3\rho_f^2 g}\right)^{1/3} = 0.0077\left[\frac{4\bar{h}L(T_g - T_w)}{h_{fg}\,\mu_f}\right]^{0.4}$$

$$\bar{h} = \frac{(0.0077)^{1/0.6}}{\left(\frac{\mu_f}{k_f^3\rho_f^2 g}\right)^{1/1.8}}\left[\frac{4L(T_g - T_w)}{h_{fg}\,\mu_f}\right]^{2/3}$$

$$\frac{\bar{h}L}{k_f} = \overline{Nu} = \frac{\frac{(0.077)^{1/0.6}L}{k_f}\left[\frac{4L(T_g - T_w)}{h_{fg}(\mu_f)}\right]^{2/3}}{\left(\frac{\mu_f}{k_f^3\rho_f^2 g}\right)^{1/1.8}}$$

$$\overline{Nu} = \frac{(0.077)^{1/0.6}L^{5/3}k_f^{2/3}[4(T_g - T_w)]^{2/3}(\rho_f^2 g)^{5/9}}{\mu_f^{1.22}\,h_{fg}^{2/3}}$$

9-2 Flat Plate

$$Re_f = \frac{4\dot{m}}{\mu_f} \qquad \dot{m} = \frac{\rho_f^2 g\,\delta^3}{3\mu_f}$$

$$\delta^3 = \left[\frac{4\mu_f k L(T_g - T_w)}{g\,h_{fg}\,\rho_f^2}\right]^{3/4}$$

$$Re_f = \frac{4\rho_f^2 g}{3\mu_f^2}\left[\frac{4\mu_f k L(T_g - T_w)}{g\,h_{fg}\,\rho_f^2}\right]^{3/4}$$

$$Re_f = 3.77\left[\frac{\rho_f^2 g\,k_f^3 L^3(T_g - T_w)^3}{\mu_f^5\,h_{fg}^3}\right]^{1/4}$$

9-3 Turbulent Film

$$\bar{h} = \frac{0.0077\, Re_f^{0.4}}{\left(\frac{\mu_f^2}{k_f^3 \rho_f^2 g}\right)^{1/3}} \qquad Re_f = \frac{4\dot{m}}{W\mu_f}$$

$$\dot{m} = \frac{\bar{h} A (T_g - T_w)}{h_{fg}} = \frac{\bar{h} L W (T_g - T_w)}{h_{fg}} \qquad \bar{h} = \frac{\dot{m} h_{fg}}{LW(T_g - T_w)}$$

$$\frac{\dot{m} h_{fg}}{LW(T_g - T_w)} = \frac{0.0077\left(\frac{4\dot{m}}{W\mu_f}\right)^{0.4}}{\left(\frac{\mu_f^2}{k_f^3 \rho_f^2 g}\right)^{1/3}}$$

$$\dot{m} = \left[\frac{0.0077\,(4)^{0.4}\, L\, W^{0.6}\, (T_g - T_w)\, k_f\, (\rho_f^2 g)^{1/3}}{\mu_f^{1.07}}\right]^{5/3}$$

9-5 $T_f = \frac{70 + 100}{2} = 85°C$ $L = 1.2$ $\rho_f = 968$ $\mu_f = 3.37 \times 10^{-4}$ $k_f = 0.674$

$h_{fg} = 2255$ KJ/Kg $\bar{h} = 1.13\left[\frac{(968)^2(9.806)(2.255\times10^6)(0.674)^3}{(1.2)(3.37\times10^{-4})(100-70)}\right]^{1/4} = 5404$ W/m²·°C

$q = (5404)(1.2)(0.3)(100 - 70) = 58359$ W

$\dot{m} = \frac{58359}{2.255 \times 10^6} = 0.026$ kg/sec $= 93.2$ kg/hr.

9-6 $L = 0.4$ m $\phi = 90 - 30 = 60°$ $T_w = 98°C$ $T_g = 100°C$ $\rho = 960$

$h_{fg} = 2.255 \times 10^6$ $k = 0.68$ $\mu = 2.82 \times 10^{-4}$

$\bar{h} = 1.13\left[\frac{(960)^2(9.806)(2.255\times10^6)(0.68)^3 \sin 60}{(0.4)(2.82\times10^{-4})(2)}\right]^{1/4} = 14152$ W/m²·°C

$q = (14152)(0.4)^2(2) = 4528$ W $\dot{m} = 2.008 \times 10^{-3}$ kg/sec

$= 7.23$ kg/hr.

9-7 $k = 0.684 \quad \mu = 3.0 \times 10^{-4} \quad \rho = 962$

$$\bar{h} = 1.13\left[\frac{(962)^2(9.8)(2.255\times10^6)(0.684)^3}{(0.5)(3\times10^{-4})(100-95)}\right]^{1/4} = 10930 \text{ W/m}^2\cdot°C$$

$$q = (10930)(0.5)^2(100-95) = 136510 \text{ W}$$

$$\dot{m} = \frac{136510}{2.255\times10^6} = 0.0606 \text{ kg/sec} = 217.9 \text{ kg/hr.}$$

9-8 $T_f = 47.5°F \quad k = 0.583 \text{ W/m}\cdot°C \quad \mu = 1.37\times10^{-3} \quad \rho = 999$

$$h = 1.13\left[\frac{(999)^2(9.8)(2.376\times10^6)(0.583)^3}{(1.5)(1.37\times10^{-3})(15)(5/9)}\right]^{1/4} = 4576 \text{ W/m}^2\cdot°C$$

$$q = (4576)(1.5)^2(15)(5/9) = 8.58\times10^4 \quad \dot{m} = q/h_{fg} = 0.0361 \frac{\text{kg}}{\text{sec}}$$

$$= 130 \text{ kg/hr}$$

9-9 $T_g = 100°F = 37.78°C \quad T_f = \frac{37.78 + 30}{2} = 33.89°C$

$\rho = 590 \quad \nu = 0.345\times10^{-6} \quad h_{fg} = 477.8 \text{ Btu/lbm} = 1111 \text{ kJ/kg}$

$k = 0.501$

$$h = 1.13\left[\frac{(590)(9.8)(1.111\times10^6)(0.501)^3}{(0.4)(0.345\times10^{-6})(7.78)}\right]^{1/4} = 5918 \text{ W/m}^2\cdot°C$$

$$q = hA\Delta T = (5918)(0.4)^2(7.78) = 7367 \text{ W}$$

$$\dot{m} = q/h_{fg} = 0.00663 \text{ kg/sec} = 2.387 \text{ kg/hr.}$$

9-10 Assume laminar condensation $T_{sat} = 328°F \quad h_{fg} = 889 \frac{\text{Btu}}{\text{lbm}}$

$P_{sat} = 100 \text{ psia} \quad T_w = 280°F \quad T_f = 304°F \quad \rho_f = 57.29 \text{ lbm/ft}^3$

$\mu = 0.44 \text{ lbm/hr ft} \quad k = 0.395 \text{ Btu/hr ft}^2 °F \quad Pr = 1.15 \quad \rho_f > \rho_v \therefore \rho_f(\rho_f - \rho_v) = \rho_f^2$

$$\bar{h} = 0.725\left[\frac{\rho_f^2 g\, h_{fg}\, k^3}{\mu d (T_g - T_w)}\right]^{1/4} = 1855 \text{ Btu/hr ft}^2 °F$$

$$\dot{m} = q/h_{fg} = \frac{\bar{h}\pi d L (T_{sat} - T_w)}{h_{fg}} \qquad \frac{\dot{m}}{L} = \frac{(1855)\pi(1)(48)}{(12)(889)} = 26.31 \text{ lbm/hr}\cdot\text{ft.}$$

9-11 $35°C = 95°F$ $\quad p_g = 0.8237$ psia $\quad p_r = (0.8)(0.8237) = 0.659$ psia

$T_g = 88°F$ $\quad h_{fg} = 1043$ Btu/lbm $= 2426$ KJ/Kg $= 31.1°C$ $\quad T_f = \frac{31.1+2}{2} = 16.55°C$

$\rho_f = 998$ $\quad \mu_f = 1.1\times10^{-3}$ $\quad k_f = 0.596$ $\quad Pr_f = 7.8$

$$\bar{h} = 0.725\left[\frac{(998)^2(9.8)(2.426\times10^6)(0.596)^3}{(1.1\times10^{-3})(0.05)(31.1-2)}\right]^{1/4} = 5425 \text{ W/m}^2\cdot°C$$

$$q = (5425)\pi(0.05)(7.5)(31.1-2) = 185970$$

$$\dot{m} = \frac{185970}{2.426\times10^6} = 0.0767 \text{ Kg/sec} = 276 \text{ Kg/hr.}$$

9-12 Assume Laminar Film & Heat Transfer Area = 1 ft²

$\bar{h}_1$: $T_f = 154°F$ $\quad \rho_f = 61.1$ $\quad \mu_f = 1.02$ $\quad k_f = 0.382$

$\bar{h}_1 = 873$ Btu/hr ft² °F

$\bar{h}_2$; $T_f = 177°F$ $\quad \rho_f = 60.6$ $\quad h_{fg} = 992.1$ $\quad \mu_f = 0.86$ $\quad k_f = 0.388$

$\bar{h}_2 = 838$ Btu/hr ft² °F

$q_1 = \bar{h}_1 A \Delta T = 1.02\times10^5$ BTU/hr

$q_2 = \bar{h}_2 A \Delta T = 1.37\times10^5$ BTU/hr

$\frac{q_2 - q_1}{q_1} = 0.344$ $\quad$ 34.4% Increase

9-13 $p_{sat} = 100$ psia $\quad h_{fg} = 888.8$ Btu/lbm $\quad T_g = 328°F$ $\quad T_w = 280°F$

$T_f = 304°F$ $\quad \rho_f = 57.17$ $\quad \mu_f = 0.45$ $\quad k_f = 0.395$

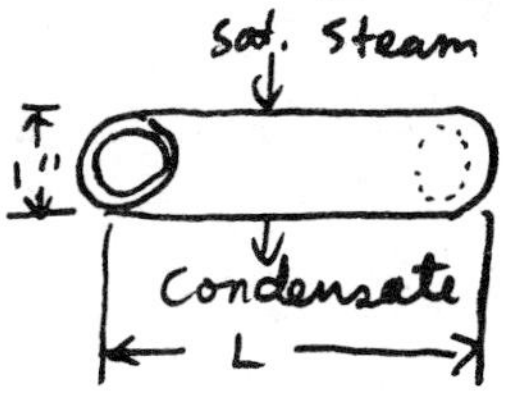

Assume Laminar Flow:

$$\bar{h} = 0.725\left[\frac{\rho^2 g\, h_{fg}\, k_f^3}{\mu_f\, d\,(T_g - T_w)}\right]^{1/4} = 1840 \text{ Btu/hr ft}^2\text{°F}$$

Check $Re_f = 73.6$

∴ Laminar

$q = \bar{h} A (T_g - T_w) = 23121$ Btu/hr $\quad \dot{m} = \frac{q}{h_{fg}} = 26.01$ lbm/hr · ft (length)

9-14 $k = 0.684$ $\mu = 3.0 \times 10^{-4}$ $\rho = 962$

$$h = 0.725\left[\frac{(962)^2(9.8)(2.255\times10^6)(0.684)^3}{(3\times10^{-4})(0.3)(5)}\right]^{1/4} = 7962 \text{ W/m}^2\cdot°C$$

$$q = h\pi dL(T_g - T_w) = (7962)\pi(0.3)(15)(5) = 5.628\times10^5 \text{ W}$$

$$\dot{m} = q/h_{fg} = 0.2496 \text{ kg/sec} = 898 \text{ kg/hr.}$$

9-16 $n = 20$ $T_f = \frac{88+100}{2} = 94°C$ $\rho_f = 963.2$ $d = 0.25 \text{ in} = 0.00635$ m

$\mu_f = 3.06\times10^{-4}$ $k_f = 0.627$ $h_{fg} = 2255$ kJ/kg

$$\bar{h} = 0.725\left[\frac{(963)^2(9.8)(2.255\times10^6)(0.627)^3}{(3.06\times10^4)(20)(0.00635)(100-88)}\right]^{1/4} = 7845 \text{ W/m}^2\cdot°C$$

$$q/L = (400)\pi(0.00635)(7845)(100-88) = 7.51\times10^5 \text{ W/m}$$

$$\dot{m} = \frac{7.51\times10^5}{2.255\times10^6} = 0.333 \text{ kg/sec} = 1200 \text{ kg/hr}\cdot\text{m}$$

9-17 $\dot{m} = 1.3$ kg/s $d = 1.25$ cm $S_p = 1.9$ cm $T_w = 93°C$

$T_g = 100°C$ n = number of rows $T_f = 96.5°C$ $h_{fg} = 2255$ kJ/kg

$\rho = 962$ $\mu = 2.96\times10^{-4}$ $k = 0.68$

$$\bar{h} = 0.725\left[\frac{(962)^2(9.8)(2.255\times10^6)(0.68)^3}{(2.96\times10^{-4})(0.0125)(7)n}\right]^{1/4} = \frac{16186}{n^{1/4}}$$

$$q = (1.3)(2.255\times10^6) = 2.932\times10^6 \text{ W} = \bar{h}A(T_g - T_w)$$

x = side of square array $3x$ = length $x = n(0.0125) + (n-1)(0.019)$

$$A = n^2\pi d(3x) = 0.00371n^3 - 0.00224n^2$$

$$2.932\times10^6 = \frac{113302}{n^{1/4}}(0.00371n^3 - 0.00224n^2)$$

By iteration $n = 25$ $x = 0.769$ $3x = 2.306$

number of tubes $= n^2 = 625$

9-18 $\dot{m} = 600$ kg/hr $= 0.1667$ kg/s $\quad q = \dot{m} h_{fg} = (0.1667)(2.255 \times 10^6)$

$q = 3.75 \times 10^5$ W $\quad T_f = 98.5\,°C \quad \rho = 960 \quad \mu = 2.82 \times 10^{-4}$

$k = 0.68$ W/m·°C $\quad n = 20 \quad d = 0.01$ m

$$\bar{h} = 0.725 \left[\frac{(960)^2(9.8)(2.255 \times 10^6)(0.68)^3}{(20)(0.01)(2.82 \times 10^{-4})(3)} \right]^{1/4} = 10123 \text{ W/m}^2 \cdot °C$$

$q = \bar{h}\, n \pi d L (T_g - T_w)$

$3.75 \times 10^5 = (10123)(400)\pi(0.01)L(3) \qquad L = 0.983$ m

9-19 $d = 0.05 \quad T_g = 100°C \quad T_w = 98°C \quad L = 1.5$ m $\quad \rho = 960 \quad k = 0.68$

$\mu = 2.82 \times 10^{-4} \quad h'_{fg} \approx h_{fg} = 2.255 \times 10^6$

$$\bar{h} = 0.555 \left[\frac{(960)^2(9.8)(0.68)^3(2.255 \times 10^6)}{(2.82 \times 10^{-4})(0.05)(2)} \right]^{1/4} = 12117 \text{ W/m}^2 \cdot °C$$

$q = (12117)\pi(0.05)(1.5)(2) = 5710$ W $\qquad \dot{m} = \frac{5710}{2.255 \times 10^6} = 2.532 \times 10^{-3}$ kg/s

$= 9.12$ kg/hr.

9-20 $\rho = 962 \quad \mu = 3.0 \times 10^{-4} \quad k = 0.68 \quad n = 10 \quad d = 1.0$ in $= 0.0254$ m

$$h = 0.725 \left[\frac{(962)^2(9.8)(2.255 \times 10^6)(0.68)^3}{(10)(0.0254)(3.0 \times 10^{-4})(5)} \right]^{1/4} = 8264 \text{ W/m}^2 \cdot °C$$

$q = (8264)(100)\pi(0.0254)(2)(0.3048)(5) = 2.01 \times 10^5$ W

$\dot{m} = q/h_{fg} = 0.0891$ kg/s $= 320.9$ kg/hr.

9-21 $T_f = 98.5°C \quad \rho = 960 \quad \mu = 2.82 \times 10^{-4} \quad k = 0.68$ W/m·°C $\quad n = 10$

$d = 0.0254$ m $\qquad h = 0.725 \left[\frac{(960)^2(9.8)(2.255 \times 10^6)(0.68)^3}{(10)(0.0254)(2.82 \times 10^{-4})(3)} \right]^{1/4} = 9526$ W/m²·°C

$q = (9526)(100)\pi(0.0254)(3)(0.3048)(3) = 2.085 \times 10^5$ W

$\dot{m} = q/h_{fg} = 0.0925$ kg/sec $= 332.9$ kg/hr.

9-23 $h_{fg} = 488.5$ Btu/lbm $= 1136$ kJ/kg $T_f = \frac{90+82}{2} = 86°F = 30°C$

$\rho = 596$ $\nu = 0.349 \times 10^{-6}$ $k = 0.507$ $n = 20$ $d = 0.00635$ m

$$h = 0.725\left[\frac{(596)^2(9.8)(1.136\times10^6)(0.507)^3}{(20)(0.00635)(90-82)(5/9)}\right]^{1/4} = 709 \text{ W/m}^2\cdot°C$$

$$q = (709)(400)\pi(0.00635)(0.3048)(90-82)(5/9) = 7664 \text{ W}$$

$$\dot{m} = q/h_{fg} = 0.00675 \text{ kg/s} = 24.3 \text{ kg/hr.}$$

9-24 $\dot{m} = 10\,000$ kg/hr $= 2.778$ kg/sec $h_{fg} = 55.93$ Btu/lbm

$h_{fg} = 130.09$ kJ/kg $q = \dot{m}h_{fg} = 361400$ W $T_f = \frac{100+90}{2} = 95°F$

$T_f = 35°C$ $\rho = 1276$ $\nu = 0.193 \times 10^{-6}$ $k = 0.07$ W/m·°C

$n = 25$ $d = 0.012$

$$\bar{h} = 0.725\left[\frac{(1276)(9.8)(1.3\times10^5)(0.07)^3}{(25)(0.0125)(10)(5/9)(0.193\times10^{-6})}\right]^{1/4} = 823.4 \text{ W/m}^2\cdot°C$$

$$q = hA\Delta T = hn\pi dL\Delta T$$

$$361400 = (823.4)(625)\pi(0.0125)L(10)(5/9)$$

$$L = 3.22 \text{ m}$$

9-25 $d = 0.012$ m $T_f = \frac{90+80}{2} = 85°F = 29.44°C$ $\rho = 1295$ $k = 0.07$

$\nu = 0.194 \times 10^{-6}$ $c = 984$ J/kg·°C $h_{fg} = 57.46$ Btu/lbm $= 133.65$ kJ/kg

$$h'_{fg} = 133.65 + (0.68)(0.984)(10)(5/9) = 137.4 \text{ kJ/kg}$$

$$\bar{h} = 0.555\left[\frac{(1295)(9.8)(0.137\times10^6)(0.07)^3}{(0.194\times10^{-6})(0.012)(10)(5/9)}\right]^{1/4} = 1446 \text{ W/m}^2\cdot°C$$

$$q/L = h\pi d\Delta T = (1446)\pi(0.012)(10)(5/9)$$

$$= 303 \text{ W/m}$$

9-26 $\Delta T_x = (107-100) = 7°C \qquad q/A = 7.96(7)^4 = 19.11\ kW/m^2$

$q = (19110)(0.3) = 5734\ W$

9-27

Surface tension of H_2O @ 212°F = 58.8 dyne/cm

(H_2O vapor, 213°F, p = 14.993)

$\pi r^2 (p_v - p_\ell) =$ pressure force $\qquad 2\pi r\sigma =$ surface tension force

H_2O @ 212°F $\qquad p_\ell = 14.696 \qquad r = \dfrac{2\sigma}{p_v - p_\ell} = 0.00226$ in.

9-29 $\Delta T_x = 117 - 100 = 17°C \qquad C_{sf} = 0.013 \qquad q/A = \dfrac{0.7\ MW}{m^2}$

$q = (0.7\times10^6)(0.3)^2 = 63000\ W$

9-33

$$Re_f = \frac{4\Gamma}{\mu_f} \qquad h_x = \left[\frac{\rho^2 g h_{fg} k^3}{4\mu x (T_g - T_w)}\right]^{1/4}$$

$$\bar{h} = \frac{1}{L}\int_0^L h_x\, dx \qquad \bar{h} = \frac{1}{L}\int_0^L \left[\frac{\rho^2 g h_{fg} k^3}{4\mu_f \Delta T}\right]^{1/4} dx$$

$$\bar{h} = \frac{4}{3}\left[\frac{L^3 \rho^2 g k^3 h_{fg}}{L^4\, 4\mu_f \Delta T}\right]^{1/4} = \frac{4}{3}\left[h_x\right]_{x=L}$$

$$\bar{h} = \Delta T L = q = h_{fg}\Gamma \qquad \Gamma = \frac{\bar{h}\Delta T L}{h_{fg}}$$

$$Re_f = \frac{4\Gamma}{\mu_f} = \frac{4\bar{h}\Delta T L}{\mu_f h_{fg}} \qquad Re^{-1/3} = \left[\frac{\mu_f h_{fg}}{4\bar{h}\Delta T L}\right]^{1/3} = \bar{h}\left[\frac{\mu_f h_{fg}}{4\bar{h}^4 \Delta T L}\right]^{1/3}$$

$$\bar{h}^4 = \left(\frac{4}{3}\right)^4 \left[\frac{\rho_f^2 g h_{fg} k^3}{4L\mu_f \Delta T}\right] \qquad \left(\frac{4}{3}\right)^{4/3} Re_f^{-1/3} = \bar{h}\left[\frac{\mu_f}{\rho_f^2 k_f^3 g}\right]^{1/3}$$

$$1.466\, Re_f^{-1/3} = \bar{h}\left[\frac{\mu_f}{\rho_f^2 k_f^3 g}\right]^{1/3}$$

9-34 $\Delta T_x = 11°C$ $\quad q/A = 7\times10^5$ Btu/hr ft^2 $= 2.21$ MW/m^2

9-35 $T_w = 110°C$ $\quad T_{b\,avg} = 96°C$ $\quad \Delta T_x = 10°C$ $\quad d = 1.25$ cm $\quad u_m = 1.2$ m/sec

Brass $C_{sf} = 0.006$ for Plat $q/A = 2.5\times10^4$ Btu/hr ft^2 $= 7.885\times10^4$ W/m^2

$$\left(q/A\right)_{Brass} = (7.885\times10^4)\left(\frac{0.013}{0.006}\right)^3 = 8.02\times10^5 \text{ W/m}^2$$

$$\left.q/L\right)_{boiling} = (8.02\times10^5)\,\pi\,(0.0125) = 31494 \text{ W/m}$$

Forced Convection

@ 96°C $\rho = 960$ $\mu = 2.96\times10^{-4}$ $k = 0.68$ $P_r = 1.83$

$$Re = \frac{(960)(1.2)(0.0125)}{2.96\times10^{-4}} = 48650 \qquad h = \frac{0.68}{0.0125}(0.019)(48650)^{0.8}(1.83)^{0.4} = 7396 \text{ W/m}^2\cdot°C$$

$$\left.q/L\right)_{conv} = (7396)\,\pi\,(0.0125)(14) = 4066 \text{ W/m} \qquad \left.q/L\right)_{total} = 31494 + 4066 =$$

$$= 35560 \text{ W/m}$$

9-36 $q = (2.3)(2.255\times10^6) = 5.187\times10^6$ J/hr $= 1441$ W

$$q/A = \frac{1441}{\pi(0.15)^2} = 20382 \text{ W/m}^2 \qquad q/A = 5.56\,(\Delta T_x)^4$$

$\Delta T_x = 7.78°C$ $\quad T_w = 107.8°C$

9-37 $\Delta T_x = 11°C$ $\quad \left.q/A\right)_{plat}$ $\quad \left.q/A\right)_{cu} = 4000$ $\frac{\text{Btu}}{\text{hr ft}^2} = 126160$ W/m^2

$$q/L = (126160)\,\pi\,(0.005) = 1982 \text{ W/m}$$

9-39 $\Delta T_x = 10°C$ $\quad p = 3$ MN/m^2 $= 435$ psia $\quad d = 0.02$ $\quad L = 1$ m

$h = 2.54\,(10)^3 e^{3/1.551} = 17576$ W/m$^2\cdot$°C $\quad q = (17576)\,\pi\,(0.02)(1)(10) = 11042$ W

h_{fg} @ 435 psia $= 771.3$ Btu/lbm $= 1.794\times10^6$ J/kg

$$\dot m = \frac{11042}{1.794\times10^6} = 6.15\times10^{-3} \text{ kg/sec} = 22.16 \text{ kg/hr.}$$

9-40 $q/A = 0.2\ MW/m^2 = 200\ kW/m^2$ $d = 0.003$ $L = 7.5\ cm$ $p = 1.6\ atm$

$h = 5.56\,(\Delta T_x)^3$ $q/A = h\,\Delta T_x$ $200 \times 10^3 = 5.56\,(\Delta T_x)^4 (1.6)^{0.4}$

$\Delta T_x = 13.14°C$ at 1.16 atm $p = 23.52$ psia $T_{sat} = 236.5°F$
$= 113.6\ °C$

$T_w = 113.6 + 13.14 = 126.7°C$

9-41 $d = 1.0\ in = 2.54\ cm$ $p = 14.7 + 5 = 19.7$ psia $\Delta T_x = 4°C$

Assume : horizontal $h = 5.56\,(4)^3 = 355.8\ W/m^2 \cdot °C$

$q/A = h\,\Delta T = (355.8)(4) = 1.423\ kW/m^2$

above relation does not apply.

$h = 1042\,(3)^{1/3} = 1654\ W/m^2 \cdot °C$ $q/A = h\,\Delta T = 6.616\ kW/m^2$

$\dot{m} = 2000\ lbm/hr = 0.255\ kg/sec$ $q = (0.255)(2.255 \times 10^6) = 6616\ A$

$A = 86.91\ m^2 = \pi d L$ $L = 1089\ m$

9-42 $\Delta T_x = 15°C$ $h/h_w = 0.83$ $(q/A)_{water} = 2 \times 10^5\ Btu/hr\,ft^2$

$(q/A)_{water} = 6.308 \times 10^5\ W/m^2$ $h_w = \dfrac{6.308 \times 10^5}{15} = 42050\ W/m^2 \cdot °C$

$h_{(glycerine)} = (0.83)(42050) = 34900\ W/m^2 \cdot °C$

9-44 $\Delta T_x = 30°F$ $C_{sf} = 0.008$ For plat $q/A = 2.5 \times 10^5 \dfrac{Btu}{hr\,ft^2}$

$q/A)_{ss} = (2.5 \times 10^5)\left(\dfrac{0.13}{0.008}\right)^3 = 1.073 \times 10^6\ Btu/hr\,ft^2$

$= 3.383\ MW/m^2$

9-46 $\Delta T_x = 232 - 212 = 20°F$

From figure $q/A = 2.3 \times 10^5$ Btu/hr ft^2

$= 7.25 \times 10^5$ W/m^2

$q = (q/A)\pi d L = (7.25\times10^5)\pi(0.001)(0.12) = 273.5$ W

9-47 $T_f = \frac{94+100}{2} = 97°C$ $d = 0.04$ m $\rho = 962$ $k = 0.68$

$\mu = 3\times10^{-4}$ $h_{fg} = 2.255\times10^6$ J/kg

$$\bar{h} = 0.725\left[\frac{(962)^2(9.8)(2.255\times10^6)(0.68)^3}{(3\times10^{-4})(0.04)(100-94)}\right]^{1/4} = 12533 \text{ W/m}^2\cdot°C$$

$q/L = h\pi d\,\Delta T = (12533)\pi(0.04)(100-94) = 9450$ W/m

9-48 $q/A)_{total} = q/A)_{FC} + q/A)_{boiling}$

$\Delta T_x = 110 - 98 = 12°C = 21.6°F$ $q/A|_{plat} = 4\times10^4 \frac{Btu}{hr\,ft^2} = 1.26\times10^5 \frac{W}{m^2}$

At 100°C $k = 0.68$ $Pr = 1.7$

$$h = \frac{k}{d}(0.019)(40000)^{0.8}(1.7)^{0.4} = 3070 \text{ W/m}^2\cdot°C$$

$$q/A|_{brass} = (1.26\times10^5)\left(\frac{C_{plat}}{C_{brass}}\right)^3 = 12.8\times10^5 \text{ W/m}^2$$

$$h_b = \frac{q/A|_{brass}}{\Delta T} = \frac{12.8\times10^5}{110-98} = 10670 \text{ W/m}^2\cdot°C$$

$h\,(total) = 10670 + 3070 = 13740$ W/m$^2\cdot$°C

9-50 $h_{fg} = 2.255 \times 10^6$ J/kg $\quad \rho_v = 1.5$ kg/m³ $\quad \rho_l = 998$ kg/m³

$\sigma = 70$ mN/m $\quad (q/A)_{max} = \frac{\pi}{24}(2.255\times10^6)(1.5)\left[\frac{(0.07)(9.8)(996)}{(1.5)^2}\right]^{1/4}\left(1+\frac{1.5}{998}\right)^{1/2}$

$(q/A)_{max} = 1.85$ MW/m²

9-52 $h_{fg} = 2.255 \times 10^6$ J/kg $\quad T_w = 813$ K $\quad T_{sat} = 373$ K

a) $T_f = \frac{540+100}{2} = 320°C = 593$ K $\quad \rho_l = 960 \quad \rho_v = 0.365$

$k_g = 0.042 \quad \mu_v = 20.6\times10^{-6} \quad c_{pv} = 2000 \quad \Delta T_x = 440$

$$h_b = 0.62\left[\frac{(0.042)^3(0.365)(960)(9.8)[2.255\times10^6 + 2000](0.4)(4000)}{(0.0125)(20.6\times10^{-6})(440)}\right]^{1/4}$$

$= 171.5$ W/m²·°C

$$h_r = \frac{(5.669\times10^{-8})(0.8)[813^4 - 373^4]}{813-373} = 43.04$$

$h = (171.5)^{4/3} h^{-1/3} + 43.04 \quad h = 204.7$ W/m²·°C

$q/A = h\Delta T_x = (204.7)(440) = 90068$ W/m²

9-53 $T_w = 280°C \quad T_{sat} = 100°C \quad V = 3$ m/sec $\quad d = 0.0004$ m

flow $= 1$ L/hr @ 25°C $= 1000\times10^{-6}\times996 = 0.996$ kg/hr $= 2.767\times10^{-4}$ kg/s

$T_f = \frac{280+100}{2} = 190°C = 463$ K $\quad h_{fg} = 2.255\times10^6 \quad \rho_L = 996$

$\rho_{vf} = 0.478 \quad c_{pv} = 1980 \quad \sigma = 58.8$ mN/m

$\lambda = 2.255\times10^6 + (1980)\left(\frac{280-100}{2}\right) = 2.433\times10^6$ J/kg

$$Q_{max}/\text{drop} = (1.83\times10^{-3})(996)(0.0004)^3(2.433\times10^6)\left[\frac{(996)^2(3)^2(0.0004)}{(0.478)(0.0588)(1)}\right]^{0.341}$$

$= 0.0156$ J/drop

9-53 contd

$$\text{Mass/drop} = (996)(4/3\,\pi)(0.0002)^3 = 3.338 \times 10^{-8}\ \text{kg}$$

$$\#\ \text{of drops} = \frac{2.767 \times 10^{-4}\ \text{kg/s}}{3.338 \times 10^{-8}\ \text{kg/drop}} = 8290\ \text{drops/sec}$$

$$Q_{max} = (0.0156)(8290) = 129.3\ \text{W}$$

9-54 $T_f = \frac{100+92}{2} = 96°C \quad d = 0.0125 \quad n = \sqrt{196} = 14$

$\rho = 961 \text{ kg/m}^3 \quad \mu = 2.96 \times 10^{-4} \quad k = 0.68$

$h_{fg} = 2255 \text{ kJ/kg}$

$$\bar{h} = 0.725 \left[\frac{(961)^2 (9.8)(2255 \times 10^3)(0.68)^3}{(2.96 \times 10^{-4})(14)(0.0125)(100-92)} \right]^{1/4}$$

$= 8088 \text{ W/m}^2\text{-}°C$

$q = (8088)(196)\pi(0.0125)(2.0)(100-92)$

$= 9.96 \times 10^5 \text{ W} = \dot{m}(2255 \times 10^3)$

$\dot{m} = 0.442 \text{ kg/s}$

9-55 $T_f = \frac{100+91}{2} = 95.5°C$

$\rho = 961 \text{ kg/m}^3 \quad \mu = 2.97 \times 10^{-4} \quad k = 0.68$

$h_{fg} = 2255 \text{ kJ/kg}$

$$1800 = \frac{4 \bar{h} L (100-91)}{2255 \times 10^3 (2.97 \times 10^{-4})}$$

$$\bar{h} L = 33487 = 1.13 \left[\frac{(961)^2 (9.8)(2255 \times 10^3)(0.68)^3}{(2.97 \times 10^{-4})(100-91)} \right]^{1/4} L^{3/4}$$

$L^{3/4} = 4.234$

$L = 6.85 \text{ m}$

$$\frac{\dot{m}}{\rho} = \frac{(1800)(2.97 \times 10^{-4})}{4} = 0.134 \text{ kg/s-m depth}$$

9-56 $C_{sf} = 0.008$ $\quad$ C_{sf} plat $= 0.013$

$\Delta T_x = 15°C$

$q/A|_{plat} = 2.7 \times 10^5$ Btu/hr-ft^2 $= 851$ Kw/m^2

$$(q/A)_{ss} = (851)\left(\frac{0.013}{0.008}\right)^3 = 3.65 \text{ MW/m}^2$$

9-58 $T_f = \dfrac{100+94}{2} = 97°C$ $\quad h_{fg} = 2255$ kJ/kg

$\rho = 961$ kg/m^3 $\quad \mu = 2.98 \times 10^{-4}$ $\quad k = 0.68$

$$\bar{h} = 0.725\left[\frac{(961)^2(9.8)(2255\times10^3)(0.68)^3}{(2.98\times10^{-4})(0.3)(100-94)}\right]^{1/4}$$

$= 7582$ W/m^2-°C

$q/L = (7582)\,\pi(0.3)(100-94) = 42877$ W/m

$= \dfrac{\dot{m}}{L}(2255 \times 10^3)$

$\dfrac{\dot{m}}{L} = 0.019$ kg/s m length

9-59 $T_f = \frac{100+93}{2} = 96.5°C \quad h_{fg} = 2255 \times 10^3 \ J/kg$

$\rho = 961 \ kg/m^3 \quad \mu = 2.96 \times 10^{-4} \quad k = 0.68$

$$\bar{h} = 1.13\left[\frac{(961)^2(9.8)(2255\times10^3)(0.68)^3}{(0.2)(2.96\times10^{-4})(100-93)}\right]^{1/4}$$

$= 12605 \ W/m^2\text{-}°C$

$q = (12605)(0.2)^2(100-93) = 3530 \ W$

$= \dot{m}(2255 \times 10^3)$

$\dot{m} = 0.00156 \ kg/s$

$$\delta = \left[\frac{(4)(2.96\times10^{-4})(0.68)(0.2)(100-93)}{(9.8)(2255\times10^3)(961)^2}\right]^{1/4}$$

$= 8.6 \times 10^{-5} \ m$

10-1 Pipe nearly constant temperature $T_w = 82°C$ $T_\infty = 30°C$

$\epsilon \approx 0.8$ $h = 1.32\left(\frac{82-30}{0.0564}\right)^{1/4} = 7.27\ W/m^2 \cdot °C$

$q_{conv} = (7.27)\pi(0.0564)(15)(82-30) = 1005\ W$

$q_{rad} = (5.669 \times 10^{-8})\pi(0.0564)(15)(0.8)[355^4 - 303^4] = 898\ W$

$q_{tot} = 1903 = (0.6)(4175)\Delta T_w$ $\Delta T_w = 0.76\ °C$

$T_{w\ exit} = 82 - 0.76 = 81.2°C$

10-3

Inside Tube $T = 473K$ $d = 0.025$ $u_m = 6\ m/sec$ $\mu = 2.58 \times 10^{-5}$

$\rho = \frac{2.07 \times 10^5}{(287)(473)} = 1.525$ $k = 0.0385$ $c_p = 1030$ $Pr = 0.681$

$\dot{m} = (1.525)\pi(0.0125)^2(6) = 4.491 \times 10^{-3}\ kg/sec$ $Re = \frac{(1.525)(6)(0.025)}{2.58 \times 10^{-5}} = 8866$

$h_i = \frac{0.0385}{0.025}(8866)^{0.8}(0.681)^{0.3} = 45.4\ W/m^2 \cdot °C$

Conduction resistance negligable.

Outside Tube $T_f \approx \frac{200+20}{2} = 110°C = 383K$ $\nu_f = 25.15 \times 10^{-6}$

$k_f = 0.0324$ $Pr_f = 0.69$ $d_o = 0.025 + 0.0016 = 0.0266$

$Re = \frac{(12)(0.0266)}{25.15 \times 10^{-6}} = 12691$ $C = 0.193$ $n = 0.618$ from Chap. 6

$h_o = \frac{(0.0324)(0.193)}{0.0266}(12691)^{0.618}(0.69)^{1/3} = 71.36\ W/m^2 \cdot °C$

Base U on A_i

$U_i = \frac{1}{\frac{1}{45.4} + \frac{0.025}{(0.0266)(71.36)}} = 28.41\ W/m^2 \cdot °C$ $A_i = \pi(0.025)(3) = 0.236\ m^2$

$C_{min} = (4.491 \times 10^{-3})(1030) = 4.626$ $C_{min}/C_{max} = 0$

$NTU = \frac{(28.41)(0.236)}{4.626} = 1.45$ $\epsilon = 0.78$ $\Delta T_i = (0.78)(200-20) = 140.4°C$

$T_{exit} = 200 - 140.4 = 59.6°C$ If reduce air flow in half,

$NTU = (2)(1.45) = 2.90$ $\epsilon = 0.95$ $\Delta T_i = (0.95)(200-20) = 171$

$T_{exit} = 200 - 171 = 29°C$

10-6 Water at 90°C : $\rho = 965$ $\mu = 3.16 \times 10^{-4}$ $k = 0.676$ $Pr = 1.96$

$$Re = \frac{(965)(4)(0.025)}{3.16 \times 10^{-4}} = 3.05 \times 10^5 \quad h_i = \frac{0.676}{0.025}(0.023)(3.05 \times 10^5)^{0.8}(1.96)^{0.3}$$

$h_i = 18590$ W/m²·°C

Engine Oil at 20°C: $\nu = 0.0009$ $k = 0.145$ $Pr = 10400$

$$D_H = 0.0375 - 0.0266 = 0.0109 \text{ m} \qquad Re = \frac{(7)(0.0109)}{0.0009} = 84.78$$

$Re\,Pr = (84.78)(10400) = 8.82 \times 10^5$ h_o is smaller compared to h_i so approximately h_o can be obtained from const. Temp. eq.

$$Re\,Pr\,\frac{d}{L} = \frac{(8.82 \times 10^5)(0.0109)}{6} = 1602.3 \qquad \nu_w @ 90°C = 0.289 \times 10^{-4}$$

$$h_o = \frac{0.145}{0.0109}(1.86)(1602.3)^{1/3}\left(\frac{9}{0.289}\right)^{0.14} = 468.6 \text{ W/m}^2\cdot°C$$

Based on A_i :

$$U_i = \frac{1}{\frac{1}{18590} + \frac{0.025}{(0.0266)(468.6)}} = 485.6 \text{ W/m}^2\cdot°C$$

10-7 $d_o = 1.315$ in $d_i = 0.957$ in $k = 43$ W/m²·°C

For 1m length: $\frac{\ln(r_o/r_i)}{2\pi k L} = 1.176 \times 10^{-3}$

$$1/h_oA_o = \frac{1}{(180)\pi(1.315)(0.0254)} = 6.05294$$

$$1/h_iA_i = \frac{1}{(65)\pi(0.957)(0.0254)} = 0.20146$$

$$UA = \frac{1}{0.05294 + 0.20146 + 1.176 \times 10^{-3}} = 3.9127$$

$$U_i = \frac{3.9127}{\pi(0.957)(0.0254)}$$

$= 51.24$ W/m²·°C

10-8 $d_o = 1.315 \quad d_i = 1.049 \text{ in} \quad k = 43 \text{ W/m}\cdot{}^\circ\text{C} \quad \bar{T}_b = \frac{80+60}{2} = 70^\circ\text{C}$

$\rho = 858 \quad c_p = 2090 \quad \nu = 0.6 \times 10^{-4} \quad k = 0.139 \quad Pr = 770$

$$Re = \frac{(5)(1.049)(0.0254)}{0.6 \times 10^{-4}} = 2220$$

Free Convection $T_f = \frac{70+20}{2} = 45^\circ\text{C}$

$$\frac{g\beta\rho^2 c_p}{\mu k} = 4.2 \times 10^{10} \qquad k = 0.64$$

$$Gr\,Pr = (4.2\times10^{10})(1.315)^3(0.0254)^3(70-20) = 7.82 \times 10^7$$

$$h_o = \frac{0.64}{(1.315)(0.0254)}(0.53)(7.82\times10^7)^{1/4} = 953 \text{ W/m}^2\cdot{}^\circ\text{C}$$

$$q = \dot{m} c_p \Delta T_b = (858)(5)\frac{\pi(1.049)^2}{4}(0.0254)^2(2090)(80-60) = 99970 \text{ W}$$

Assume inside barely turbulent:

$$h_i = \frac{0.139}{(1.049)(0.0254)}(0.023)(2220)^{0.8}(770)^{0.3} = 419 \text{ W/m}^2\cdot{}^\circ\text{C}$$

Neglect conduction resistance

$$U = \frac{1}{1/h_i + 1/h_o} = 291 \text{ W/m}^2\cdot{}^\circ\text{C}$$

$$q = U\pi d L \Delta T \qquad d = \frac{1.049 + 1.315}{2} = 1.18 \text{ in}$$

$$L = \frac{99970}{(291)\pi(1.18)(0.0254)(70-20)} = 72.8 \text{ m}$$

10-9 a)

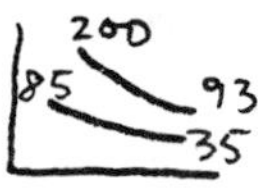

$$\Delta T_m = \frac{(200-85)-(93-35)}{\ln\left(\frac{115}{58}\right)} = 83.3^\circ\text{C} \qquad P = 0.303$$

$$R = \frac{200-93}{85-35} = 2.14 \qquad F = 0.92$$

$$q = (2.5)(4175)(85-35) = (180)A(0.92)(83.3) = 521.8 \text{ kW} \qquad A = 37.8 \text{ m}^2$$

(b) $m_g c_g (200-93) = (2.5)(4175)(85-35) \qquad m_g c_g = 4877 = C_{min}$

$$C_{min}/C_{max} = \frac{4877}{10437} = 0.467 \qquad \epsilon = \frac{200-93}{200-35} = 0.648$$

$$\frac{AU}{C_{min}} = 1.4 \qquad A = \frac{(1.4)(4877)}{180} = 37.9 \text{ m}^2$$

10-12

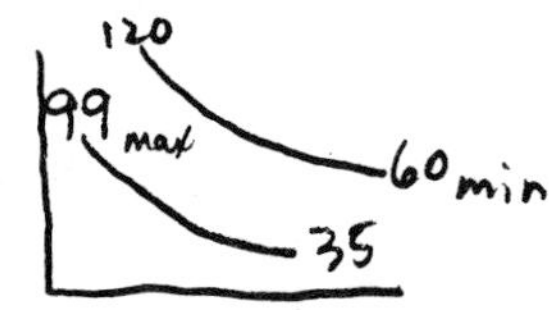

$\dot{m}_w = 230$ kg/hr $A = 1.4\ m^2$ $c_o = 2100$ $U = 2800$

$m_w C_w = \frac{(230)(4175)}{3600} = 266.7$ Assume water is min. fluid

$U/C_{min} = \frac{(280)(1.4)}{266.7} = 1.47$ $\epsilon = \frac{99-35}{120-35} = 0.753$

$C_{min}/C_{max} = 0.125$ $C_{max} = \frac{266.7}{0.125} = 2133.6 = M_o c_o$

$M_o = \frac{2133.6}{2100} = 1.016$ kg/s $= 3658$ kg/hr.

10-13 At 40 mph $q = 60\,000$ Btu/hr $U = 35$ Btu/hr ft² °F

$\Delta T = 10$°F $U = k_1 v^{0.7} = 0.0066\, v^{0.7}$ $\dot{m}_{air} = K_2 v = 0.118\, v$

At 40 mph = 211200 ft/hr $\epsilon_a = \frac{\Delta T_a}{T_f - T_{a1}} = 0.182$ $C_{min}/C_{max} = 0$

NTU = 0.20 $A = \left(\frac{NTU}{C_{min}}\right)/U = 34.3\ ft^2$

At 30 mph = 158000 ft/hr $U = 28.5$ $\dot{m} = 18700$ $C_{min} = 4490$

$NTU_{max} = 0.218$ $\epsilon = 0.185 = \frac{T_{ea} - 95}{55}$, $T_{ea} = 105.2$°F

$q = \dot{m}_a c_{pa} \Delta T_a = 58000$ Btu/hr 3.3% red.

At 20 mph = 105600 ft/hr $U = 21.6$ $\dot{m}_a = 12450$ $C_{min} = 2990$

$NTU_{max} = 0.248$ $\epsilon = 0.20 = \frac{T_{ea} - 95}{55}$ $T_{ea} = 106$°F

$q = 32900$ Btu/hr 45.2% reduction

At 10 mph = 52800 ft/hr $U = 13.3$ $\dot{m}_a = 6230$ $C_{min} = 1495$

$NTU_{max} = 0.305$ $\epsilon = 0.26$ $T_{ea} = 109.3$°F $q = 21400$ Btu/hr 64.4% Red.

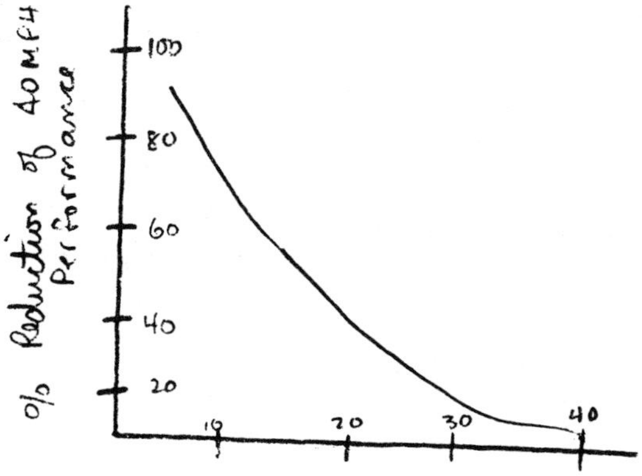

10-14 $A = 4.64\ m^2$ $U = 280$ $\dot{m}_a = 0.45\ kg/sec$ $c_a = 1009$ $c_w = 4175$

$q = (\dot{m}c)_w(93-20) = (0.45)(1009)(260-T_e)$

$q = (280)(4.64)\dfrac{(260-93)-(T_e-20)}{\ln\left(\dfrac{260-93}{T_e-20}\right)}$ By Iteration: $T_e = 45.4°C$

$\dot{m}_w = \dfrac{(0.45)(1009)(260-45.4)}{(4175)(93-20)} = 0.32\ kg/sec$

10-15 $q = (0.6)(4175)(90-35) = (0.9)(2100)(175-T_{e_o}) = 137775\ W$

$T_{e_o} = 102°C$ $\Delta T_m = \dfrac{85-67}{\ln(85/67)} = 75.64$ $A = \dfrac{137775}{(425)(75.64)} = 4.286\ m^2$

Oil is min Fluid: $\epsilon = \dfrac{175-102}{175-35} = 0.521$

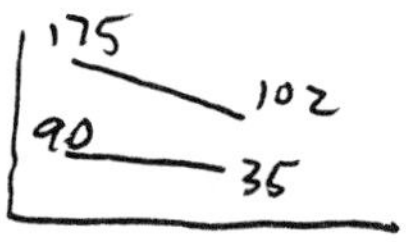

10-16 $p = 345\ kN/m^2 = 50\ psia$ $T_{sat} = 281°F = 138°C$

$\Delta T_m = \dfrac{53-39}{\ln(53/39)} = 45.64$ $q = (7.5)(4175)(99-85) = 438375\ W$

$q = (2800)A(45.64)$ $A = 3.43\ m^2 = (30)\pi(0.025)(2L)$ $L = 0.728\ m$

10-17 $R_f = 0.0005 = \dfrac{1}{U_{dirty}} - \dfrac{1}{U_{clean}}$ $U_{dirty} = 1167\ W/m^2 \cdot °C$

$q = (1167)(3.43)(45.64) = 182636\ W$ $\Delta T_w = \dfrac{182636}{(7.5)(4175)} = 5.83°C$

$T_{w_e} = 99 - 5.83 = 93.2°C$

10-19 Glycol $c_p = 2742$ $q = \dfrac{4500}{3600}(2742)(140-80) = 205650\ W$

$q = \dot{m}_w(4175)(85-35)$ $\dot{m}_w = 0.985\ kg/sec$ $U = 850\ W/m^2 \cdot °C$ $C_w = 4112$

$C_g = \dfrac{(4500)(2742)}{3600} = 3428$ $C_{min}/C_{max} = 0.834$ $\epsilon = \dfrac{140-80}{140-35} = 0.571$

$(U/C_{min})A = 1.3$ $A = 5.24\ m^2$

10-20 $UA/C_{min} = 2.6$ $C_{min}/C_{max} = 0.417$ $\epsilon = 0.82$

$\Delta T_g = (0.82)(140-35) = 86.1°C$ $q = (86.1)\left(\frac{3428}{2}\right) = 147575$ W

reduction of 28.2% in q. $\Delta T_w = \frac{147575}{4112} = 35.9°C$ $T_{w_e} = 49.1°C$

10-21 $C_{min} = C_g = 4877$ $C_{max} = C_w = 10437/2 = 5219$ $\frac{AU}{C_{min}} = 1.4$

$C_{min}/C_{max} = (0.467)(2) = 0.934$ $\epsilon = 0.55 = \frac{\Delta T_g}{200-35}$ $\Delta T_g = 90.75$ °C

$q = (4877)(90.75) = 442.6$ kW % reduction = 15%

10-22 $\Delta T_m = 83.3°C$ (a) $P = 0.303$ $R = 2.14$ $F = 0.86$

$521.8 \times 10^3 = (180) A (0.86)(83.3)$ $A = 40.5 \text{ m}^2$

(b) $C_{min}/C_{max} = 0.467$ $\epsilon = 0.648$ $UA/C_{min} = 1.55$

$A = \frac{(1.5)(4877)}{180} = 40.6 \text{ m}^2$

10-23 $AU/C_{min} = 1.5$ $C_{min}/C_{max} = 0.934$ $\epsilon = 0.53$ $\Delta T_g = 87.45$

$q = (4877)(87.45) = 426.5$ kW % reduction $= \frac{521.8 - 426.5}{521.8} = 18.3\%$

10-24

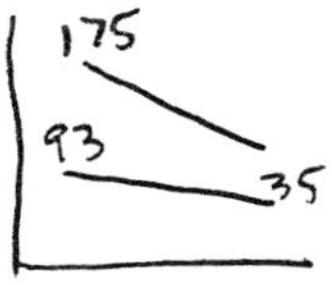

$M_o C_o = \frac{(230)(2100)}{3600} = 134.17$ $M_w C_w = \frac{(230)(4175)}{3600} = 266.7$

$\Delta T_o = (93-35)\left(\frac{266.7}{134.17}\right) = 115.3$ $T_{o\,out} = 175 - 115.3$
$= 59.7°C$

$\Delta T_m = \frac{82 - 24.7}{\ln\left(\frac{82}{24.7}\right)} = 47.75$ °C $q = (266.7)(93-35) = 15469$ W

Ex. 1 $A = \frac{15469}{(570)(47.75)} = 0.568 \text{ m}^2$ not large enough

Ex 2 $A = \frac{15469}{(370)(47.75)} = 0.876 \text{ m}^2$ 0.94 m^2 is large enough

10-25 $p = 83\ kN/m^2 = 12\ psia$ $\quad T_{sat} = 204\ °F = 95.6\ °C$

$h_{fg} = 976\ Btu/lbm = 2.27 \times 10^6\ J/kg$ $\quad q = (2.27\times10^6)(0.76/60) = 28753\ W$

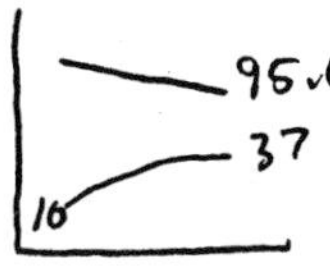

$$\Delta T_m = \frac{85.6 - 38.6}{\ln\left(\frac{85.6}{38.6}\right)} = 59\ °C$$

$$A = \frac{q}{U\Delta T_m} = \frac{28753}{(3400)(59)} = 0.143\ m^2$$

10-26 $28753 = (3400)(0.143)\Delta T_m$ $\quad \Delta T_m = 59°C$

$$\Delta T_m = \frac{(95.6 - 30) - (95.6 - T_{we})}{\ln\left(\frac{65.6}{95.6 - T_{we}}\right)} = 59$$ By iteration $T_{we} = 42.6\ °C$

$28153 = q = \dot{m}_w(4175)(42.6 - 30)$ $\quad \dot{m}_w = 0.5466\ kg/sec$

$$\dot{m}_w = \frac{28753}{(4175)(57-10)} = 0.1465\ kg/sec \qquad \%\ increase = \frac{0.5466 - 0.1465}{0.1465}$$

% increase = 273 %

10-27

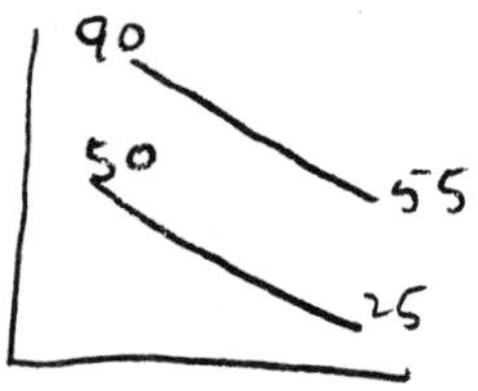

$$\Delta T_m = \frac{40 - 30}{\ln\left(\frac{40}{30}\right)} = 34.76\ °C$$

$$A = \frac{q}{U\Delta T_m} = \frac{29000}{(340)(34.76)} = 2.45\ m^2$$

10-28

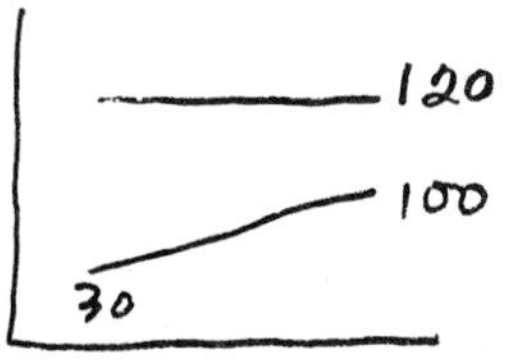

$$\Delta T_m = \frac{90 - 20}{\ln\left(\frac{90}{20}\right)} = 46.54\ °C$$

$$q = m_w c_w \Delta T_w = (2.5)(4180)(100 - 30)$$
$$= 7.315 \times 10^5\ W$$

$$A = \frac{q}{U\Delta T_m} = \frac{7.315\times10^5}{(2000)(46.54)} = 7.86\ m^2$$

10-29 $0.0002 = \frac{1}{U} - \frac{1}{2000}$ $U = 143 \text{ W/m}^2\cdot°C$

$C_w = C_{min} = (2.5)(4180) = 10450$ $NTU = \frac{UA}{C_{min}} = \frac{(143)(7.86)}{10450} = 0.1076$

$\epsilon = 1 - e^{-0.1076} = 0.1020 = \frac{\Delta T_w}{120-30}$ $\Delta T_w = 9.18°C$

$T_{w\,exit} = 30 + 9.18 = 39.18°C$

10-30 $C = 1.0$ $C_{min} = C_{max} = (0.5)(1006) = 503$ $n = 0.903$

$N = \frac{(40)(20)}{503} = 1.59$ $\epsilon = 0.567$ $\Delta T = (0.567)(400-20) = 215.4°C$

$T_{ce} = 215.4 + 20 = 235.4°C$ $T_{he} = 400 - 215.4 = 184.6°C$

10-31 $C_a = 1005$ $\rho_a = \frac{1.013 \times 10^5}{(287)(283)} = 1.247 \text{ kg/m}^3$

$\dot{m}_a = (1500)(1.247) = 1871 \text{ kg/min} = 31.19 \text{ kg/sec}$

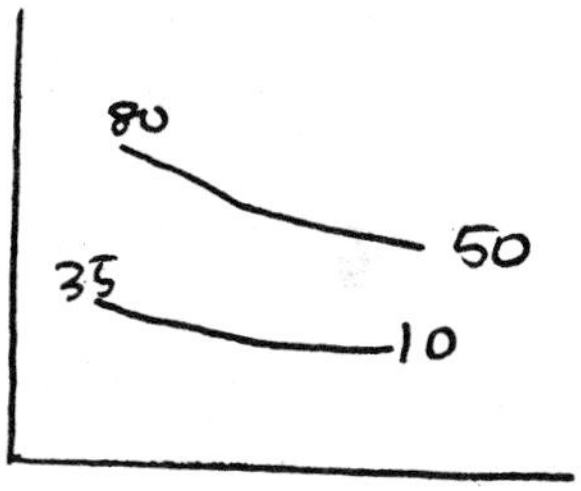

Both fluids unmixed

$\Delta T_m = \frac{45-40}{\ln(45/40)} = 42.45°C$

$R = \frac{80-50}{35-10} = 1.2$ $P = \frac{35-10}{80-10} = 0.357$ $F = 0.96$

$q = m_a C_a \Delta T_o = (31.19)(1005)(35-10) = 7.84 \times 10^5 \text{ W}$

$A = q/UF\Delta T_m = \frac{7.84 \times 10^5}{(50)(0.96)(42.45)} = 384.6 \text{ m}^2$

10-32 Both fluids unmixed

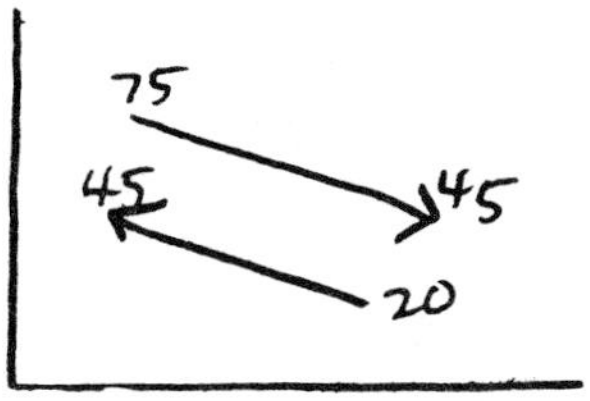

$\Delta T_m = \frac{30-25}{\ln(30/25)} = 27.42°C$

$R = \frac{75-45}{45-20} = 1.2$ $P = \frac{45-20}{75-20} = 0.455$

$F = 0.91$

$A = q/UF\Delta T_m = \frac{(100{,}000)/3.413}{(50)(0.91)(27.42)} = 23.49 \text{ m}^2$

10-33 $C_o = 1920 \quad c_w = 4180 \quad C_o = \frac{(95)(1920)}{60} = 3040$ (min)

$C_w = \frac{(55)(4180)}{60} = 3832 \qquad C_{min}/C_{max} = 0.793$

$NTU = \frac{(250)(14)}{3040} = 1.151 \qquad \epsilon = 0.5 = \frac{\Delta T_o}{120-30} \qquad \Delta T_o = 45°C$

$T_{oe} = 120 - 45 = 75°C \qquad q = C_o \, \Delta T_o = (3040)(45) = 1.368 \times 10^5 \text{ W}$

$\Delta T_w = 45\left(\frac{3040}{3832}\right) = 35.7°C \qquad T_{we} = 30 + 35.7 = 65.7°C$

10-34 h (steam) ~ 9000 W/m$^2\cdot$°C

$U = \frac{1}{\frac{1}{h_o} + \frac{1}{h_i}} = \frac{1}{\frac{1}{250} + \frac{1}{9000}} = 243 \text{ W/m}^2\cdot°C$

$C_w = C_{min} = 6967 \qquad \epsilon = \frac{175-50}{250-50} = 0.625$

$N = -\ln(1-\epsilon) = 0.98 = UA/C_{min} \qquad A = \frac{(0.98)(6967)}{243} = 28.1 \text{ m}^2$

For reduced water flow by 60%

$C_{min} = (0.4)(6967) = 2787 \qquad N = \frac{(243)(28.1)}{2787} = 2.45$

$\epsilon = 1 - e^{-2.45} = 0.914 = \Delta T_w / 250-50 \qquad \Delta T_w = 183°C$

$T_{w\,exit} = 50 + 183 = 233°C$

10-35 $c_w = (5)(4180) = 20900 \qquad c_a = 4800$

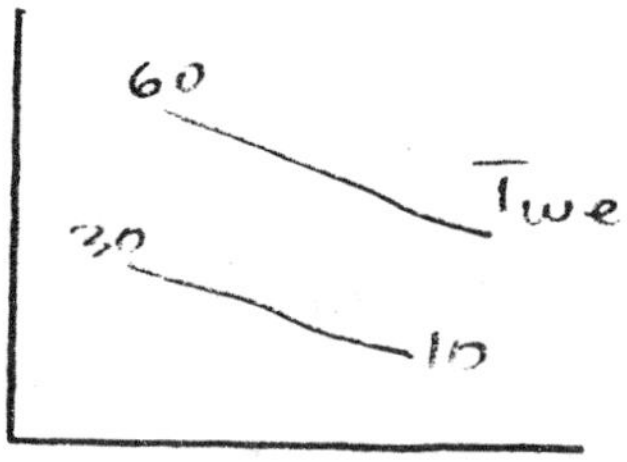

$q = 20900(60 - T_{we}) = (4800)(\dot{m}_a)(30-10)$

$20900(60 - T_{we}) = UA\,\Delta T_m = (800)(30)\frac{30 - T_{we} + 10}{\ln\left(\frac{30}{T_{we}-10}\right)}$

Solve by iteration: $T_{we} = 31°C$

$\dot{m}_a = \frac{(20900)(60-31)}{(4800)(20)} = 6.31$ kg/sec

10-36 $U = 150\ W/m^2 \cdot °C$ $A = 30\ m^2$ $\epsilon = \frac{\Delta T_a}{100-10} = 0.85$

$\Delta T_a = 76.5°C$ $T_{ae} = 76.5 + 10 = 86.5\ °C$

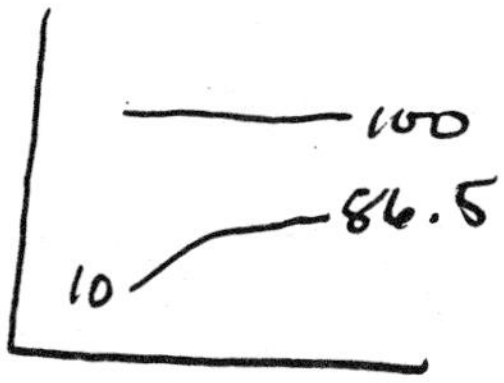

$$\Delta T_m = \frac{90 - 13.5}{\ln\left(\frac{90}{13.5}\right)} = 40.32°C$$

$q = UA\,\Delta T_m = (150)(30)(40.32) = 1.81 \times 10^5\ W$

10-37 $0.0002 + 0.0004 = \frac{1}{U} - \frac{1}{150}$ $U = 137.6\ W/m^2 \cdot °C$

$\frac{A(\text{dirty})}{30} = \frac{150}{137.6}$ $A(\text{dirty}) = 32.7\ m^2$ 9% increase

10-38 $N = (2)(1.59) = 3.18$ $n = 0.775$ $\epsilon = 0.693$ $T_{ce} = 283.3°C$

$T_{he} = 136.7°C$ $\Delta T = (0.693)(400-20) = 263.3$

$N = \frac{1}{2}(1.59) = 0.795$ $n = 1.052$ $\epsilon = 0.416$ $\Delta T = 158.2$

$T_{ce} = 178.2°C$ $T_{he} = 241.8\ °C$

10-39 $C_a = \frac{65}{60}(1006) = 1090\ W/°C$ $q = (1090)(45-30) = 16350\ W$

$U = 52$ $A = 8.0$ $T_{w,i} = 90°C$ Assume air min. fluid

$\epsilon = \frac{45-30}{90-30} = 0.25$ $N = \frac{(52)(8)}{1090} = 0.382$

Unable to match on graph so C_w is min. Iterative: $C = C_w/C_a$

C	N	n	ϵ_c	ΔT_w	$\Delta T_a = C\Delta T_w$	$\Delta T_a - 15$
0.6	0.636	1.104	0.405	24.3	14.6	−0.42
0.48	0.795	1.052	0.48	28.8	13.82	−1.17
0.50	0.764		0.4664	27.98	13.99	−1.008
0.7	0.546	1.143	0.3574	21.44	15.01	+0.01

$T_{we} = 90 - 21.44 = 68.6°C$

10-41 $m_h = 2.6\ kg/s \quad m_c = 1.3\ kg/s \quad \Delta T_h = \frac{1}{2}\Delta T_c = \frac{1}{2}(32-4) = 14$

$$\Delta T_m = \frac{67-81}{\ln\left(\frac{67}{81}\right)} = 73.78$$

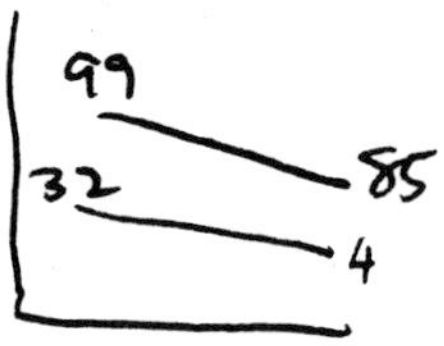

$$q = (1.3)(4175)(32-4) = (830)A(73.78)$$

$$A = 2.48\ m^2 \qquad \epsilon = \frac{32-4}{99-4} = 29.5\%$$

10-42 $\epsilon_c = \dfrac{T_{c2} - T_{c1}}{T_{h1} - T_{c1}} \qquad dq = U\,dA\,(T_h - T_c) = m_c c_c\, dT_c$

$$\frac{dT_c}{T_h - T_c} = \frac{U\,dA}{\dot{m}_c c_c} \qquad -\ln(T_h - T_c)\Big]_1^2 = \frac{UA}{\dot{m}_c c_c}\Big]_0^A$$

$$\frac{T_h - T_{c2}}{T_h - T_{c1}} = e^{-UA/\dot{m}_c c_c} \qquad 1 - \frac{T_{c2} - T_{c1}}{T_h - T_{c1}} = e^{-UA/\dot{m}_c c_c}$$

$$\epsilon = \frac{T_{c2} - T_{c1}}{T_h - T_c} = 1 - e^{-UA/\dot{m}_c c_c}$$

10-43 $\Delta T_w = 75 - 30 = 45°C \qquad \Delta T_o = 48 - 25 = 23°C$

water is minimum fluid. $\epsilon = \dfrac{\Delta T_w}{\Delta T_{max}} = \dfrac{45}{75-25} = 0.9$

10-44 Sat steam: $p_s = 100$ psia $T_s = 328°F \quad h = 1000\ Btu/hr\,ft^2\,°F$

$h_{fg} = 888.8\ Btu/lbm$ neglect wall resistance.

CO_2: $\dot{m}_c = 3600\ lbm/hr \quad p_c = 15$ psia $T_{c2} = 70°F \quad T_{c1} = 200°F$

$c_{pc} = 0.208\ Btu/lbm\,°F \qquad \rho_c = 0.1106\ lbm/ft^3$ @ $90°F \qquad \frac{S_0}{d} = \frac{S_n}{d} = 3/2$

$C = 0.25 \quad n = 0.62 \qquad Nu = \frac{hd}{k} = C\left(\frac{u_{max} d}{\nu}\right)^n$

$T_f = 231°F \qquad u_\infty = \frac{mc}{\rho A} \qquad \rho = 0.0874 \qquad u_{max} = u_\infty\left(\frac{S_n}{S_n - d}\right)$

$k = 0.0134 \qquad Re_d = \frac{\rho u_m d}{\mu} \qquad h_c = \frac{k}{d}\, C\,(Re)^n$

$q_c = UAF(\Delta T_m) \quad F = 1$

10-44 contd. Assume $L = 3.25'' = 0.271$ ft $\quad UA = \dfrac{1}{\dfrac{1}{h_i A_i} + \dfrac{1}{h_o A_o}}$

$A = 0.1775$ ft^2 $\quad \Delta T_m = \dfrac{(T_{h2} - T_{h1}) - (T_{h1} - T_{c1})}{\ln\left(\dfrac{T_{h2} - T_{h1}}{T_{h1} - T_{c1}}\right)}$

$u_\infty = 199000$ ft/hr $\quad u_{max} = 597000$ ft/hr

$Re_d = 24200 \quad h = 83.5 \quad UA(\Delta T_m) = \dot{m} c_p (T_2 - T_1)_{CO_2}$

$UA = \dfrac{4.98}{t} \quad L = 107.5 \quad \dfrac{107.5}{400} = 0.268 = 3.22 \text{ in} \approx 3.25''$

this assumption was o.k.

10-47

$m_o(2100)(138-93) = (2.5)(4175)(65-25)$

$m_o = 4.417$ kg/sec $\quad U = 450 \quad A_1 = A_2 \quad$ Assume oil is min. fluid

$\epsilon_1 = \dfrac{138 - T_1}{138 - 25} \quad \epsilon_2 = \dfrac{138 - T_2}{138 - 50} \quad C_{max_1} = (2.5)(4175) = 10438$

$C_{max_2} = (1.88)(4175) = 7849 \quad \dfrac{NTU_1}{NTU_2} = \dfrac{C_{min_1}}{C_{min_2}} = \dfrac{m_{o2}}{m_{o1}} = \dfrac{4.417}{m_{o1}} - 1$

$(10438)(50-25) = C_{min_1}(138 - T_1) = 260950$

$(7849)(65-50) = C_{min_2}(138 - T_2) = 117735$

By Iteration: $\quad m_{o1} = 3.638 \quad T_1 = 103.8°C \quad \epsilon_1 = 0.3023$

$C_1 = 0.732 \quad N_1 = 0.4099$

$$A_1 = A_2 = \frac{(0.4099)(3.638)(2100)}{450} = 6.96 \text{ m}^2$$

10-48 Water is min fluid $\quad F = 1 \quad A_1 = 1.535$ m^2

$(\Delta T_m)_1 = \dfrac{(138-25) - (138-50)}{\ln\left(\frac{113}{88}\right)} = 99.98°C \quad A_2 = 0.863$ m^2

$(\Delta T_m)_2 = \dfrac{(138-50) - (138-65)}{\ln(88/73)} = 80.27°C$

$q_1 = (2.5)(4175)(50-25) = 1700 A_1(99.98) = 260.9$ kW

$q_2 = (1.88)(4175)(65-50) = 1700 A_2(80.27) = 117.7$ kW

$m_{s1} = q_1/h_{fg} = 0.116$ kg/sec $\quad m_{s2} = q_2/h_{fg} = 0.052$ kg/sec

10-51 $0.004 = \frac{1}{U} - \frac{1}{340}$ $U = 144$ $A = (2.45)(\frac{340}{144}) = 5.78\ m^2$

with original area, Oil is min fluid $c = 1900$

$C_o = \frac{29000}{90-55} = 828.6$ $C_w = \frac{29000}{50-25} = 1160$

$C_{min}/C_{max} = 0.714$ $NTU = \frac{(144)(2.45)}{828.6} = 0.426$

$\epsilon = 0.29 = \frac{\Delta T_o}{90-25}$ $\Delta T_o = 18.85°C$ $q = (828.6)(18.85) = 15619\ W$

reduced by 46%

10-52

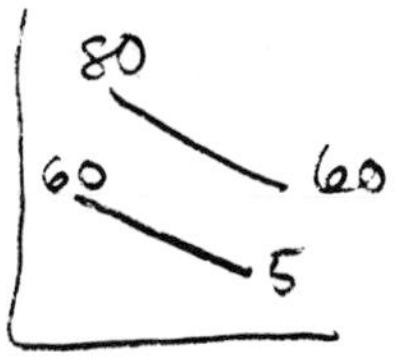

$q = 60\ kW$ $U = 1100\ W/m^2 \cdot °C$

Cold fluid is minimum fluid $\epsilon = \frac{60-5}{80-5} = 0.733$

$\frac{C_{min}}{C_{max}} = \frac{20}{55} = 0.364$ $NTU = 1.9$

$C_{min} = \frac{60000}{60-5} = 1091$ $A = \frac{(1091)(1.9)}{1100} = 1.88\ m^2$

11-53 $NTU = 1.9$ $\frac{C_{min}}{C_{max}} = (2)(0.364) = 0.728$

$\epsilon = 0.69 = \frac{\Delta T_{min}}{80-5}$ $\Delta T_{min} = 51.75\ °C$

$q = (51.75)(1091) = 56460\ W$ only reduced by 6%

10-55 $29000 = C_a(25-15) = C_w(70-40)$ $C_w = 966.7$

$\frac{C_w}{C_a} = \frac{C_{min}}{C_{max}} = \frac{10}{30} = 0.333$ $\epsilon = \frac{30}{70-15} = 0.545$

$\frac{UA}{C_{min}} = 1.0$ $A = \frac{(1.0)(966.7)}{45} = 21.5\ m^2$

10-56 $C_w = 9667/3 = 322.2$ $\qquad \frac{C_{min}}{C_{max}} = 0.333/3 = 0.111$

$NTU = \frac{UA}{C_{min}} = \frac{(45)(21.5)}{322.2} = 3.0$ $\qquad \epsilon = 0.92 = \frac{\Delta T_w}{70-15}$

$\Delta T_w = 50.6\ °C$ $\qquad q = C_w \Delta T_w = 16303\ W$

10-57 For parallel flow assume $\dot{m}_c = \dot{m}_h$

$q = \dot{m}_h c_{ph}(T_{h_1} - T_{h_2}) = \dot{m}_c c_{pc}(T_{c_1} - T_{c_2})$ $\qquad c_{ph} = c_{pc}$ $\quad \dot{m}_h = \dot{m}_c$

$$\frac{T_{h2} - T_{c2}}{T_{h1} - T_{c1}} = \exp UA\left[\frac{1}{\dot{m}_h c_h} + \frac{1}{\dot{m}_h c_h}\right] = \exp\left[\frac{-UA}{\dot{m}_c c_{pc}}\left(1 + \frac{\dot{m}_c c_{pc}}{\dot{m}_h c_{ph}}\right)\right]$$

$$\frac{T_{h2} - T_{c2}}{T_{h1} - T_{c1}} = \exp\left[-2\frac{UA}{\dot{m}_c c_{pc}}\right]$$

consider the cold fluid: $\epsilon = \frac{T_{c2} - T_{c1}}{T_{h1} - T_{c1}}$

$$\frac{T_{h2} - T_{c2}}{T_{h1} - T_{c1}} = \frac{T_{h1} - 2T_{c2} + T_{c1}}{T_{h1} - T_{c1}} \qquad T_{h2} = T_{h1} + (T_{c1} - T_{c2})$$

$$\frac{(T_{h1} - T_{c1}) + (T_{c1} - T_{c2}) + (T_{c1} - T_{c2})}{T_{h1} - T_{c1}} = 1 - 2\epsilon$$

$$\epsilon = \frac{1 - e^{-2UA}}{2}$$

For counter flow:

$\dot{m}_h c_h dT_h = \dot{m}_c c_c dT_c = dq$

$dq = U dA (T_h - T_c)$

$q = UA(T_h - T_c)$

$$\epsilon = \frac{q(act)}{q_{max}} = \frac{UA(T_h - T_c)}{\dot{m}c(T_{h1} - T_{h2})}$$

$$\epsilon = NTU\frac{(T_{h1} - T_{c1})}{(T_{h1} - T_{c2})}$$

$$\epsilon = NTU\left[1 + \frac{T_{c2} - T_{c1}}{T_{h1} - T_{c2}}\right]$$

$$\epsilon = NTU[1 - \epsilon]$$

$$\epsilon = \frac{NTU}{1 + NTU}$$

10-58 $C_w(90-55) = C_o(50-25)$ $\frac{C_w}{C_o} = \frac{25}{35}$ $C_w = C_{min}$

$\epsilon = \frac{90-55}{90-25} = 53.8\%$

10-59 $C_g = (3428)(0.6) = 2056.8$ $C_{min}/C_{max} = \frac{2057}{4112} = 0.500$

$\frac{UA}{C_{min}} = \frac{(850)(5.24)}{2057} = 2.17$ $\epsilon = 0.77$

$q = (2057)(0.77)(140 - T_{wi}) = (4112)(85 - T_{wi})$ $T_{wi} = 50.54\,°C$

Preheater heats water from 35°C to 50.54°C

$q = (4112)(50.54 - 35) = 63900\ W$

$\Delta T_m = \frac{115 - 99.5}{\ln\left(\frac{115}{99.5}\right)} = 107\,°C$ $A = \frac{63900}{(2000)(107)} = 0.299\ m^2$

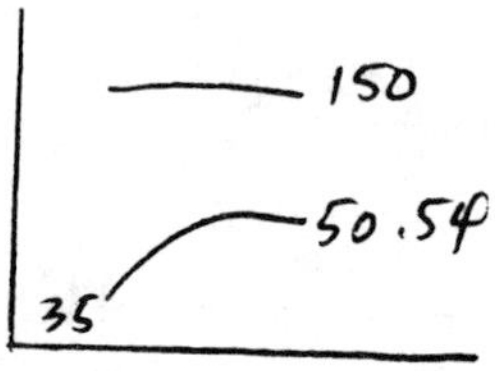

@ $T = 150°C = 302°F$ $h_{fg} = 910\ Btu/lbm$

$h_{fg} = 2.117 \times 10^6\ J/kg$ $m_s = \frac{63900}{2.117 \times 10^6} = 0.03\ kg/sec$

10-61 $m_a = 5.0$ kg/sec $A = 110\ m^2$ $U = 50\ W/m^2 \cdot °C$

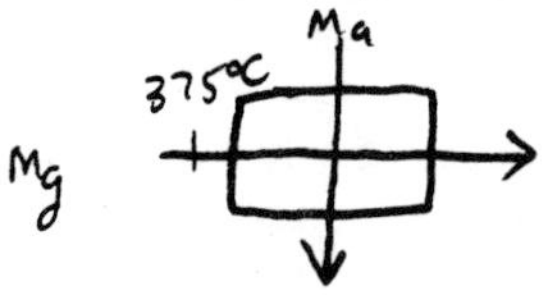

$m_g = 5\ kg/s$

(a) Both unmixed $\frac{C_{min}}{C_{max}} = 1.0$ $NTU = \frac{(110)(50)}{(5)(1005)} = 1.095$

$\epsilon = 0.49$ $\Delta T = (0.49)(375 - 27) = 170.5$

$T_{a2} = 170.5 + 27 = 197.5\,°C$

$T_{g2} = 375 - 170.5 = 204.5\,°C$

$q = (5025)(170.5) = 8.57 \times 10^5\ W$

(b) $\epsilon = 0.48$ one fluid mixed

about same results

10-62 Both unmixed $h_a = 174$ $h_s = 5000$ $U = 168$ $C = 0$

$\text{Volume} = (0.5)(0.4) = 0.2\ m^3 = 7.063\ ft^3$ $\text{Area} = (7.063)(229) = 1617\ ft^2$

$A = 150.3\ m^2$ $C_{min} = (1.1774)(0.5)(15)(1006) = 8883\ W/°C$

$$N = \frac{(168)(150.3)}{8883} = 2.843 \qquad \epsilon = 1 - e^{-2.843} = 0.942$$

$\Delta T_a = (0.942)(373 - 300) = 68.7°C$ $T_{a2} = 368.7K = 95.7°C$

$q = (8883)(68.7) = 6.1 \times 10^5\ W$

10-63 $m_w = 30\ kg/s$ $C_w = (30)(4174) = 125220$ $T_{w1} = 20°C$

$T_{w2} = 40°C$ $T_{oi} = 200°C$ $U = 275$ $q = (125220)(40-20) = 2.504 \times 10^6\ W$

$$N = \frac{1}{C-1} \ln\left(\frac{\epsilon - 1}{C\epsilon - 1}\right)$$

T_{oe}	ΔT_o	C_o	C	ϵ	NTU
190	10	250440	0.5	0.1111	0.1212
180	20	125220	1.0	0.1111	0.125
140	60	41740	0.333	0.333	0.432
80	120	20870	0.167	0.667	1.177

10-64 $m_w = 50\ kg/s$ $T_{w1} = 60°C$ $T_{w2} = 90°C$ $U = 4500$

$T_s = 200°C$ $C_w = (50)(4174) = 208700$ $C = 0$

$$\epsilon = \frac{30}{200-60} = 0.2143 \qquad N = -\ln(1 - 0.2143) = 0.2412$$

$$A = \frac{(0.2412)(208700)}{4500} = 11.18\ m^2 \qquad \epsilon = \frac{T_{w2} - 60}{T_s - 60} = 0.2143$$

$T_{w2} = 0.2143\ T_s + 47.1$

10-65 (a) $q = (5.0)(4180)(80-30) = 1.045 \times 10^6$

$$\epsilon = \frac{80-30}{100-30} = 0.7143 \qquad N = -\ln(1-\epsilon) = 1.253$$

$$A = \frac{(1.253)(5)(4180)}{900} = 29.09 \text{ m}^2$$

(b) $$N = \frac{(900)(29.09)}{(1.3)(4180)} = 4.818$$

$$\epsilon = 1 - e^{-N} = 0.992$$

$$\Delta T_w = (0.992)(100-30) = 69.4°C$$

$$T_{we} = 69.4 + 30 = 99.4 °C$$

10-68

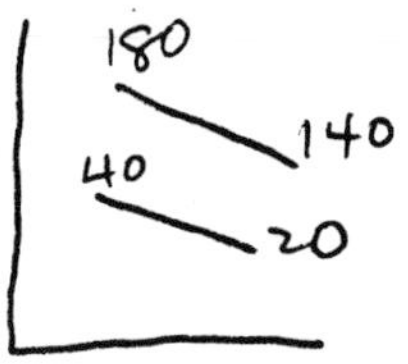

$U = 130 \text{ W/m}^2 \cdot °C \qquad c_o = 2100 \text{ J/kg °C}$

$\dot{m}_w = 3.0 \text{ kg/sec} \qquad C_w = (3)(4180) = 12540$

$q = C_w \Delta T_w = C_o \Delta T_o = 2.508 \times 10^5 \text{ W}$

$$C_o = (12540)\frac{(40-20)}{180-140} = 6270 \qquad \frac{C_{min}}{C_{max}} = 0.5$$

$$\epsilon = \frac{180-140}{180-120} = 0.25 \qquad NTU = 0.35 \quad A = \frac{(0.35)(6270)}{130} = 16.88 \text{ m}^2$$

if water flow cut in half $C_w = 6270$

$q = (2.508 \times 10^5)(0.5) = 1.254 \times 10^5 \text{ W}$

Oil flow will be reduced so it will still be min. fluid.

$$q = C_o(180 - T_{o2}) = (130)(16.88)\frac{(80-40)-(T_{o2}-20)}{\ln\left(\frac{140}{T_{o2}-20}\right)}$$

Solve by iteration: $T_{o2} = 36°C$

$C_o(180-36) = 1.254 \times 10^5 \qquad C_o = 870.8$

Flow is reduced by 86%

10-69 $\dot{m}_a = 0.8$ kg/sec $T_{a1} = 85°F$ $T_{a2} = 45°F$ $T_{w1} = 37.4°F$

$U = 55\ W/m^2 \cdot °C$ $\dot{m}_w = 0.75$ kg/sec $C_a = (0.8)(1005) = 804$

$C_w = (0.75)(4180) = 3135$ $\epsilon = \frac{85-45}{85-37.4} = 0.84$

$\frac{C_{min}}{C_{max}} = \frac{804}{3135} = 0.256$ $NTU = 2.4$ $A = \frac{(2.4)(804)}{55} = 35.08\ m^2$

Water cut in half: $C_{min}/C_{max} = (2)(0.256) = 0.512$

$NTU = 2.4$ $\epsilon = 0.76 = \Delta T_a / 85-37.4$ $\Delta T_a = 36.18°F$

$\%\ \text{reduction} = \left(1 - \frac{36.18}{40}\right) \times 100 = 9.6\ \%$

10-70 $\epsilon = \frac{85-45}{85-35} = 0.8$ $NTU = -\ln(1-\epsilon) = 1.609$

$A = \frac{(1.609)(804)}{125} = 10.35\ m^2$ Reduce air flow by 1/3

$NTU = \frac{1.609}{2/3} = 2.414$ $\epsilon = 1 - e^{-N} = 0.91 = \Delta T_a / 85-35$

$\Delta T_a = 45.5°F$

Initial $q = (804)(85-45)(5/9) = 17867\ W$

Reduced $q = (804)(2/3)(45.5)(5/9) = 13549\ W$

Reduced by 24%

10-71 $M_o = 4000$ kg/hr $= 1.111$ kg/sec $\quad c_o = 2000$ J/kg·°C

$C_o = C_{min} = (2000)(1.111) = 2222$

$\epsilon = \frac{\Delta T_o}{\Delta T_{max}} = \frac{80-40}{100-40} = 0.667 \quad N = -\ln(1-\epsilon) = 1.099 = \frac{UA}{C_{min}}$

$$A = \frac{(1.099)(2222)}{1200} = 2.034\ m^2$$

when oil flow cut in half:

$NTU = (2)(1.099) = 2.198 \quad \epsilon = 1 - e^{-N} = 0.889 = \frac{\Delta T_o}{100-40}$

$\Delta T_o = 53.34°C \quad q = C_o \Delta T_o = \frac{2222}{2}(53.34) = 59260\ W$

$= \dot{m}_s h_{fg}$

$$\dot{m}_s = \frac{59260}{2.255 \times 10^6} = 0.0262\ \text{kg/sec}$$

10-72 $\epsilon = 0.8 = \frac{\Delta T_w}{100-50} \quad \Delta T_w = 32°C \quad N = -\ln(1-\epsilon) = 1.609$

$$A = \frac{(1.609)(5)(4180)}{1200} = 28.03\ m^2$$

10-73 Reduce water flow in half $\quad NTU = \frac{1.609}{0.5} = 3.218$

$\epsilon = 1 - e^{-N} = 0.96 = \frac{\Delta T_w}{100-50} \quad \Delta T_w = 48°C$

$T_{we} = 48 + 50 = 98°C \quad q = C_w \Delta T_w = (5)(0.5)(4180)(48)$

$= 5.016 \times 10^5\ W$

10-74 $\Delta T_w = 180 - 125 = 55$ $\Delta T_a = 115 - 45 = 70$

Air is minimum fluid: $\epsilon = \frac{70}{180-45} = 0.519$

10-75 $q = \dot{m}_w c_w (55) = \dot{m}_a c_a (70)$

$\dot{m}_w = 150$ kg/min $= 2.5$ kg/sec

$q = (2.5)(4180)(55)(5/9) = 3.193 \times 10^5$ W

$\frac{C_a}{C_w} = \frac{55}{70} = 0.786$ $NTU = 1.2 = \frac{UA}{C_a}$

$A = \frac{(1.2)(2.5)(4180)(0.786)}{50} = 197\ m^2$

10-76 $\dot{m}_g = 10$ kg/s $\dot{m}_w = 15$ kg/sec $c_g = 2420$

$c_w = 4180$ kJ/kg·°C $C_g = 24200$ $C_w = 62700$

$\frac{C_{min}}{C_{max}} = 0.386$ Glycol is minimum fluid

$\epsilon = \frac{40-20}{70-20} = 0.4$ $NTU = 0.6 = \frac{UA}{C_{min}}$

$A = \frac{(0.6)(24200)}{40} = 363\ m^2$

10-77 $C_w = (10)(4180) = 41800$ $U = 35$ W/m²·°C

$\frac{C_{min}}{C_{max}} = \frac{24200}{41800} = 0.579$ $NTU = \frac{(35)(363)}{24200} = 0.525$

$\epsilon = 0.38 = \Delta T_g / 70-20$ $\Delta T_g = 19°C$

$t_g = 20 + 19 = 39°C$

10-78 Water is minimum fluid

$q = 800 \times 10^6 = C_w(30-25)$ $C_w = 1.6 \times 10^8$ $U = 2000\ \mathrm{W/m^2 \cdot {}^\circ C}$

$\epsilon = \dfrac{\Delta T_w}{\Delta T_{max}} = \dfrac{30-25}{100-25} = 0.06667$ $N = -\ln(1-\epsilon) = 0.069 = \dfrac{UA}{C_{min}}$

$$A = \frac{(0.069)(1.6 \times 10^8)}{2000} = 5519\ \mathrm{m^2}$$

10-79 Water flow in half $C_w = 0.8 \times 10^8$ $NTU = 0.138$

$\epsilon = 1 - e^{-N} = 0.1289 = \Delta T_w / 100-25$ $\Delta T_w = 9.67°C$

$q = C_w \Delta T_w = 7.73 \times 10^8\ \mathrm{W} = \dot{m}_s h_{fg}$ $h_{fg} = 2.255 \times 10^6\ \mathrm{J/kg}$

$$\dot{m}_s = \frac{7.73 \times 10^8}{2.25 \times 10^6} = 343\ \mathrm{kg/sec} = 1.235 \times 10^6\ \mathrm{kg/hr.}$$

10-80 $T_{g_1} = 20°C$ $T_{g_2} = 60°C$ $c_g = 2474$ $\dot{m}_g = 1.2\ \mathrm{kg/s}$

$T_{w_1} = 90°C$ $T_{w_2} = 50°C$ $U = 600\ \mathrm{W/m^2 \cdot {}^\circ C}$

$C_g = (1.2)(2474) = 2969 = C_w$ $\dfrac{C_{min}}{C_{max}} = 1.0$

$\epsilon = \dfrac{40}{90-20} = 0.571$

$NTU = 1.4 = UA/C_{min}$

$$A = \frac{(1.4)(2969)}{600} = 6.93\ \mathrm{m^2}$$

10-82 $\rho = 1.1774$ kg/m³ $c_p = 1005$ J/kg °C $Pr = 0.708$

$\mu = 1.846 \times 10^{-5}$ $\sigma = 0.697 = \frac{A_c}{A}$ $D_h = 0.0118$ ft $= 0.0036$ m

Frontal Area $= (0.3)(0.6) = 0.18$ m² $A_c = (0.697)(0.18) = 0.1254$ m²

$$G = \frac{\rho u_\infty A}{A_c} = \frac{(1.1774)(10)}{0.67} = 17.57 \text{ kg/m}^2\cdot\text{s}$$

$$Re \frac{D_h G}{\mu} = \frac{(0.0036)(17.57)}{1.846 \times 10^{-5}} = 3426 \quad St\, Pr^{2/3} = 0.004 = \frac{h}{G c_p} Pr^{2/3}$$

$$h = (0.004)(17.57)(1005)(0.708)^{-2/3} = 88.09 \text{ W/m}^2\cdot°\text{C}$$

This coefficient is controlling

$$\text{Volume} = (0.3)^2(0.6) = 0.054 \text{ m}^3$$

$$\text{Heat Transfer Area} = (0.054)(\alpha) = (0.054)\left(229 \frac{\text{ft}^2}{\text{ft}^3} \times 3.28 \text{ ft/m}\right)$$

$$= 40.56 \text{ m}^2$$

$$q = h A \left(100 - \frac{27}{2} - \frac{T_2}{2}\right) = \dot{m} c_p (T_2 - 27)$$

$$(88.9)(40.56)\left(100 - \frac{27}{2} - \frac{T_2}{2}\right) = (1.1774)(10)(0.18)(1005)(T_2 - 27)$$

$$T_2 = 93.93°\text{C}$$

$$q = 1.43 \times 10^5 \text{ W} = \dot{m}_s h_{fg} \quad h_{fg} = 2.255 \times 10^6 \text{ J/kg}$$

$$\dot{m}_s = q/h_{fg} = 0.63 \text{ kg/s}$$

10-85

$c_g = 2382\ J/kg\text{-}^\circ C$

$q = m_w c_w \Delta T_w = m_g c_g \Delta T_g$

$= (0.5)(4180)(80-60) = (0.7)(2382)\Delta T_g = 41800\ W$

$\Delta T_g = 25.1\ ^\circ C \qquad = UA\Delta T_w$

$\epsilon = \dfrac{25.1}{80-20} = 0.418$

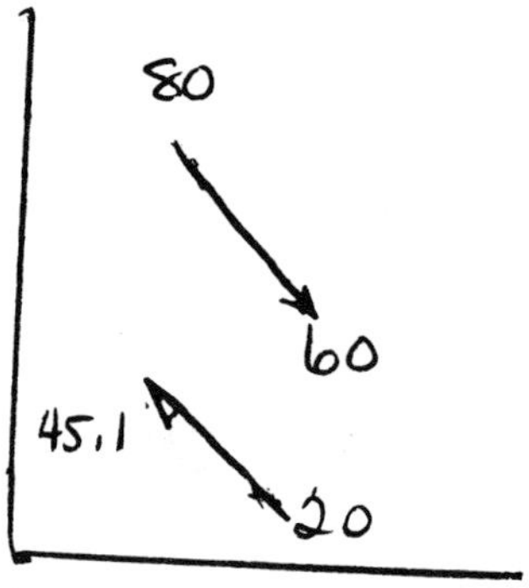

Assume counterflow

$\Delta T_m = \dfrac{34.9-40}{\ln \frac{34.9}{40}} = 37.4\ ^\circ C$

$A = \dfrac{41800}{(1000)(37.4)} = 1.12\ m^2$

10-86

$q = M_w C_w \Delta T_w = M_o C_o \Delta T_o$

$= (0.6)(4180)(50-20) = (1.2)(2100)(100-T_{oe}) = 75240\ W$

$\Delta T_o = 29.86$

$T_{oe} = 70.14^\circ C$

water is min fluid and mixed
Oil is max fluid and unmixed

$\epsilon = \dfrac{30}{100-20} = 0.375 \qquad C_{min}/C_{max} = \dfrac{29.86}{30} = 0.995$

$N = \dfrac{-1}{0.995} \ln\left[1 + 0.995 \ln(1-0.375)\right] = 0.634 = \dfrac{UA}{C_{min}}$

$A = \dfrac{(0.634)(0.6)(4180)}{250} = 0.636\ m^2$

10-87 Water is min fluid and unmixed
Oil is max fluid and mixed

$$N = -\ln\left[1 + \frac{1}{0.995}\ln(1-0.995(0.375))\right]$$

$$= 0.469$$

$$A = 0.470\ m^2$$

10-88

$$q = m_w C_w \Delta T_w = m_o C_o \Delta T_o$$

$$= (2)(4180)(70-10) = (3)(2100)(120-T_{oe}) = 501600\ \text{W}$$

$$\Delta T_o = 79.62$$

$$T_{oe} = 40.38°C$$

Oil is min fluid

$$C = \frac{60}{79.62} = 0.7536$$

$$\epsilon = \frac{79.62}{120-10} = 0.7238, \quad n = 3$$

$$\epsilon_p = \frac{[(1-\epsilon C)/(1-\epsilon)]^{1/n} - 1}{[(1-\epsilon C)/(1-\epsilon)]^{1/n} - C}$$ (see Prob. 10-95)

$$= \frac{\left[\frac{1-(0.7278)(0.7536)}{1-0.7238}\right]^{1/3} - 1}{[\quad]^{1/3} - C} = \frac{1.1806 - 1}{1.1806 - 0.7536} = 0.423$$

From Fig. 10-16, N (1 shell pass) = 0.75

N(total) = (3)(0.75) = 2.25

$$2.25 = \frac{UA}{(3)(2100)} \qquad A = \frac{(2.25)(3)(2100)}{300} = 47.25\ m^2$$

10-89

$C_w = (1)(4180) = 4180$

$C_o = (3)(2100) = 6300$

$C_{min}/C_{max} = 4180/6300 = 0.663$

$$NTU = \frac{(300)(47.25)}{4180} = 3.391$$

$N(1\ \text{Shell}) = 1.13$ $\quad \epsilon_p = 0.54$

From Fig. 10-16

From Table 10-3

$$\epsilon = \frac{\left[\frac{1-(0.54)(0.663)}{1-0.54}\right]^3 - 1}{\left[\quad\right]^3 - 0.663} = 0.836$$

$$= \frac{\Delta T_w}{120-10} \qquad \Delta T_w = 91.97°C$$

$T_{we} = 10 + 91.97 = 101.97°C$

From energy balance: $\Delta T_o = 61.02\ °C$

$T_{oe} = 120 - 61.02 = 58.98\ °C$

$q = 3.844 \times 10^5$ W

10-90

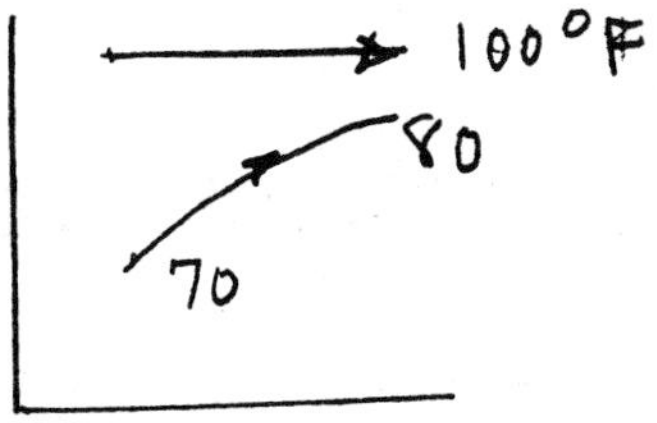

$q = (1800)(56) = 1{,}008 \times 10^5$ Btu/hr $= 29534$ W

$= UA\,\Delta T_m$

$$\Delta T_m = \frac{30-20}{\ln\frac{30}{20}} = 24.66\ °F = 13.7\ °C$$

$$A = \frac{29534}{(13.7)(700)} = 3.08\ m^2$$

$$\epsilon = \frac{10}{100-70} = 1/3 \qquad C = 0$$

$$N = 0.0405 = -\ln(1-\epsilon)$$

10-91 Reduce $\dot{m}_w$ in half

$$N = \frac{(0.405)}{1/2} = 0.81$$

$$\epsilon = 1 - e^{-N} = 1 - \exp(-0.81) = 0.555 = \frac{\Delta T_w}{100-70}$$

$$\Delta T_w = 16.65\ °F$$

$$\frac{q(1/2)}{q(\text{full})} = \frac{(\tfrac{1}{2})(16.65)}{(1)(10)} = 0.833$$

So, condensation of freon reduced by 12.7%

10-92

$$m_h C_h = m_c C_c = (3)(4180) = 12540$$

$C = 1.0$ $n = 4$

$$\epsilon = \frac{20}{80-10} = 0.2857$$

Using Eq. in Table 10-3 for $C = 1$

$$\epsilon[1+(n-1)\epsilon_p] = n\epsilon_p$$

$$\epsilon_p = \frac{\epsilon}{n - \epsilon(n-1)} = \frac{0.2857}{4-(3)(0.2857)} = 0.0909$$

From Table 10-4

$$N = \frac{1}{\sqrt{2}} \ln\left[\frac{22-1-1-\sqrt{2}}{22-1-1+\sqrt{2}}\right]$$

$$= 0.1002 = \frac{UA}{C_{min}} = \frac{1000\,A}{(3)(4180)}$$

$A = 1.256\ m^2$

$A_{(total)} = (4)(1.256) = 5.024\ m^2$

10-93

New $C_{min} = (1.5)(4180) = 6270$

$C = 0.5$

One shell pass $N = (2)(0.1002) = 0.2004$

From Table 10-3

$$\epsilon_p = 2\left\{1 + 0.5 + 1.118 \times \frac{1 + e(-0.224)}{1 - e(-0.224)}\right\}^{-1}$$

$= 0.173$

Also from Table 10-3

$$\epsilon = \frac{\left[\frac{1 - 0.0865}{1 - 0.173}\right]^4 - 1}{\left[\quad\right]^4 - C} = 0.4943 = \frac{\Delta T_w}{80 - 10}$$

$\Delta T_w = 34.6°C$

$q = (1.5)(4180)(34.6) = 2.156 \times 10^5$ W

For Prob. 10-92 $q = (3)(4180)(20) = 2.508 \times 10^5$W

14% reduction in heat transfer

10-94 $C = 0.5$ and 1.0

Take N (1 shell) $= 1.0$

Fig. 10-16 $\epsilon_p = 0.525$ @ $C = 0.5$

$\epsilon_p = 0.45$ @ $C = 1.0$

Fig. 10-17 @ $N = 2$, $\epsilon = 0.76$ @ $C = 0.5$

$\epsilon = 0.62$ @ $C = 1.0$

at $C = 0.5$

$$\epsilon = \frac{\left[\frac{1-0.2625}{1-0.525}\right]^2 - 1}{\left[\quad\right]^2 - 0.5} = 0.738$$

at $C = 1.0$

$$\epsilon = \frac{(2)(0.45)}{1+(2-1)(0.45)} = 0.621$$

Good agreement, considering accuracy of reading figures.

10-95

Solve third from last equation of Table 10-3 for ϵ_p.

10-96 $C_o = (7.0)(2100) = 14700$ $T_{wi} = 20°C$

$C_{mixed} = C_W = (3.5)(4180) = 14630$ $T_{oi} = 100°C$

$\epsilon = 0.6 = \dfrac{\Delta T_w}{100-20}$ $\Delta T_w = 48°C$

$q = (14630)(48) = (14700)\Delta T_o$ $\Delta T_o = 47.8°C$

$T_{we} = 68°C$ $T_{oe} = 52.3°C$

$C_{mixed}/C_{unmixed} = 14630/14700 = 0.995$

$N = 2.5 = \dfrac{UA}{14630}$ $UA = 36575$

10-97 $C_W = (1.0)(4180) = 4180$

$NTU = \dfrac{(2500)(0.8)}{4180} = 0.4785$

$\epsilon = 1 - e^{-0.4785} = 0.38 = \dfrac{\Delta T_w}{100-20}$

$\Delta T_w = 30.4°C$ $T_{we} = 20 + 30.4 = 50.4°C$

10-98

$C_w = (1.5)(4180) = 6270$

$C_g = (3.0)(2474) = 7422$

$q = (6270)(50-20) = 7422\ \Delta T_g = 1.881 \times 10^5\ \text{W}$

$\Delta T_g = 25.34\ °C$

$$\epsilon = \frac{\Delta T_w}{80-20} = \frac{30}{60} = 0.5$$

$C = 6270/7422 = 0.845\ ,\ n = 4$

From Prob. 10-95

$$\epsilon_p = \frac{\left[\dfrac{1-0.5(0.845)}{1-0.5}\right]^{1/4} - 1}{\left[\quad\right]^{1/4} - 0.845} = 0.1914$$

From Table 10-4, $(1+c^2)^{1/2} = 1.3092$

$$N_p = \frac{-1}{1.3092} \ln\left[\frac{\frac{2}{0.1914} - 1 - 0.845 - 1.3092}{\frac{2}{0.1914} - 1 - 0.845 + 1.3092}\right] = 0.234$$

$$N(\text{total}) = (4)(0.234) = 0.937 = \frac{(900)\,A}{6270}$$

$A = 6.53\ m^2$

10-99 $C_g = 7422/2 = 3711 = C_{min}$

$$N_p = \frac{(900)(6.53)}{(371)\ (4)} = 0.3959$$

$$C = 3711/6270 = 0.592$$

$$(1+c^2)^{1/2} = 1.162 \qquad N(1+c^2)^{1/2} = 0.46$$

From Table 10-3

$$\epsilon_p = 2\left\{1 + 0.592 + 1.162 \times \frac{1+e^{-0.46}}{1-e^{-0.46}}\right\}^{-1}$$

$$= 0.297$$

From Table 10-4

$$\epsilon = \frac{\left[\frac{1-0.297(0.592)}{1-0.297}\right]^4 - 1}{\left[\quad\right]^4 - 0.592} = 0.685 = \frac{\Delta T_g}{80-20}$$

$$\Delta T_g = 41.13°C \qquad \Delta T_w = (41.13)(3711)/6270$$

$$= 24.34°C$$

$$T_{we} = 20 + 24.34 = 44.34°C$$

10-100

Oil min fluid

$$\epsilon = \frac{60-30}{110-30} = 0.375$$

$$C = \frac{20}{30} = 0.667$$

From Prob. 10-95, $n = 3$

$$\epsilon_p = \frac{\left[\frac{1-0.375(0.667)}{1-0.375}\right]^{1/3} - 1}{\left[\quad\right]^{1/3} - 0.667} = 0.158$$

10-101

$$q = (3)(4180)(20) = C_o(30) \qquad C_o = C_{min} = 8360$$

From Table 10-4, $(1+c^2)^{1/2} = 1.202$

$$N_p = \frac{-1}{1.202} \times \ln\left[\frac{\frac{2}{0.158} - 1 - 0.667 - 1.202}{\frac{2}{0.158} - 1 - 0.667 + 1.202}\right] = 0.183$$

$$N(total) = (3)(0.183) = 0.548 = \frac{230A}{8360}$$

$$A = 19.92 \text{ m}^2$$

11-1

$\dot{m} = -D\frac{dc}{dx}$ $D = \frac{-\dot{m}}{dc/dx}$ $\dot{m} \approx \bar{v} \sim T^{1/2}$ $\frac{dc}{dx} \sim p \sim \frac{1}{T}$

$\therefore D \sim \frac{T^{1/2}}{1/T} \sim T^{3/2}$

11-2

$V(C_6H_6) = 6(14.8+3.7) - 15 = 96$ $V_{air} = 29.9$

$T = 25°C = 298°K$ $M(C_6H_6) = 78$ $M(air) = 28.9$

$$D = \frac{(435.7)(298)^{3/2}}{(1.0132\times10^5)\left[(96)^{1/3}+(29.9)^{1/3}\right]^2}\left[\frac{1}{78}+\frac{1}{289}\right]^{1/2}$$

$D = 0.082\ cm^2/sec$

11-3

$T_\infty = 25°C$ $u_\infty = 1.5\ m/sec$ $L = 30cm$

°F	T_w °C	h_{fg} $\times10^{-6}$	C_w	T_f °K	ν $\times10^6$	k	c_p	Pr	ρ_f
59	115	2.47	0.0128	293	14.81	0.026	1005	0.69	1.2
95	35	2.42	0.039	303	15.98	0.0265	↓	↓	1.17
113	45	2.40	0.065	308	16.49	0.027	↓	↓	1.15
149	65	2.34	0.164	318	17.51	0.0275	↓	↓	1.11

11-3 cont'd

$$\bar{h} = \frac{k}{L}(0.664)\, Re_L^{1/2}\, Pr^{1/3}$$

$$h_D = \frac{\bar{h}}{\rho_f c_p}\left(\frac{Pr}{Sc}\right)^{2/3} \qquad \left(\frac{Pr}{Sc}\right)^{2/3} = \left(\frac{0.69}{0.6}\right)^{2/3} = 1.098$$

$$Pr^{1/3} = 0.884$$

T_W	Re_L	$\bar{h}$	$h_D \times 10^3$	q_{conv}	q_{mass}	q'_{tot} W
15	30375	8.87	8.08	-7.983	22.99	15.01
35	28160	8.7	8.12	+7.83	68.97	76.8
45	27289	8.73	8.29	15.71	116.39	132.1
65	25700	8.63	8.49	31.07	293.23	324.3

$$q = \bar{h}A(T_w - T_\infty) + h_D A (C_w - C_\infty)\, h_{fg}$$

$$q = q_{conv} + q_{mass}$$

11-4

$h_{fg} = 3.77 \times 10^5$ J/kg $\quad T_w = 26°C = 299°K$

$p = 13.3$ kN/m² $\quad T_\infty - T_w = \frac{h_D}{h}(C_w - C_\infty) h_{fg}$

$$C_w = \frac{(13300)(78)}{(8316)(299)} = 0.417 \quad S_c = 1.76$$

$P_r = 0.7$ $\quad$ Assume $T_f \approx 70°C = 343°K$

$$\rho_f = \frac{1.0132 \times 10^5}{(287)(343)} = 1.029 \quad \rho_f = \frac{1.0132 \times 10^5}{(287)(343)} = 1.029$$

$$\frac{h_D}{h} = \frac{1}{(1.029)(1009)}\left(\frac{0.7}{1.76}\right)^{2/3} = 5.209 \times 10^{-4}$$

$$T_\infty - T_w = (5.209 \times 10^{-4})(0.417)(3.77 \times 10^5) = 81.89°C \quad T_\infty = 107.89°C$$

Repeat with ρ_f at $T_f = \frac{108+26}{2} = 67°C = 340°K$. Close enough.

11-5

$h_{fg} = 2.45 \times 10^6$ J/kg $\quad u_m = 3$ m/sec $\quad \rho = \frac{1.0132 \times 10^5}{(287)(298)} = 1.185$

$\mu = 1.98 \times 10^{-5}$ $\quad k = 0.026$ $\quad P_r = 0.7$ $\quad S_c = 0.6$

$D = 0.256$ cm²/sec $= 2.56 \times 10^{-5}$ m²/sec $\quad Re = \frac{(1.185)(0.05)(3)}{1.98 \times 10^{-5}} = 8977$

11-5 cont'd

$$h_D = \frac{2.56\times10^{-5}}{0.05}(0.023)(8977)^{0.83}(0.6)^{0.44} = 0.018$$

C_W @ 25°C = 0.0229 $\quad \dot{m}_w = (0.018)\pi(0.05)(3)\left[0.0229 - \frac{C_{W\,exit}}{2}\right]$

$\dot{m}_W$ = (air volume flow) $C_{W\,exit}$ $\quad$ air volume flow $= \frac{\pi d^2}{4} u_m = \frac{\pi(0.05)^2}{4}(3)$

air volume flow = 5.89×10^{-3} m³/sec $\quad$ Solving for C_W exit:

$C_{W\,exit}$ = 0.0192 kg/m³

11-6

p_{W_1} = 0.4593 psia $\quad$ T = 25°C = 298°K $\quad p_{W_2} = (0.4593/2) = 0.2297$ psia

D = 0.256 cm²/sec = 2.56×10^{-5} m²/sec $\quad p_{a_1}$ = 14.696 − 0.4593 = 14.237 psia

$p_{a_1} = 9.8155\times10^4$ N/m² $\quad p_{a_2}$ = 14.696 − 0.2297 = 14.466 psia = 9.9736×10^4 N/m²

$$m_W = \frac{(2.56\times10^{-5})(1.0132\times10^5)\pi(0.075)^2(18)}{(8316)(298)(0.15)}\ln\left(\frac{9.9736}{9.8155}\right) = 3.55\times10^{-8}\ \text{kg/sec}$$

= 0.128 g/hr.

11-7

D = 0.088 cm²/sec = 8.8×10^{-6} m²/sec $\quad p_{a_1}$ = 101.32 − 13.3 = 88.02 kN/m²

p_{a_2} = 101.32 − 0 = 101.32 kN/m²

$$m_B = \frac{(8.8\times10^{-6})(1.0132\times10^5)(78)\pi(0.00625)^2}{(8316)(299)(0.15)}\ln\left(\frac{101.32}{88.02}\right) = 3.22\times10^{-9}\ \text{kg/sec}$$

= 0.0116 g/hr.

11-8

$T_f = \frac{25+0}{2} = 12.5°C = 285.5°K \quad \rho = 1.34 \quad \mu = 1.64\times10^{-5} \quad k = 0.0235$

$Pr = 0.71 \quad C_p = 1009 \quad Re = \frac{(1.34)(1.5)(0.3)}{1.64\times10^{-5}} = 36768$

$h = \frac{(0.0235)}{0.3}(0.664)(36768)^{1/2}(0.71)^{1/3} = 8.9\ W/m^2\text{-}°C \qquad Sc = 0.6$

$h_D = \frac{8.9}{(1.34)(1009)}\left(\frac{0.71}{0.6}\right)^{2/3} = 7.36\times10^{-3} \quad @\ 0°C \quad C_w = 4.845\times10^{-3}$

$m_w = (7.36\times10^{-3})(0.3)^2(4.845\times10^{-3} - 0) = 3.21\times10^{-6}\ kg/sec = 0.0116\ kg/hr.$

11-9

$T_w = 32°C = 305°K = 89.6°F \quad C_w = 0.0342 \quad h_{fg} = 2.42\times10^6\ J/kg$

$Sc = 0.6 \quad Pr = 0.7 \quad$ assume $T_f \approx 350°K \quad \rho_f = 0.998 \quad C_p = 1009$

$T_\infty - T_w = \frac{(0.0342-0)(2.42\times10^6)}{(0.998)(1009)}\left(\frac{0.7}{0.6}\right)^{2/3} = 91°C \qquad T_\infty = 32+91 = 123°C$

$T_f = \frac{123+32}{2} = 77.5°C = 350.5°K$

11-10

$T_\infty = 115\ °F$

$u_\infty = 10\ mph$

$= 52800\ ft/hr.$

$L = 1\ ft$

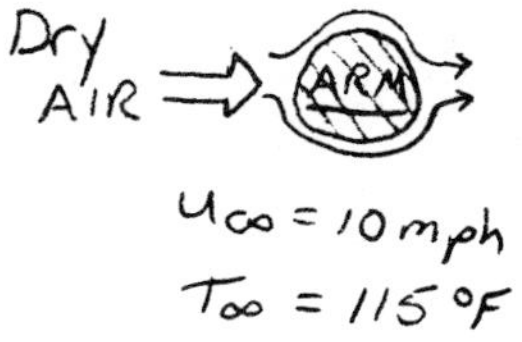

$q_{sun} = 350\,(A_{new})\,\epsilon_1$

$\epsilon_1 = 1.0$

11-10 cont'd

Neglect internal heat generation

$$q_{rad}\big|_{sun} + q_{conv}\big|_{air} = q_{rad}\big|_{arm} + q_{evap}\big|_{arm}$$

$$q_{rad}\big|_{sun} = 350\, dL = 116.5 \text{ Btu/hr.}$$

$$q_{conv} = hA\,(T_\infty - T_w)$$

$$q_{rad}\big|_{arm} = \epsilon \sigma A T_w^4$$

$$q_{evap} = h_D A (C_w - C_\infty) h_{fg}\ , \quad C_\infty = 0$$

$$350 \frac{A}{\pi} + hA(T_\infty - T_w) - \epsilon\sigma A T_w^4 - h_D A C_w h_{fg} = 0$$

$$h = \frac{k_f}{d}\, c\left(\frac{\rho u_\infty d}{\mu}\right)^n$$

assume $T_w = 58°F$

$\therefore T_f = 86.5\ °F$

$\rho = .0728$; $d = .333$ ft ; $\mu = .045$; $K = .0154$

$$Re_d = \frac{(.0708)(52800)(.333)}{.046} = 28{,}300$$

$c = 0.174$, $\quad m = 0.618$

$h = 4.53$ Btu/hr·ft²·°F

$$\frac{h}{h_D} = \rho c_p \left(\frac{Sc}{Pr}\right)^{2/3} \qquad \therefore h_D = 289 \qquad c_p = .24 \quad \frac{Sc}{Pr} = .85$$

11-10 cont'd

$$\frac{350}{\pi} + 4.53\,(T_\infty - T_w) - \epsilon\sigma T_w^4 - 289\, c_w h_{fg} = 0$$

$$111 + 256 - 123 - 237 = -3 \approx 0$$

close enough

$T_w = 58°F$

The above calculation of the radiation from the arm is not strictly correct because the effective radiation temp-ezature of the surroundings is not given. Certain view factors would also need to be known. When the calculation is repeated neglecting the radiation from the arm the result is:

$T_w = 67°F$

11-11

$$h(T_\infty - T_w) = h_D c_w h_{fg} + \sigma(T_w^4 - T_s^4) \qquad T_\infty = 43°C \quad T_s = 10°C$$

$L = 30\text{cm}$ $\quad u_\infty = 12\text{ m/sec}$ $\quad Pr = 0.71$ $\quad Sc = 0.6$ $\quad$ assume $T_f \approx 300°K$

$\rho_f = 1.177$ $\quad \mu_f = 1.983 \times 10^{-5}$ $\quad k_f = 0.02624$ $\quad c_p = 1006$ $\quad h_{fg} = 2.42 \times 10^6$

$$Re = \frac{(1.177)(12)(0.3)}{1.983 \times 10^{-5}} = 2.14 \times 10^5 \qquad \bar{h} = \frac{(0.02624)(0.664)}{0.3}(2.14 \times 10^5)^{1/2}(0.71)^{1/3}$$

$$\bar{h} = 23.95\ W/m^2 \cdot °C$$

$$h_D = \frac{23.95}{(1.177)(1.006)}\left(\frac{0.71}{0.6}\right)^{2/3} = 0.0226 \qquad \text{By iteration } T_w = 14.1°C$$

11-12

$$M_W = h_D C_W A = \frac{[h(T_\infty - T_W) - \sigma(T_W^4 - T_S^4)]A}{h_{fg}}$$

$$M_W = \frac{(0.091)[(23.95)(43-14.1) - (5.669\times10^{-8})(287.1^4 - 283^4)]}{2.42\times10^6}$$

$M_W = 2.49\times10^{-5}$ kg/sec $\quad M_W = 0.09$ kg/hr.

11-13

$T_\infty = 115\ ^\circ F$

$u_\infty = 10 mph = 52800$ ft/hr

$$\dot{q}_{gen}V + q_{rad/sun} + q_{conv/air} = q_{rad/arm} + q_{evap/arm}$$

$$\dot{q}V = \frac{(180)(\pi)(d^2)(L)}{(4)} = 15.7\ \frac{BTU}{hr}$$

$$\frac{A\ 15.7}{A} + 350\frac{A}{\pi} + hA(T_\infty - T_w) - \epsilon\sigma A T_w^4 - h_D A C_w h_{fg} = 0$$

$$h = \frac{k_f}{d} C\left(\frac{\rho u_\infty d}{\mu}\right)^n$$

$$\frac{h}{h_D} = \rho c_p \left(\frac{Sc}{Pr}\right)^{2/3}$$

11-13 cont'd

Assume value of T_w, evaluate properties at film temperature.

Solve for h, h_o & Re_d

$C = 0.174$ $\quad \frac{Sc}{Pr} = 0.85$

$n = 0.618$

$h_{fg} = 1057$ BTU/lbm

$C_w = 0.000976$ lbm/ft^3

By trial & error:

$T_w = 51.785$ °F

11-16

$T_\infty = 65°C \quad L = 0.3\,m \quad u_\infty = 6\,m/sec \quad T_w = 38°C = 311°K$

$T_f = \frac{38+65}{2} = 51.5°C = 324.5°K \quad \rho_f = 1.088 \quad \mu_f = 2.03\times10^{-5}$

$C_p = 1008 \quad k_f = 0.0281 \quad Sc = 0.6 \quad Pr = 0.7 \quad Re = \frac{(1088)(6)(0.3)}{2.03\times10^{-5}}$

$Re = 96472 \quad \bar{h} = \frac{0.0281}{0.3}(0.664)(96472)^{1/2}(0.7)^{1/3} = 17.5\ W/m^2\text{-}°C$

$h_D = \frac{17.15}{(1.088)(1008)}\left(\frac{0.7}{0.6}\right)^{2/3} = 0.0173$

$hA(T_\infty - T_w) + \sigma A(T_\infty^4 - T_w^4) + m_w i_w(25°C) = h_D A C_w h_{fg}$

i_w is neg. compared to h_{fg}. m_w is determined by energy balance:

$h_{fg} = 2.41\times10^6$ @ 38°C

11-16 cont'd

$$M_w = \frac{(0.09)\left[(17.5)(65-38) + (5.669 \times 10^{-8})(338^4 - 311^4)\right]}{2.41 \times 10^6}$$

$$M_w = 2.51 \times 10^{-5} \text{ kg/sec} = 0.09 \text{ kg/hr}$$

12-1

$$\frac{v_w}{u_\infty}\sqrt{Re_x} = 0.1 \qquad Re_x = 10^5 \qquad r = 0.82 = \frac{T_{aw} - T_\infty}{T_0 - T_\infty}$$

$T_\infty = -40°C = 233°K \qquad T_0 = (233)\left[1 + (0.2)(4)^2\right] = 978.6\ °K$

$T_{aw} = 844.4\ °K = 571.4°C$

12-2

$$\frac{Nu_x}{\sqrt{Re_x}} = 0.24 \quad \text{for zero injection} \quad \frac{Nu}{\sqrt{Re_x}} = 0.29$$

$$\text{reduction} = \frac{0.29 - 0.24}{0.29} = 17.2\%$$

12-3

$$\frac{v_w}{u_\infty}\sqrt{Re_x} = 0.5 \qquad T_\infty = 540°C = 813°K \qquad T_w = 150°C$$

$$c_p = 1009 \qquad \frac{Nu_x}{\sqrt{Re_x}} = 0.3 \quad r = 0.7 \quad T_0 - T_\infty = \frac{(600)^2}{(2)(1009)} = 178°C$$

$T_{aw} = 540 + (0.7)(178) = 665°C = 938°K$

$$T_r = \frac{938 + 813}{2} = 875°K \qquad \rho = 0.404 \quad \mu = 3.83 \times 10^{-5} \quad k = 0.061$$

12-3 cont'd

$$Re_x = \frac{(0.404)(0.0075)(600)}{3.83 \times 10^{-5}} = 47467 \qquad Nu_x = (0.3)(47467)^{1/2} = 65.36$$

$$h_x = \frac{(65.36)(0.061)}{0.0075} = 531.6 \text{ W/m}^2\text{-°C}$$

$$q/A = h_x \Delta T = (531.6)(150-665) = -273.8 \text{ kW/m}^2$$

12-6

$T_\infty = -50°C = 223°K \quad p = 10^{-6}$ atm $\quad a = [(1.4)(287)(223)]^{1/2} = 306$ m/sec

$u_\infty = (5)(306) = 1530$ m/sec $\quad \mu = 1.41 \times 10^{-5} \quad c_p = 1006$

$$\rho = \frac{0.10132}{(287)(223)} = 1.58 \times 10^{-6}$$

$$Re = \frac{(1.58 \times 10^{-6})(1530)(0.025 \times 10^{-3})}{1.41 \times 10^{-5}} = 4.29 \times 10^{-3}$$

Free Molecule range $\quad s = (1.4/2)^{1/2}(5) = 4.183$

$\frac{\gamma+1}{\gamma} r = 2.05 \qquad \frac{1}{\alpha}\frac{\gamma}{\gamma+1} St = 0.16 \qquad r = 1.196 \qquad St = 0.247$

$T_0 = (233)[1 + (0.2)(5)^2] = 1398°K \quad T_{aw} = 1626°K$

$h = (1.58 \times 10^{-6})(1006)(1530)(0.247) = 0.601$ W/m²-°C

$(0.601)(1626 - T_{rw}) = (5.669 \times 10^{-8})(0.4)(T_{rw}^4 - 223^4)$

$T_{rw} = 429°K = 156°C$

12-9

$T_\infty = -60°C = 213°K \quad M = 5.0 \quad L = 0.3 \quad T_{rs} = -70°C = 203°K$

$\alpha = 0.88 \quad p = 10^{-10}\ atm \quad a = [(1.4)(287)(213)]^{1/2} = 292\ m/sec$

$u_\infty = (5)(292) = 1463\ m/sec \quad \mu = 1.37\times 10^{-5} \quad c_p = 1005$

$\rho = \dfrac{1.0132\times 10^{-5}}{(287)(213)} = 1.66\times 10^{-10} \qquad Re = \dfrac{(1.66\times10^{-10})(1463)(0.3)}{1.37\times 10^{-5}} = 5.31\times 10^{-3}$

Free molecule flow $\quad S = (0.7)^{1/2}(5) = 4.183 \quad \dfrac{\gamma+1}{\gamma} r = 2.0$

$\dfrac{1}{\alpha}\dfrac{\gamma}{\gamma+1} St = 0.04 \qquad r = 1.167 \quad St = 0.0603 \quad T_0 = (213)[1+(0.2)(25)]$

$T_0 = 1278°K \quad T_{aw} = 1456°K \quad h = (1.66\times10^{-10})(1005)(1463)(0.0603)$

$h = 1.472\times 10^{-5}$

For $\epsilon = 0.3 \quad (1.472\times10^{5})(1456 - T_{rw}) = (5.669\times10^{-8})(0.3)(T_{rw}^4 - 203^4)$

$T_{rw} \approx 203°K$

12-11

$p = 10^{-5}\ atm = 1.0132\ N/m^2 \quad \alpha_1 = 0.87 \quad \alpha_2 = 0.95 \quad \epsilon_1 = 0.08 \quad \epsilon_2 = 0.23$

$T_1 = 70°C = 343°K \quad T_2 = 4°C = 277°K \quad L = 1.3\ mm \quad Pr = 0.7 \quad \gamma = 1.4$

$k = 0.027 \quad T_m = 310°K$

$\lambda = \left(\dfrac{(2.27\times10^{-5})(310)}{1.0132}\right) = 7.13\times10^{-3}\ m = 7.13\ mm$

$\Delta T_1 = \left(\dfrac{2-0.87}{0.87}\right)\left(\dfrac{2.8}{2.4}\right)\left(\dfrac{7.13\times10^{-3}}{0.7}\right)\left(\dfrac{70-4-\Delta T_1-\Delta T_2}{1.3\times10^{-3}}\right) \qquad \Delta T_1 = 34.11\ °C$

12-11 contd.

$$\Delta T_2 = \left(\frac{2-0.95}{0.95}\right)\left(\frac{2.8}{2.4}\right)\left(\frac{7.13\times10^{-3}}{0.7}\right)\left(\frac{70-4-\Delta T_1-\Delta T_2}{1.3\times10^{-3}}\right) \qquad \Delta T_2 = 29.02°C$$

Conduction:

$$q/A = k\,\frac{T_1 - T_2 - \Delta T_1 - \Delta T_2}{L} = \frac{(0.027)(70-4-34.11-29.02)}{0.013} = 5.96\ \frac{W}{m^2}$$

Radiation

$$q/A = \frac{(5.669\times10^{-8})(343^4 - 277^4)}{\frac{1}{0.8} + \frac{1}{0.23} - 1} = 28.45\ W/m^2$$

q/A total $= 34.41\ W/m^2$

12-12

$q = 200{,}000$ Btu/hr $= 58599$ W axial heat flux $= 295\ W/cm^2$

$$A = \frac{58599}{295} = 198.6\ cm^2$$

12-13

$$\frac{\delta_o}{\delta_m} = 0.5 \qquad N = 0.27$$

$$\frac{Nu_m}{Nu_o} = \left[\frac{(5.33)(0.27)(1)}{1-e^{-5.33(0.27)(1)}}\right]^{1/2} = 1.373 \qquad 37\%\ \text{increase}$$

2-14

$$a = [(1.4)(287)(233)]^{1/2} = 306 \text{ m/sec}$$

$$u_\infty = (4)(306) = 1224 \text{ m/sec} \qquad v_w = \frac{(0.1)(1224)}{(10^5)^{1/2}} = 0.387 \text{ m/sec.}$$